AF753535

MATHEMATICS RESEARCH DEVELOPMENTS

UNDERSTANDING INTEGRO-DIFFERENTIAL EQUATIONS

Mathematics Research Developments

Additional books in this series can be found on Nova's website under the Series tab.

Additional e-books in this series can be found on Nova's website under the e-book tab.

MATHEMATICS RESEARCH DEVELOPMENTS

UNDERSTANDING INTEGRO-DIFFERENTIAL EQUATIONS

J. Vasundhara Devi, PhD
Zahia Drici, PhD
Farzana A. McRae, PhD
EDITORS

nova
science publishers
www.novapublishers.com

Copyright © 2023 by Nova Science Publishers, Inc.
DOI: https://doi.org/10.52305/NRLJ6556

All rights reserved. No part of this book may be reproduced, stored in a retrieval system or transmitted in any form or by any means: electronic, electrostatic, magnetic, tape, mechanical photocopying, recording or otherwise without the written permission of the Publisher.

We have partnered with Copyright Clearance Center to make it easy for you to obtain permissions to reuse content from this publication. Simply navigate to this publication's page on Nova's website and locate the "Get Permission" button below the title description. This button is linked directly to the title's permission page on copyright.com. Alternatively, you can visit copyright.com and search by title, ISBN, or ISSN.

For further questions about using the service on copyright.com, please contact:
Copyright Clearance Center
Phone: +1-(978) 750-8400 Fax: +1-(978) 750-4470 E-mail: info@copyright.com.

NOTICE TO THE READER

The Publisher has taken reasonable care in the preparation of this book, but makes no expressed or implied warranty of any kind and assumes no responsibility for any errors or omissions. No liability is assumed for incidental or consequential damages in connection with or arising out of information contained in this book. The Publisher shall not be liable for any special, consequential, or exemplary damages resulting, in whole or in part, from the readers' use of, or reliance upon, this material. Any parts of this book based on government reports are so indicated and copyright is claimed for those parts to the extent applicable to compilations of such works.

Independent verification should be sought for any data, advice or recommendations contained in this book. In addition, no responsibility is assumed by the Publisher for any injury and/or damage to persons or property arising from any methods, products, instructions, ideas or otherwise contained in this publication.

This publication is designed to provide accurate and authoritative information with regard to the subject matter covered herein. It is sold with the clear understanding that the Publisher is not engaged in rendering legal or any other professional services. If legal or any other expert assistance is required, the services of a competent person should be sought. FROM A DECLARATION OF PARTICIPANTS JOINTLY ADOPTED BY A COMMITTEE OF THE AMERICAN BAR ASSOCIATION AND A COMMITTEE OF PUBLISHERS.

Additional color graphics may be available in the e-book version of this book.

Library of Congress Cataloging-in-Publication Data

ISBN: 979-8-89113-040-1

Published by Nova Science Publishers, Inc. † New York

Contents

PREFACE

This book presents an overview of the theory of Integro-Differential Equations (InDEs). Beginning with the history of the inception of InDEs, the existence and uniqueness results for the initial value problem of InDEs is given. Next, stability concepts are discussed, including Lyapunov stability and Ulam-Rassias stability. Numerical techniques and operational matrix methods are also introduced. A specific study of second order InDEs of Fredholm InDes in complex plane is presented and an application to Finance is given.

In: Understanding Integro-Differential Equations ISBN: 979-8-89113-040-1
Editors: J. Vasundhara Devi et al. © 2023 Nova Science Publishers, Inc.

Chapter 1

An Overview of Existence Results for Some Integro-Differential Equations

CH. V. Sreedhar *
I. S. N. R. G. Bharat
and J. Vasundhara Devi

Department of Mathematics, GVP-Prof. V. Lakshmikantham
Institute for Advanced Studies,
Gayatri Vidya Parishad College of Engineering(Autonomous),
Visakhapatnam, AP, India

Abstract

Starting with a brief sketch of the history of integro-differential equations, we proceed to give existence and uniqueness results for nonlinear Volterra integro-differential equations. Next we present monotone iterative technique for integro-differential equations with retardation and anticipation, periodic boundary-value problems of integro-differential equations, and higher order integro differential equations. Then, we give quasilinearization for periodic boundary-value problems of integro-differential equations, integro-differential equations with retardation and anticipation.

Keywords: integral equation, integro-differential equation, initial value problem, periodic boundary value problem, lower and upper solutions, existence

*Corresponding Author's Email: chaduvulasreedhar@gvpce.ac.in.

and uniqueness, monotone iterative technique, quasilinearization

AMS Subject Classification: 45D05, 45B05, 45J05, 45J99, 47G20

1. Introduction

Modeling of physical phenomenon is necessary to understand and to predict the behavior of the phenomenon in future. Modeling involves developing mathematical equations which are obtained through the laws of nature. We illustrate the process through the following example.

Electronic or magnetic polarization of substances depends on the electro magnetic field at the moment. Interestingly, for some other substances the electronic or magnetic polarization depends both on the present moment and also on the history of the electromagnetic state of the matter in all the past moments (this is called hysteresis). When these physical facts are translated into mathematical equations, the result is integral equation and if dynamics is also involved we get an integro-differential equation [1].

The theory of integro-differential equations (InDEs) can be traced to the works of Abel, Lotka, Fredholm, Malthus, Verhulst and Volterra. Abel's work on the tautochrome problem is one of the first problems involving an InDE. Fredhlom also worked on InDEs and Volterra worked with a particular type of Fredholm integro-differential equation (FInDE). To solve an InDE, it is either transformed into a system of differential equations or integral equations and the solutions of these systems are solutions of InDE [2].

As the classification of InDEs is parallel to that of integral equations, we begin with the classification of integral equations and proceed with the classification of InDEs.

In this chapter we concentrate on the theoretical aspects of solving nonlinear InDEs, particularly of Volterra type. Of the many types of existence results available, we focus on the constructive methods of monotone iterative technique (MIT) [3] and quasilinearization [4]. An obvious advantage of these techniques is that the solution is obtained in a sector, that is, a closed set. Further, the iterates are solutions of the corresponding linear integro-differential equations and are either increasing or decreasing in nature and these iterates converge to a solution of the considered problem.

We begin with existence results obtained by using fixed point theorems. We first present the MIT for InDEs involving retardation and anticipation. This result includes the MIT of an InDE and MIT for an InDE involving retardation as special cases. Next, we present MIT developed for Periodic Boundary Value Problem (PBVP) from [5]. In this paper, the authors made a simple choice of initial value as the boundary value of the previous iterate, while constructing the iterates, which guaranteed the existence of the solutions in a sector. Thus one need not prove an existence result separately. Further, we develop MIT for fourth order nonlinear InDE. To develop the MIT, we need to use the maximum principle which is established in [6]. There are some papers involving the concept of lower and upper solutions with reverse inequalities. We introduce this concept from [7].

Next, we proceed to present results obtained using the method of quasilinearization for InDEs involving retardation and anticipation and describe the method of generalized quasilinearization for PBVP of InDE [8].

2. Integral Equations

The term, integral equation, was first used by Du Bois-Reymond. The development of the theory of integral equations is one of the most powerful tools in both pure and applied mathematics. It has many applications in physical problems [9].

Definition 1. The general form of linear integral equation is given below

$$\xi(z)\,h(z) = u(z) + \lambda \int_{q(z)}^{r(z)} H(z,t)\,h(t)\,dt, \tag{1}$$

where $q(z)$ and $r(z)$ are the limits of integration, λ is a non-zero parameter, $H(z,t)$ is the kernel of the integral equation and is a function of z, $t \in [t_0, T]$.

If the unknown function $h(z)$ involved in the integral equation is nonlinear, then such an integral equation is called a nonlinear integral equation.

3. Classification of Integral Equations

In this section we present various types of integral equations that are used frequently in applications. Integral equations can be broadly categorized under two heads as Fredholm integral equations and Volterra integral equations. The classification mainly depends on the limits of integration. In Fredholm integral equation, the integral is over the whole interval of definition. If the upper limit of the integral is "t", then such an integral equation is called Volterra integral equation. The various types of integral equations are briefly given below.

3.1. Fredholm Integral Equation

In the integral equation (1), if the limits of the integration $q(z)$ and $r(z)$ are constants, then such an equation is called a Fredholm integral equation. The general form of a Fredholm integral equation is

$$\xi(z)h(z) = u(z) + \lambda \int_a^b H(z,t)h(t)\,dt, \tag{2}$$

where a and b are the limits of the integration, λ is a non-zero parameter, $H(z,t)$ is the kernel of the integral equation, and is a function of $z,t \in [t_0,\ T]$.

Example 1. $f(z) = e^z + \int_0^1 (zt) f(t)\,dt$ is a Fredholm integral equation.

3.1.1. Fredholm Integral Equation of First Kind

In the Fredholm integral equation (2) if $\xi(z) \equiv 0$ and the unknown function $h(z)$ appears only under the integral sign then such an equation is called an Fredholm integral equation of first kind. The general form of a Fredholm integral equation of first kind is

$$u(z) + \lambda \int_a^b H(z,t)h(t)\,dt = 0. \tag{3}$$

Example 2. In $e^z = \int_0^1 (zt)u(t)\,dt,$ the unknown function $u(t)$ appears only inside the integral sign and both the limits of the integration are constants, hence (3) is an example of a Fredholm integral equation of first kind.

3.1.2. Fredholm Integral Equation of Second Kind

If $\xi(z) \equiv 1$ and the unknown function $h(x)$ appears both inside and outside the integral sign in Fredholm integral equation (2) then such an equation is called a Fredholm integral equation of second kind. The general form of a Fredholm integral equation of second kind is

$$h(z) = u(z) + \lambda \int_a^b H(z,t)\, h(t)\, dt. \tag{4}$$

Example 3. The equation $u(z) = \sin(z) + \int_0^1 (zt)\, u(t)\, dt$, is an example of a Fredholm integral equation of second kind.

3.2. Volterra Integral Equation

The Volterra integral equation is an equation which contains a fixed lower limit and a variable upper limit for the integral. The general form of a Volterra integral equation is given by

$$\xi(z)\, h(z) = u(z) + \lambda \int_a^z H(z,t)\, h(t)\, dt. \tag{5}$$

Example 4. The equation $f(z) = e^z + \int_0^z (zt)\, f(t)\, dt$. It is an example of Volterra integral equation (5).

3.2.1. Volterra Integral Equation of First Kind

If $\xi(z) \equiv 0$, in the Volterra integral equation (5) then such an equation is called a Volterra integral equation of first kind.

Example 5. An example of Volterra integral equation of first kind is given by $e^z + \int_0^z (zt)\, f(t)\, dt = 0$.

3.2.2. Volterra Integral Equation of Second Kind

If $\xi(z) \equiv 1$, in Volterra integral equation (5), then we get a Volterra integral equation of second kind. The general form of Volterra integral equation of second kind is given by

$h(z) = u(z) + \lambda \int_a^z H(z,t)\, h(t)\, dt.$

Example 6. $f(z) = sinz + \int_0^z (z^2 t)\, f(t)\, dt$, is a Volterra integral equation of second kind.

4. Integro-Differential Equations

An InDE is an equation in the form

$$z^{(n)}(t) = c(t, z(t)) + \int_{q(t)}^{r(t)} H(t,s)\, z(s)\, ds. \tag{6}$$

Example 7. $z'' = 5z + 6\int_0^t 4z(\tau)\sin(t)\, d\tau.$

4.1. Classification of Integro-Differential Equations

In this section we discuss various types of InDEs that are frequently used as mathematical tools. Parallel to the classification of integral equations, InDEs are also divided into two types, namely, Fredholm integro-differential equations and Volterra integro-differential equations. These equations are further subdivided into various types, which are briefly given below.

4.1.1. Fredholm Integro-Differential Equation

In InDE (6), if the limits of integration are constants, then such an equation is called a Fredholm integro-differential equation. The Fredholm integro-differential equation has the form

$$z^{(n)}(t) = c(t, z(t)) + \lambda \int_a^b H(t,s) z(s) ds, \ n \geqslant 1. \tag{7}$$

Example 8. $z''(t) + z'(t) = z - \cos z - \int_0^{\frac{\pi}{2}} z(s) ds.$

4.1.2. Fredholm Integro-Differential Equation of Second Kind

If the kernel involved in the Fredholm InDE (7) is a separable kernel, that is, $H(t,s) = \sum_{k=1}^{n} h_k(t) r_k(s)$, then such an equation is called a Fredholm integro-differential equation of second kind.

4.1.3. Volterra Integro-Differential Equation

In InDE (6) if the upper limit of integration is variable, then such an equation is called a Volterra integro-differential equation. The Volterra integro-differential equation has the form

$$z^{(n)}(t) = c(t, z(t)) + \lambda \int_a^t H(t,s) z(s) ds, \ n \geqslant 1. \tag{8}$$

Example 9. $z'''(t) = 1 + \int_0^t tz(s) ds.$

4.1.4. Volterra-Fredhlom Integro-Differential Equation

The Volterra-Fredhlom integral equation is an equation which contains integrals with fixed limits and with variable limits. An example of Volterra-Fredholm integro-differential equation is

$$z^{(n)}(t) = u(t) + \lambda_1 \int_a^t H_1(z,t) h(t) dt + \lambda_2 \int_a^b H_2(z,t) h(t) dt. \tag{9}$$

5. Existence Using Fixed Point Theorems

In this section, we state global existence and uniqueness results for nonlinear Volterra InDEs using fixed point theorems and the comparison method with a scalar integro-differential equation.

5.1. Existence Result Using Tychonoff Fixed Point Theorem

Consider the IVP of InDE given by,

$$z'(t) = c(t, z(t)) + \int_{t_0}^{t} H(t, s, z(s))\, ds, \tag{10}$$

$$z(t_0) = z_0. \tag{11}$$

We state the following theorem dealing with the existence of solutions using Tychonoff fixed point theorem, see [10].

Theorem 1. *Assume that*

(H_1) $c \in C[\mathbb{R}_+ \times \mathbb{R}^n, \mathbb{R}^n]$, $l \in C[\mathbb{R}_+^2, \mathbb{R}_+]$, $l(t,u)$ *is monotone nondecreasing in u, for each* $t \in J$, *and* $|c(t,z)| \leqslant l(t, |z|)$, $(t,z) \in \mathbb{R}_+ \times \mathbb{R}^n$;

(H_2) $H \in C[\mathbb{R}_+^2 \times \mathbb{R}^n, \mathbb{R}^n]$, $G \in C[\mathbb{R}_+^2, \mathbb{R}_+]$, $G(t,s,u)$ *is monotone nondecreasing in u for each* $(t,s) \in \mathbb{R}_+^2$; *and*

(H_3) *for each* $u_0 > 0$, *the scalar integro-differential equation* $u'(t) = l(t, u(t)) + \int_{t_0}^{t} G(t, s, u(s))\, ds$, $u(t_0) = u_0$ *has a solution* $u(t)$ *existing for* $t \geqslant t_0$;

(H_4) $\int_{s}^{t} |H(\sigma, s, x(s))|\, d\sigma \leqslant N$ *for* $t, s \in \mathbb{R}_+$, $z \in C[\mathbb{R}_+, \mathbb{R}^n]$.

Then, for every $z_0 \in \mathbb{R}^n$ *such that* $|z_0| \leqslant u_0$, *there exists a solution* $z(t)$ *of IVP of InDE* (10) *and* (11).

5.2. Existence Using Schauder Fixed Point Theorem

The following theorem deals with the existence of solutions using Schauder fixed point theorem, for details see [10].

Theorem 2. *Assume that*

(H_1) $c \in C[J \times \mathbb{R}^n, \mathbb{R}^n]$, $H \in C[J \times J \times \mathbb{R}^n, \mathbb{R}^n]$, *and* $\int_s^t |H(\sigma, s, z(s))| d\sigma \leqslant N$,
for $t_0 \leqslant s \leqslant t \leqslant t_0 + a$,
$z \in \Omega = \{\phi \in C[J, R^n] : \phi(t_0) = z_0, \ |\phi(t) - z_0| \leqslant b\}$.

Then, the IVP of InDE (10) *and* (11) *possesses at least one solution* $z(t)$ *on* $t_0 \leqslant s \leqslant t \leqslant t_0 + \alpha$, *for some* $0 < \alpha < a$.

6. Monotone Iterative Technique

The method of monotone iterative technique (MIT), together with the method of lower and upper solutions, is an important tools that yields sequences of monotone iterates that converge to the maximal and minimal solution of the considered problem. This method is a flexible mechanism and a constructive technique to obtain the existence of solutions in a closed sector. In this section we will consider some InDEs and develop MIT.

Comparison results play a pivotal role in studying nonlinear InDEs. A complicated InDE can be studied through a much simpler scalar InDE using the comparison principle. We present the following comparison result using Dini derivatives [10, 11]. This theorem plays an important role in the qualitative theory of InDEs.

Theorem 3. *Assume that* $l \in C\left[\mathbb{R}_+^2, \mathbb{R}\right]$, $H \in C\left[\mathbb{R}_+^3, \mathbb{R}\right]$, $H(t, s, u)$ *is nondecreasing in* u *for each* (t, s) *and for* $t \geqslant t_0$,

$$D_{-}m(t) \geqslant l(t, m(t)) + \int_{t_0}^{t} H(t, s, m(s)) ds,$$

where $m \in C[\mathbb{R}_+, \mathbb{R}]$ *and* $D_- m(t) = \lim\inf_{h \to 0^-} h^{-1}[m(t+h) - m(t)]$. *Suppose that* $\gamma(t)$ *is the maximal solution of*

$$u'(t) = l(t, u(t)) + \int_{t_0}^{t} H(t, s, u(s))\, ds,\ u(t_0) = u_0 \geqslant 0,$$

existing on $[t_0, \infty)$. *Then,* $m(t) \leqslant \gamma(t),\ t \geqslant t_0,$ *provided* $m(t_0) \leqslant u_0$.

As finding the solutions of scalar integro-differential equations is also difficult, the above comparison theorems are not helpful in many situations. The previous theorems require knowledge of these solutions, which are not always. Thus, we present a comparison theorem which uses a scalar differential equation to estimate the nonlinear InDE. This result is an adaptation of results in [12] and is given in [10].

We first state a lemma that is essential to prove the comparison theorem.

Lemma 4. *Let* $l_0, l \in C[\mathbb{R}_+^2, \mathbb{R}]$ *satisfy*

$$l_0(t, u) \leqslant l(t, u),\ (t, u) \in \mathbb{R}_+^2.$$

Then, the right maximal solution $\gamma(t, t_0, u_0)$ *of*

$$u' = l(t, u), \quad u(t_0) = u_0,$$

and the left maximal solution $\eta(t, T, v_0)$ *of*

$$u' = l_0(t, u),\ u(T) = v_0 \geqslant 0,$$

satisfy the relation,

$$\gamma(t, t_0, u_0) \leqslant \eta(t, T, v_0),\ t \in [t_0, T],$$

whenever

$$\gamma(T, t_0, u_0) \leqslant v_0.$$

Theorem 5. *Let* $m \in C[\mathbb{R}_+, \mathbb{R}_+],\ l \in C[\mathbb{R}_+^2, \mathbb{R}],\ G \in C[\mathbb{R}_+^3, \mathbb{R}]$ *and*

$$D_- m(t) \geqslant l(t, m(t)) + \int_{t_0}^{t} G(t, s, m(s))\, ds,\ t \in I_0,$$

where $I_0 = \{t \geqslant t_0 : m(s) \leqslant \eta(s,t,m(t)),\ t_0 \leqslant s \leqslant t\}$, $\eta(t,T,v_0)$ *being the left maximal solution of* $u' = l_0(t,u),\ u(T) = v_0$, *existing on* $[t_0,T]$, *assume that*

$$l_0(t,u) \leqslant F(t,u,t_0),$$

where

$$F(t,u,t_0) = l(t,u) + \int_{t_0}^{t} G(t,s,u(s))\,ds,$$

and $\gamma(t)$ *is the maximal solution of*

$$u' = F(t,u,t_0),\quad u(t_0) = u_0,$$

existing on $[t_0,\infty)$, *then,* $m(t_0) \leqslant u_0$ *implies* $m(t) \leqslant \gamma(t),\ t \geqslant t_0$.

6.1. InDE with Retardation and Anticipation

The past history or memory and anticipation play a vital role in physical phenomena. Hence, we present the MIT for InDEs with retardation and anticipation (InDE with R and A). Consider the nonlinear InDE given by,

$$z' = h(t,z,Bz,z_t,z^t),\quad t \in J = [t_0,\ T], \tag{12}$$

$$z_{t_0} = \phi_0,\quad z^T = \psi_0, \tag{13}$$

where $\phi_0 \in C_1,\ \psi_0 \in C_2$ and $h \in C[J \times \mathbb{R} \times \mathbb{R} \times C_1 \times C_2,\ \mathbb{R}]$ and

$$Bz(t) = \int_{t_0}^{t} H(t,s)z(s)\,ds, \qquad t \in J,$$

$H \in C[J \times J, \mathbb{R}_+]$, $C_1 = C[[-h_1,0],\mathbb{R}]$, $C_2 = C[[0,h_2],\mathbb{R}]$ and the term $z_t = x_t(s) = z(t+s),\ s \in [-h_1,0]$, represents retardation or delay and the term $z^t = z^t(\sigma) = z(t+\sigma),\ \sigma \in [0,\ h_2]$ represents anticipation.

We begin by defining the natural lower and upper solutions for InDE with R and A (12) and (13).

Definition 2. *(i) A function $v_0 \in C^1[[t_0 - h_1, T + h_2], \mathbb{R}]$ is said to be a natural lower solution of InDE with R and A(12) and (13) if*

$$v_0' \leq h(t, v_0, Bv_0, v_{0t}, v_0^t)$$

where $v_{0t_0} = \phi_1, \; v_0^T = \psi_1, \; v_{0t_0} \leq \phi_0$ *and* $v_0^T \leq \psi_0$.

(ii) A function $w_0 \in C^1[[t_0 - h_1, T + h_2], \mathbb{R}]$ is said to be a natural upper solution of InDE with R and A (12) and (13) if

$$w_0' \geq h(t, w_0, Bw_0, w_{0t}, w_0^t)$$

where $w_{0t_0} \geq \phi_0$ *and* $w_0^T \geq \psi_0$ *and* $w_{0t_0} = \phi_2, \; w_0^T = \psi_2$. *Note that* $\phi_1, \phi_2 \in C_1, \psi_1, \psi_2 \in C_2$ *and* $\phi_1 \leq \phi_0 \leq \phi_2, \; \psi_1 \leq \psi_0 \leq \psi_2$.

Next, we introduce the concept of maximal and minimal solutions for InDE with R and A (12) and (13).

Definition 3. *(i) Let $r_1(t)$ be a solution of InDE with R and A (12) and (13), then $r_1(t)$ is said to be a maximal solution of (12) and (13) if for every solution $z(t)$ of (12) and (13) existing on $[t_0 - h_1, T + h_2]$, the inequality*

$$z(t) \leq r_1(t)$$

holds for $t \in [t_0 - h_1, T + h_2]$.

(ii) Let $\rho_1(t)$ be a solution of InDE with R and A (12) and (13), then $\rho_1(t)$ is said to be a minimal solution of (12) and (13) if for every solution $z(t)$ of (12) and (13) existing on $[t_0 - h_1, T + h_2]$, the inequality

$$\rho_1(t) \leq z(t)$$

holds for $t \in [t_0 - h_1, T + h_2]$.

Definition 4. *(i) A function $v_0 \in C^1[I, \mathbb{R}]$ is said to be a coupled lower solution of InDE with R and A (12) and (13) if*

$$v_0' \leq h(t, v_0, Bv_0, v_{0t}, w_0^t),$$

where $v_{0t_0} = \phi_1, v_0^T = \psi_1, \; v_{0t_0} \leq \phi_0$ *and* $v_0^T \leq \psi_0$;

(ii) A function $w_0 \in C^1[I, \mathbb{R}]$ is said to be a coupled upper solution of InDE with R and A (12) and (13) if

$$w_0' \geq h(t, w_0, Bw_0, w_{0t}, v_0^t),$$

where $w_{0t_0} = \phi_2,\ w_0^T = \psi_2,\ \ w_{0t_0} \geq \phi_0$ *and* $w_0^T \geq \psi_0$.
Note that $\phi_1,\ \phi_2 \in C_1,\ \psi_1,\ \psi_2 \in C_2$ *and* $\phi_1 \leq \phi_0 \leq \phi_2,\ \psi_1 \leq \psi_0 \leq \psi_2$.

The following result is from [13] and is used to develop the monotone iterates corresponding to (12) and (13).

Lemma 6. *Let* $k \in C[[t_0 - h_1, T], \mathbb{R}] \bigcap C^1[J, \mathbb{R}]$ *be continuously differentiable on* $J = [t_0, T]$ *and*

$$k'(t) \leq -L_1 k(t) - L_2 Bk(t) - L_3 \int_{-h_1}^{0} k_t(s)ds \ \text{ on } J, \tag{14}$$

where
where $L_1,\ L_2,\ L_3 \geq 0,\ Bk(t) = \int_{t_0}^{t} H(t,s)k(s)ds, \qquad k_1 = \max_{t,s\in J}\ H(t,s).$
Suppose further that either

A) $k(t_0) \leq k_{t_0}(s) \leq 0,\ \ s \in [-h_1, 0]$ *and*
$$[L_1 + L_2 k_1 (T - t_0) + L_3 h_1](T - t_0) \leq 1,$$
or
B) $k_{t_0}(s) \leq 0,\ s \in [-h_1, 0],\ k \in C^1[[t_0 - h_1,\ T], \mathbb{R}]$ *and* $k'(t) \leq \frac{\lambda}{T - t_0 + h_1}$,

where $t \in [t_0 - h_1, t_0],\ \min_{[t_0 - h_1, t_0]}\ k(s) = -\lambda,\ \lambda \geq 0$ *and*
$$[L_1 + L_2 k_1 (T - t_0) + L_3 h_1](T - t_0 + h_1) \leq 1. \tag{15}$$

Then, $k(t) \leq 0$ *on J.*

In order to develop MIT for InDE with R and A (12) and (13), we need the following assumptions relative to (12) and (13).

(H_1) Let $v_0, w_0 \in C^1[I, \mathbb{R}]$, satisfying

$$v_0' \leq h(t, v_0, Bv_0, v_{ot}, v_0^t),$$

$$v_{0t_0} = \phi_1,\ \ v_0^T = \psi_1,$$

$$w_0' \geq h(t, w_0, Bw_0, w_{ot}, w_0^t),$$

$w_{0t_0} = \phi_2,\ w_0^T = \psi_2$, such that $\phi_1 \le \phi_0 \le \phi_2,\ \psi_1 \le \psi_0 \le \psi_2,\ v_0 \le w_0$ on $I = [t_0 - h_1, T + h_2]$, $\phi_1,\ \phi_2 \in C_1 = C[[-h_1, 0], \mathbb{R}]$ and $\psi_1,\ \psi_2 \in C_2 = C[[0, h_2], \mathbb{R}]$;

(H_2) $h(t,x,\xi,\phi,\psi)$ is nonincreasing in ψ for each (t,x,ξ,ϕ);

(H_3) $h(t,x,\xi_1,\phi_1,\psi) - h(t,y,\xi_2,\phi_2,\psi)$

$$\ge -L_1(x-y) - L_2(\xi_1 - \xi_2) - L_3 \int_{-h_1}^{0} (\phi_1 - \phi_2)(s)ds$$

with $[L_1 + L_2 k_1(T - t_0) + L_3 h_1](T - t_0) \le 1$ and $L(T - t_0) > \frac{1}{2}$, whenever $v_0 \le y \le x \le w_0$, $Bv_0 \le \xi_2 \le \xi_1 \le Bw_0$, $k_1 = \max_{t,s \in J}, H(t,s)$ $L_1, L_2, L_3 \ge 0$, $v_{0t} \le \phi_2 \le \phi_1 \le w_{ot}$, $\psi \in C_2$, $\phi_1,\ \phi_2 \in C_1$;

(H_4) $v_{0t_0} - \phi_0,\ \phi_0 - w_{0t_0}$ satisfy the assumptions of Lemma 6;

(H_5) $h(t,x,\xi_1,\phi_1,\psi_1) - h(t,y,\xi_2,\phi_2,\psi_2)$

$$\le -O_1(x-y) + O_2(\xi_1 - \xi_2) + O_3 \int_{-h_1}^{0} (\phi_1 - \phi_2)(s)ds$$
$$+ O_4 \int_{0}^{h_2} (\psi_1 - \psi_2)(s)ds,$$

where $v_0 \le y \le x \le w_0$, $Bv_0 \le \xi_2 \le \xi_1 \le Bw_0$, $v_{0t_0} \le \phi_2 \le \phi_1 \le w_{0t_0}$, $v_0^t \le \psi_2 \le \psi_1 \le w_0^t$, $L_1, L_2, L_3, L_4 \ge 0$ and $[O_2 k_1 T + O_3 h_1 + O_4 h_2] < O_1$.

We now state the following theorem dealing with the monotone iterative technique for InDE involving retardation and anticipation.

Theorem 7. Suppose that the assumptions (H_1) to (H_5) are satisfied. Then there exist monotone sequences $\{v_n\}$ and $\{w_n\}$ that converge uniformly to a solution of InDE with R and A (12) and (13) on I.

The uniqueness of the solution for InDE with R and A (12) and (13) can be obtained by using the following lemma .

Lemma 8. *Let $k \in C^1[[t_0 - h_1, T + h_2], \mathbb{R}]$ be continuously differentiable on $J = [t_0, T]$ and*

$$k'(t) \leq -L_1 p(t) + L_2 Bk(t) + L_3 \int_{-h_1}^{0} k_t(s)ds + L_4 \int_{0}^{h_2} k^t(s)ds,$$

on J, where

$$Bk(t) = \int_{t_0}^{t} H(t,s)k(s)ds, \quad k_1 = \max_{t,s\in J} H(t,s),$$

and $L_1, L_2, L_3, L_4 > 0$ satisfying $[L_2 k_1 (T - t_0) + L_3 h_1 + L_4 h_2] < L_1$, $k_{t_0}(s) \leq 0,\ k^T \leq 0$. Then, $k(t) \leq 0$ on J.

Now we state the criteria for obtaining the unique solution of InDE with R and A (12) and (13).

Theorem 9. Suppose that (H_1) to (H_5) are satisfied. If in addition,

(H_6) $h(t,x,\xi_1,\phi_1,\psi_2) - h(t,y,\xi_2,\phi_2,\psi_1)$

$$\leq -L_1(x-y) + L_2(\xi_1 - \xi_2) + L_3 \int_{-h_1}^{0} (\phi_1 - \phi_2)(s)ds + L_4 \int_{0}^{h_2} (\psi_1 - \psi_2)(s)ds,$$

where $v_0 \leq y \leq x \leq w_0$, $Bv_0 \leq \xi_2 \leq \xi_1 \leq Bw_0$, $v_{0t_0} \leq \phi_2 \leq \phi_1 \leq w_{0t_0}$, $v_0^t \leq \psi_2 \leq \psi_1 \leq w_0^t$, $L_1, L_2, L_3, L_4 \geq 0$, $[L_2 k_1 T + L_3 h_1 + L_4 h_2] < L_1$.

Then, $\rho = x = r$ is the unique solution for InDE with R and A (12) and (13) on J.

Remark 1. Observe that if h is restricted to $h \in C[J \times \mathbb{R} \times \mathbb{R} \times C_1,\ \mathbb{R}]$ then the InDE with R and A (12) and (13) is reduced to an InDE with retardation. If $h \in C[J \times \mathbb{R} \times \mathbb{R},\ \mathbb{R}]$ then the system InDE with R and A (12) and (13) is reduced to an ordinary InDE. In both the cases the above theorem holds with suitable modifications.

6.2. PBVP of InDE

We present below MIT developed for PBVP of InDE using coupled upper and lower solutions of Type I. This result is interesting due to the following reason.

The authors in [5] considered the iterates of the linear InDEs as an IVP with a clever choice of the initial value as the boundary value of the previous iterate. This choice resulted in obtaining the existence of solution. Thus there was no need to prove the existence of solution separately.

$$u'(t) = h(t,u,Bu) + l(t,u,Bu),\ t \in J = [0,T],\ u(0) = u(T), \tag{16}$$

where $h, l \in C[J \times \mathbb{R} \times \mathbb{R}, \mathbb{R}]$, $Bu(t) = \int_0^t H(t,s)u(s)\,ds$, and $H \in C[J \times \mathbb{R}, \mathbb{R}^+]$ To develop the MIT, the following definitions and comparison result are required.

Definition 5. *Functions $v_0, w_0 \in C^1[J,\mathbb{R}]$ are said to be coupled lower and upper solutions of Type-I for PBVP of InDE* (16) *if*

$$v_0'(t) \leqslant h(t, v_0, Bv_0) + l(t, w_0, Bw_0),\ t \in J,\ v_0(0) \leqslant v_0(T);$$

and

$$w_0'(t) \geqslant h(t, w_0, Bw_0) + l(t, v_0, Bv_0),\ t \in J,\ w_0(0) \geqslant w_0(T).$$

Definition 6. *Two functions $v_0, w_0 \in C^1[J,\mathbb{R}]$ are said to be coupled lower and upper solutions of Type-II for PBVP of InDE* (16) *if*

$$v_0'(t) \leqslant h(t, w_0, Bw_0) + l(t, v_0, Bv_0),\ t \in J,\ v_0(0) \leqslant v_0(T);$$

and

$$w_0'(t) \geqslant h(t, v_0, Bv_0) + l(t, w_0, Bw_0),\ t \in J,\ w_0(0) \geqslant w_0(T).$$

Lemma 10. Let $q \in C^1[J,\mathbb{R}]$ be such that $q' \leqslant -Mq + NBq$ and $q(0) \leqslant q(T)$, where constants $M > 0$, $\mathrm{N} > 0$, and $k_0 = \max\{H(t,s) : (t,s) \in J \times J\}$ satisfy $Nk_0T < M$. Then, $q(t) \leqslant 0$ for $t \in J$.

Using the above definitions, the comparison result and the fact that the corresponding IVP of linear integro-differential equations have unique solutions. We now state the MIT in this setup.

Theorem 11. *Suppose that*

1. *v_0, w_0 are coupled lower and upper solutions of Type-I relative to the PBVP of InDE* (16) *on J;*
2. *The forcing functions $h(t,u,Bu)$ and $l(t,u,Bu)$ are nondecreasing and nonincreasing respectively in the last two variables.*

Then, the iterative scheme

$$v_n{}'(t) = h(t, v_{n-1}, Bv_{n-1}) + l(t, w_{n-1}, Bw_{n-1}),\ t \in J,\ v_n(0) = v_{n-1}(T);$$

and

$$w_n{}'(t) = h(t, w_{n-1}, Bw_{n-1}) + l(t, v_{n-1}, Bv_{n-1}),\ t \in J,\ v_n(0) = v_{n-1}(T);$$

yields two monotone sequences $\{v_n\}$ and $\{w_n\}$ such that $v_0 \leqslant v_1 \leqslant v_2 \leqslant \ldots \leqslant v_k \leqslant \ldots \leqslant w_k \leqslant \ldots \leqslant w_2 \leqslant w_1 \leqslant w_0$ on J. Further $v_n \to \rho$ and $w_n \to r$, where ρ and r are coupled minimal and maximal solutions of the PBVP of InDE (16) *on J with $\rho \leq r$ on J. Also,*

$$\rho'(t) \leqslant h(t, \rho, B\rho) + l(t, r, Br),\ t \in J,\ \rho(0) = \rho(T);$$

and

$$r'(t) \geqslant h(t, r, Br) + l(t, \rho, B\rho),\ t \in J,\ r(0) = r(T).$$

The following theorem deals with the uniqueness of the solutions of PBVP of InDE (16) on J.

Theorem 12. *If, in addition to the assumptions of Theorem* 11, *the forcing functions $h(t,u,Bu)$ and $l(t,u,Bu)$ satisfy*

$$h(t, u_1, Bu_1) - h(t, u_2, Bu_2) \leqslant M_1(u_1 - u_2) + N_1 B(u_1 - u_2),\ t \in J;$$

$$l(t, u_1, Bu_1) - l(t, u_2, Bu_2) \leqslant M_2(u_1 - u_2) + N_2 B(u_1 - u_2),\ t \in J.$$

whenever $u_1 \leqslant u_2$, $M_i > 0$, $N_i > 0$, for $i = 1, 2$, and $M_1 > M_2$. Then, $\rho = r$ on J. Consequently PBVP of InDE (16) *has a unique solution on J.*

In place of natural upper and lower solutions of PBVP of InDE (16), if coupled lower and upper solutions of Type-II are considered then the iterative scheme guarantees coupled, intertwining, alternating sequences that converge uniformly and monotonically to the coupled minimal and maximal solutions of PBVP of InDE (16) on J. We state this result from [5]

Theorem 13. *Let the hypothesis of Theorem* 11, *hold. Then, for any solution* $u(t)$ *for PBVP of InDE* (16) *with* $v_0(t) \leqslant u(t) \leqslant w_0(t)$ *on* J, *the iterative scheme*

$$v_n{}'(t) = h(t, w_{n-1}, Bw_{n-1}) + l(t, v_{n-1}, Bv_{n-1}),\ t \in J,\ v_n(0) = v_{n-1}(T);$$

and

$$w_n{}'(t) = h(t, v_{n-1}, Bv_{n-1}) + l(t, w_{n-1}, Bw_{n-1}),\ t \in J,\ v_n(0) = v_{n-1}(T);$$

yields the intertwined alternating sequences $\{v_{2n}, w_{2n+1}\}, \{w_{2n}, v_{2n+1}\}$ *satisfying* $v_0 \leqslant w_1 \leqslant v_2 \leqslant \ldots \leqslant v_{2n} \leqslant w_{2n+1} \leqslant u \leqslant v_{2n+1} \leqslant w_{2n} \ldots \leqslant\leqslant w_2 \leqslant v_1 \leqslant w_0$ *on* J, *for every* $n \geq 1$. *Moreover the sequences* $\{v_{2n}, w_{2n+1}\}, \{w_{2n}, v_{2n+1}\}$ *converge to* ρ *and* r *respectively on* J, *where* ρ *and* r *are coupled minimal and maximal solutions for PBVP of InDE* (16), *with* $\rho \leq r$ *on* J.

The same technique was extended for nonlinear boundary value problems with casual operators in [14].

6.3. MIT for Higher Order Integro-Differential Equations

We consider the nonlinear fourth-order InDE given by

$$y^{(4)}(x) = C(x, y(x), Ty(x)), \tag{17}$$

with the Navier boundary conditions

$$y(0) = y(1) = y''(0) = y''(1) = 0, \tag{18}$$

where $C \in C([0,1] \times \mathbb{R}^2, \mathbb{R})$, $Ty(x) = \int_0^1 h(x,t)y(t)\,dt$, $h \in C([0,1] \times [0,1], \mathbb{R})$. The BVP of higher-order InDE (17) and (18) is a generalization of the linear fourth-order problem.

$$y^{(4)}(x) + \lambda y(x) - \mu \int_0^1 h(x,t)y(t)\,dt = p(x),\ \ x \in (0,1), \tag{19}$$

$$y(0) = y(1) = y''(0) = y''(1) = 0, \tag{20}$$

where λ and μ are constants, $p \in C[0,1]$.

It is known that for general second-order DEs with various types of boundary conditions, the existence of lower and upper solutions satisfying $v_0 \leq w_0$ is sufficient for existence of a solution in a sector. But in [15] there is a counter-example to show that this result does not hold for fourth-order DEs. The application of the method lower and upper solutions for boundary value problems of fourth order DEs is highly dependent on the conclusion of the maximum principle for the corresponding linear operators.

In [6], in view of this observation, they developed first the maximum principle and then the MIT for linear higher- order InDE (19), (20) and then the MIT has been developed. We proceed to present the established results from [6]. Consider the nonlocal nonhomogeneous boundary value problem

$$y^{(4)}(x) + \lambda y(x) - \mu \int_0^1 h(x,t) y(t)\, dt = p(x), \quad x \in (0,1), \tag{21}$$

$$y(0) = M, y(1) = N, y''(0) = P, y''(1) = Q. \tag{22}$$

Theorem 14. *Assume that*

(H_1) $0 < \lambda \leqslant c_1 \approx 950.8843$ *and*

(H_2) $\|h\|_\infty = \max\{h(x,t)\} < \frac{1}{|\mu| \max_{x\in[0,1]} \int_0^1 G(x,s)ds}$ where $(x,t) \in [0,1]\times[0,1]$, *then for any* $p \in C[0,1]$ *then non homogeneous BVP of InDE* (21), (22) *has a unique solution.*

We now state the maximum principle, which is essential to show that the lower and upper solutions v_0, w_0 are such that $v_0 \leq w_0$.

Theorem 15. *Assume that* $0 < \lambda \leqslant c_1 \approx 950.8843$, $\mu h(x,t) \geq 0$ *and* $\|h\|_\infty = \max\{h(x,t)\} < \frac{1}{|\mu| \max_{x\in[0,1]} \int_0^1 G(x,s)ds}$ where$(x,t) \in [0,1]\times[0,1]$, *then*

1. *The nonhomogeneous BVP of InDE* (21), (22) *has a unique solution* $y \geq 0$, *if* $p(x) \geq 0$, $M \geq 0$, $N \geq 0$, $P \leq 0$, $Q \leq 0$ $\forall x \in [0,1]$,
2. *The non homogeneous BVP of InDE* (21), (22) *has a unique solution* $y \leq 0$, *if* $p(x) \geq 0$, $M \leq 0$, $N \leq 0$, $P \geq 0$, $Q \geq 0$ $\forall x \in [0,1]$.

To develop the monotone iterative technique the following definitions are necessary.

Definition 7. *The function* $v_0 \in C^4[0,1]$ *is said to be a lower solution for the BVP of higher order InDE* (17), (18) *if*

$$v_0^{(4)}(x) \leqslant C(x, v_0(x), Tv_0(x)),\ x \in (0,1) \tag{23}$$

$$v_0(0) \leqslant 0,\ v_0(1) \leqslant 0,\ v_0''(0) \geqslant 0,\ v_0''(1) \geqslant 0, \tag{24}$$

where $Tv_0(x) = \int_0^1 h(x,t) v_0(t)\, dt$

Definition 8. *The function* $w_0 \in C^4[0,1]$ *is said to be an upper solution for the BVP of higher-order InDE* (17), (18) *if*

$$w_0^{(4)}(x) \geqslant C(x, w_0(x), Tw_0(x)),\ x \in (0,1), \tag{25}$$

$$w_0(0) \geqslant 0,\ w_0(1) \geqslant 0,\ w_0(0) \leqslant 0,\ w_0''(1) \leqslant 0, \tag{26}$$

where $Tw_0(x) = \int_0^1 h(x,t) w_0(t)\, dt$.

We now state the MIT from [6],

Theorem 16. *Let* v_0, w_0 *be lower and upper solutions for the nonhomogeneous BVP of higher-order InDE* (17), (18) *which satisfy* $v_0(x) \leqslant w_0(x)$ *for* $x \in [0,1]$ *and* $C \in C([0,1] \times \mathbb{R}^2, \mathbb{R})$, $h \geqslant 0$. *If there exist two constants* $\lambda, \mu > 0$ *satisfying* $\lambda < c_1 \approx 125.137$ *and* $\|h\|_\infty < \frac{1}{|\mu| \max\limits_{x\in[0,1]} \int_0^1 G(x,s)ds}$ *such that*

$$C(x, r_1, s_1) - C(x, r_2, s_2) \geqslant -\lambda (r_1 - r_2) + \mu (s_1 - s_2), \tag{27}$$

for $v_0(x) \leqslant r_1 \leqslant r_2 \leqslant w_0(x)$ and

$$\int_0^1 h(x,t) v_0(t)\, dt \leqslant s_1 \leqslant s_2 \leqslant \int_0^1 h(x,t) w_0(t)\, dt,\ x \in [0,1],$$

then, by the maximum principle the iterative sequences $\{v_n\}$ and $\{w_n\}$ with the initial functions $y_0 = v_0 = w_0$, satisfy $v_{n-1} \leqslant v_n \leqslant w_n \leqslant w_{n-1}$ and converge uniformly to the extremal solutions of BVP of higher-order InDE (17), (18) in $[v_0, w_0]$.

In [7], the authors using the method of upper and lower solutions together with the monotone iterative technique, developed the existence of maximal and minimal solutions under reversed ordering condition.

Consider the class of integro-differential equations with integral boundary conditions

$$x'(t) = C(t, x(t), Hx(t), Sx(t)), \; t \in J, \tag{28}$$

$$x(0) + \lambda \int_0^T x(s)\, ds = x(T), \tag{29}$$

where $J = [0,T]$, $T > 0$, $C \in C(J \times \mathbb{R}^3, \mathbb{R})$, $H(t,s) \in C(D, \mathbb{R}^+)$, $Sx(t) = \int_0^t h(t,s)x(s)\, ds$, where $h(t,s) \in C(J \times J, \mathbb{R}^+)$, $\mathbb{R}^+ = [0, +\infty)$, $\lambda \geqslant 0.]$ and $D = \{(t,s) \in \mathbb{R}^2, 0 \leqslant s \leqslant t \leqslant T\}$.

Note that if $\lambda = 0$, in (29), then class of InDE (28), (29) is a PBVP of InDE.

We state the following lemmas from [7].

Lemma 17. *Assume that there exist* $P > 0$, Q, $R \geqslant 0$ *such that*

$$x'(t) \geqslant Px(t) + Q \int_0^t k(t,s)x(s)\, ds + R \int_0^T h(t,s)x(s)\, ds, \; t \in J,$$

$$x(0) \geqslant x(T) \frac{e^{PT}(Qk_0 + Rh_0)\left(1 - e^{-PT}\right)\left(e^{PT} - 1\right)}{P^2} < 1.$$

Then, $x(t) \leq 0$, *where* $k_0 = \max\{k(t,s) : (t,s) \in D\}$, $h_0 = \max\{k(t,s); (t,s) \in J \times J\}$.

Lemma 18. *Assume that* $P > 0$, Q, $R \geqslant 0$, $\sigma(t) \in C[J, \mathbb{R}]$ *and* $\frac{(Qk_0 + Rh_0)T}{P} < 1$, *then, the equation*

$$x'(t) - Px(t) - Q \int_0^t R(t,s)x(s)\, ds - L \int_0^T h(t,s)x(s)\, ds = \sigma(t), \; t \in J$$

$$x(0)+d=x(T),\ d\in\mathbb{R},$$

has a unique solution.

The following definitions deal with lower and upper solutions of class of InDE with integral boundary condition (28), (29).

Definition 9. *The functions $v_0, w_0 \in C'[J,R]$ are said to be the lower and upper solutions of class of InDE with integral boundary condition* (28), (29), *respectively if*

$$v_0'(t) \leqslant C(t,x(t),Hx(t),Sx(t)),\ t\in J$$

$$v_0(0)+\lambda\int_0^T x(s)\,ds \leqslant v_0(T),$$

$$w_0'(t) \geqslant C(t,x(t),Hx(t),Sx(t)),\ t\in J,$$

$$w_0(0)+\lambda\int_0^T x(s)\,ds \geqslant w_0(T).$$

The following assumptions are necessary to develop the MIT for InDE with integral boundary condition (28), (29).

(H_1) $v_0, w_0 \in C^1[J,R]$ are lower and upper solutions of class of InDE with integral boundary conditions (28), (29), respectively then $w_0(t) \leqslant v_0(t)$ for $t\in J$;

(H_2) $C\in C\left(J\times\mathbb{R}^3,\mathbb{R}\right)$;

(H_3) *there exist* $P>0,\ Q,\ R\geqslant 0$ such that $C(t,m_2,n_2,l_2)-C(t,m_1,n_1,l_1)$

$$\leqslant -P(m_2-m_1)+Q(n_2-n_1)+R(l_2-l_1)$$

if $w_0\leqslant m_1\leqslant m_2\leqslant v_0$, and $Kw_0\leqslant Kn_1\leqslant Kn_2\leqslant v_0$, $Sw_0\leqslant Sl_1\leqslant Sl_2\leqslant Sv_0,\ t\in J$;

(H_4) $\frac{e^{PT}(Qk_0+Rh_0)\left(1-e^{-PT}\right)\left(e^{PT}-1\right)}{P^2}\leqslant 1$

(H_5) $\frac{(Qk_0+Rh_0)T}{P}<1$ where $k_0=\max\left\{K(t,s):(t,s)\in D\right\},$

$$h_0 = \max\{S(t,s);(t,s) \in J \times J\}$$

The following lemma deals with lower and upper solutions satisfying reverse inequalities.

Lemma 19. *Assume that* $(H_1)-(H_5)$ *hold if*

$$y'(t) - Py(t) - Q\int_0^t K(t,s)y(s)ds - R\int_0^T S(t,s)y(s)ds =$$

$$C(t, v_0(t), Hv_0(t), Sv_0(t)) - Pv_0(t) - Q\int_0^t K(t,s)v_0(s)ds$$

$$-R\int_0^T S(t,s)v_0(s)ds, t \in J$$

$$y(0) + \lambda\int_0^T v_0(s)ds = y(T)$$

and

$$z'(t) - Pz(t) - Q\int_0^t K(t,s)z(s)ds - R\int_0^T S(t,s)z(s)ds =$$

$$C(t, w_0(t), Kw_0(t), Sw(t)) - Pw_0(t) - Q\int_0^t K(t,s)w_0(s)ds$$

$$-R\int_0^T S(t,s)w_0(s)ds, t \in J$$

$$z(0) + \lambda\int_0^T w_0(s)ds = z(T)$$

then, $w_0(t) \leqslant z(t) \leqslant y(t) \leqslant v_0(t),\ t \in J$

Using the reverse inequalities for lower and upper solutions, the MIT is developed for InDE with integral boundary conditions (28), (29). This result is stated below without proof [7].

Theorem 20. *Suppose that* $(H_1)-(H_5)$ *hold. Then, there exist monotone sequences* $\{v_n\}$, $\{w_n\}$ *that converge uniformly and monotonically to* ρ, *and* r *respectively on* J. *Moreover,* ρ *and* r *are maximal and minimal solutions of InDE with integral boundary conditions* (28), (29) *in* $[w_0, v_0] = \{g \in C'(J,R) : w_0 \leqslant g \leqslant v_0\}$

7. Quasilinearization

Quasilinearization is another constructive iterative technique yielding quadratic convergence. This method coupled with the method of lower and upper solutions yields a unique solution of the considered problem.

The quasilinearization technique was developed using coupled lower and upper solutions for an IVP of InDE and then this result was used to obtain existence and uniqueness of solutions for the corresponding PBVP of an InDE given by

$$z' = h_1(t,z,Bz) + h_2(t,z,Bz), \tag{30}$$

$$h(0) = h(T). \tag{31}$$

In order to develop the quasilinearization technique for the PBVP of InDE (30) and (31), the quasilinearization technique for the following initial value problem of an InDE was first obtained.

$$z' = h_1(t,z,Bz) + h_2(t,z,Bz), \tag{32}$$

$$h(0) = h_0. \tag{33}$$

where $h_1, h_2 \in C[I \times \mathbb{R}^n \times \mathbb{R}^n, \mathbb{R}^n]$, $Bx(t) = \int_0^t H(t,s)x(s)ds$, with $H \in C[I \times I, \mathbb{R}_+]$ and $I = [0,T]$.

Before developing the quasiliearization technique for the IVP of an InDE (32) and (33), we present the following basic definitions.

Definition 10. *Let* $v_0, w_0 \in C^1[I, \mathbb{R}^n]$. *Then,* v_0, w_0 *are said to be*
(a) natural lower and upper solutions of PBVP of InDE (32) *and* (33) *if*

$$\left.\begin{aligned} v' &\leq h_1(t,v_0,Bv_0) + h_2(t,v_0,Bv_0), \;\; v_0(0) \leq z_0, \\ w_0' &\geq h_1(t,w_0,Bw_0) + h_2(t,w_0,Bw_0), \;\; w_0(0) \geq z_0, \;\; t \in I; \end{aligned}\right\} \tag{34}$$

(*b*) *coupled lower and upper solutions of Type I for PBVP of InDE* (32) *and* (33) *if*

$$\left.\begin{aligned} v_0' &\le h_1(t,v_0,Bv_0)+h_2(t,w_0,Bw_0),\ \ v_0(0)\le z_0,\\ w_0' &\ge h_1(t,w_0,Bw_0)+h_2(t,v_0,Bv_0),\ \ w_0(0)\ge z_0,\ \ t\in I; \end{aligned}\right\} \tag{35}$$

(*c*) *coupled lower and upper solutions of Type II for PBVP of InDE* (32) *and* (33) *if*

$$\left.\begin{aligned} v_0' &\le h_1(t,w_0,Bw_0)+h_2(t,v_0,Bv_0),\ \ v_0(0)\le z_0,\\ w_0' &\ge h_1(t,v_0,Bv_0)+h_2(t,w_0,Bw_0),\ \ w_0(0)\ge z_0,\ \ t\in I; \end{aligned}\right\} \tag{36}$$

(*d*) *coupled lower and upper solutions of Type III for PBVP of InDE* (32) *and* (33) *if*

$$\left.\begin{aligned} v_0' &\le h_1(t,w_0,Bw_0)+h_2(t,w_0,Bw_0),\ \ v_0(0)\le z_0,\\ w_0' &\ge h_1(t,v_0,Bv_0)+h_2(t,v_0,Bv_0),\ \ w_0(0)\ge z_0,\ \ t\in I. \end{aligned}\right\} \tag{37}$$

We observe that the lower and upper solutions defined by (34) and (37) reduce to (35) and (36) when $v(t)\le w(t)$, $t\in I$. This is due to the fact that $h_1(t,z,\xi)$ is nondecreasing in z and y and $h_2(t,z,\xi)$ is nonincreasing in z and y for each $t\in I$, and hence it is sufficient to investigate the cases (35) and (36).

We begin with a comparison result, which will be useful in establishing quasilinearization.

Lemma 21. Let $q\in C^1[I,R]$, where $I=[0,T]$ is such that

$$q\,'(t)\le -Mq(t)-NBq(t)\ \ on\ I,\ \ q(0)\le 0, \tag{38}$$

where $M>0$, $N\ge 0$ are constants such that

$$Nk_1T(e^{MT}-1)\le M, \tag{39}$$

where $k_1 = max_{t\in I}\,H(t,s)$. Then, $q(t)\le 0$ on I.

The following assumptions are useful for developing the quasilinearization technique for PBVP of InDE (30) and (31).

(H_1) (i) The second-order Frechet derivatives of $h_1(t,z,\xi)$, $h_2(t,z,\xi)$ with respect to all variables exist and are bounded;

(ii) $h_1(t,z,\xi)$ is convex in z, ξ ;

(iii) $h_{1z}(t,z,\xi)$ is nondecreasing in ξ for each (t,z);

(iv) h_1 is nondecreasing function in z, ξ for each $t \in I$ and h_2 is nonincreasing function in z, ξ for each $t \in I$.

(H_2) (i) $-M_1 \leq h_{1z}(t,z,\xi) \leq -M, \quad 0 < M < M_1$;

(ii) $-M_2 \leq h_{1\xi}(t,z,\xi) \leq -N, \quad 0 < N < M_2$;

(iii) $Nk_1T < M$; where $M > 0, \ N \geq 0$.

(H_3) v_0, w_0 are coupled lower and upper solutions IVP of InDE (32) and (33).

(H_4) (i) $h_1(t,z,Bz) \geq h_1(t,y,By) + f_{1z}(t,y,By)(z-y) + h_{1\xi}(t,y,By)(Bx-By)$;

(ii) $|h_{1z}(t,z,Bz) - h_{1y}(t,y,By)| \leq L_1(z-y) + M_1(Bz-By), \quad L_1, M_1 \geq 0$

Theorem 22. *Suppose that the assumptions (H_1) to (H_4) are satisfied. Then, there exists a monotone sequence $\{v_n\}$, such that $v_n \to \rho$, as $n \to \infty$ uniformly and monotonically to the unique solution ρ of the IVP of InDE (32) and (33)on I and the convergence is quadratic.*

Next, an existence and uniqueness result is obtained for an PBVP of InDE.

$$z' = h_1(t,z,Bz) + h_2(t,z,Bz), \tag{40}$$

$$z(0) = z(T), \tag{41}$$

where $h_1, h_2 \in C[I \times \mathbb{R}^n \times \mathbb{R}^n, \mathbb{R}^n]$, $Bx(t) = \int_0^t H(t,s)x(s)ds$, and $K \in C[I \times I, \mathbb{R}_+]$, $I = [0,T]$.

The following definitions give the various types of lower and upper solutions corresponding to the PBVP of InDE (40) and (41).

Definition 11. *Let $v_0, w_0 \in C^1[I, \mathbb{R}^n]$. Then, v_0, w_0 are said to be*
(i) natural lower and upper solutions for PBVP of InDE (40) *and* (41) *if*

$$\left.\begin{array}{r} v_0' \leq h_1(t, v_0, Bv_0) + h_2(t, v_0, Bv_0), \;\; v_0(0) \leq v_0(T), \\ w_0' \geq h_1(t, w_0, Bw_0) + h_2(t, w_0, Bw_0), \;\; w_0(0) \geq w_0(T), \;\; t \in I; \end{array}\right\} \tag{42}$$

(ii) coupled lower and upper solutions of Type III for PBVP of InDE (40) *and* (41) *if*

$$\left.\begin{array}{l} v_0' \leq h_1(t, w_0, Bw_0) + h_2(t, w_0, Bw_0), \;\; v_0(0) \leq v_0(T), \\ w_0' \geq h_1(t, v_0, Bv_0) + h_2(t, v_0, Bv_0), \;\; w_0(0) \geq w_0(T), \;\; t \in I. \end{array}\right\} \tag{43}$$

Now we present the quasilinearization technique to establish the existence of a unique solution of the PBVP of InDE (40) and (41). It is interesting to observe that the proof of the theorem follows the technique developed in [5].

Theorem 23. *Suppose that the assumptions of Theorem* 22 *are satisfied. Then, there exists monotone sequence $\{v_n\}$, such that $v_n \to \rho$, as $n \to \infty$ uniformly and monotonically to the unique solution ρ for PBVP of InDE* (40) *and* (41) *on I and the convergence is quadratic.*

In [9] the quasilinearization is developed for the integro-differential equation with retardation and anticipation given by

$$z' = h(t, z, Bz, z_t, z^t), \quad t \in J = [0, T], \tag{44}$$

$$z_0 = \phi_0, \;\; z^T = \psi_0, \tag{45}$$

where $\phi_0 \in C_1$, $\psi_0 \subset C_2$, $h \in C[J \times \mathbb{R} \times \mathbb{R} \times C_1 \times C_2, \; \mathbb{R}]$. and

$$Bz(t) = \int_{t_0}^{t} H(t,s) z(s) ds$$

where $H \in C[J \times J, \mathbb{R}_+]$, $C_1 = C[[-h_1, 0], \mathbb{R}]$ and $C_2 = C[[0, h_2], \mathbb{R}]$.

In order to state an existence and uniqueness result using quasilinearization we list the following assumptions relative to InDE with R and A (44) and (45) for convenience.
(H_1)(i) All second order frechet derivatives of $h(t, z, \xi, \phi, \psi)$ exist and

are bounded;

(ii) $h(t,z,\xi,\phi,\psi)$ is convex in z, ξ, ϕ and is concave in ψ;

(iii) $h_z(t,z,\xi,\phi,\psi)$ is nondecreasing in ξ for each (t,z,ϕ,ψ) and is nondecreasing in ϕ for each (t,z,ξ,ψ) and nondecreasing in ψ for each (t,z,ξ,ϕ);

(iv) $h_\xi(t,z,\xi,\phi,\psi)$ is nondecreasing in z for each (t,ξ,ϕ,ψ) and is nondecreasing in ϕ for each (t,z,ξ,ψ) and is nondecreasing in ψ for each (t,z,ξ,ϕ);

(v) $h_\phi(t,z,\xi,\phi,\psi)$ is nondecreasing in z for each (t,ξ,ϕ,ψ) and is nondecreasing in ξ for each (t,z,ϕ,ψ) and is idependent of ψ for each (t,z,ξ,ϕ);

(vi) $h_\psi(t,z,\xi,\phi,\psi)$ is nondecreasing in z for each (t,ξ,ϕ,ψ) and is nondecreasing in ξ for each (t,z,ϕ,ψ) and is idependent of ϕ for each (t,z,ξ,ψ);

(H_2) (i) $-M_1 \leq h_z(t,z,\xi,\phi,\psi) \leq -N_1, \quad 0 < N_1 < M_1;$

(ii) $-M_2 \leq h_\xi(t,z,\xi,\phi,\psi) \leq N_2, \quad 0 < N_2 < M_2;$

(iii) $-M_3 \int_{-h_1}^{0} \eta_1(s)ds \leq h_\phi(t,z,\xi,\phi,\psi)\eta_1 \leq N_3 \int_{-h_1}^{0} \eta_1(s)ds, \quad 0 < N_3 < M_3,$ where $\eta_1 \in C_1$;

(iv) $0 \leq h_\psi(t,z,\xi,\phi,\psi)\eta_2 \leq N_4 \int_0^{h_2} \eta_2(s)ds$, where $\eta_2 \in C_2, \ N_4 > 0;$

(v) $[N_2 k_1 T + N_3 h_1 + N_4 h_2] < N_1;$

(H_3) $v_0,\ w_0$ are natural lower and upper solutions of InDE with R and A (44) and (45) ,that is, $v_0,\ w_0 \in C^1[J,R]$, where $J = [0,T]$

$$v_0' \leq h(t, v_0, Bv_0, v_{ot}, {v_0}^t) \quad v_{00} = \phi_1, \ \ {v_0}^T = \psi_1,$$

$$w_0' \geq h(t, w_0, Bw_0, w_{ot}, {w_0}^t) \quad w_{00} = \phi_2, \ \ {w_0}^T = \psi_2,$$

where $v_0 \leq w_0,\ Bv_0 \leq Bw_0,\ t \in J$ and $\phi_1, \phi_2 \in C_1,\ \psi_1, \psi_2 \in C_2,$

such that $\phi_1 \leq \phi_0 \leq \phi_2$ and $\psi_1 \leq \psi_0 \leq \psi_2$;

(H_4) $v_{00} - \phi_0$, $\phi_0 - w_{00}$ satisfy the assumptions of Lemma (6).

Theorem 24. *Suppose that the assumptions* (H_1) *to* (H_4) *are satisfied. Then, there exist monotone sequences* $\{v_n\}$, $\{w_n\}$, *which converge uniformly on* J *to a unique solution of InDE with R and A* (44), (45) *and the convergence is quadratic.*

Conclusion

In this chapter we briefly described different types of integro-differential equations, and presented some criteria for the existence and uniqueness of the solution of a nonlinear Volterra integro-differential equation. Further we discussed the monotone iterative technique for integro-differential equations with retardation and anticipation, periodic boundary-value problems of integro-differential equations, and higher order integro differential equations. Next, quasilinearization for periodic boundary-value problems of integro-differential equations, and integro-differential equations with retardation and anticipation were given.

References

[1] Vito Volterra, *Theory of Functionals and of Integral and Integro-differential equations* , Dover Publications, New York, 1959.

[2] Addul-Majid Wazwaz, *Linear and Nonlinear Integral Equations Methods and Applications,* Higher Education Press, Beijing and Springer-Verlag Berlin Heidelberg Dordrecht, London, New York, 2011.

[3] G. S. Ladde, V. Lakshmikantham and A. S. Vatsala, *Monotone Iterative Technique for Nonlinear Differential Equations,* Pitman Publishing INC., 1985.

[4] V. Lakshmikantham, A. S. Vatsala, *Generalized Quasilinearization for Nonlinear Problems* , Kluwer Academic Publishers, Netherlands, 1998.

[5] S. G. Pandit, D. H. Dezerna, J. O. Adeyeye, Periodic boundary value problems for nonlinear integro differential equations, *Proceedings of Neural, Parallel, and Scientific Computations* , 4 (2010) 316-320.

[6] Jinxiang Wang, *Monotone iterative technique for nonlinear fourth order integro-differential equations*, arXiv:2003.04697v1 [Math. CA], pg. 1-13, 2020.

[7] Guoping Chen, Jianhua Shen,Monotone iterative technique for a class of integro-differential equations with integral boundary conditions, *International Journal of Math. Analysis*, Vol. 1, No.3, 2007, 113-122.

[8] Ch. V. Sreedhar, J.Vasundhara Devi, Generalized Quasilinearization using coupled lower and upper solutions of periodic boundary value problem of an integro-differential equation, *European Journal of Pure and Applied Mathematics*, Vol. 12, No. 4, 2019, 1662-1675.

[9] J.Vasundhara Devi, Ch. V. Sreedhar, Quasilinearization for integro-differential equations with retardation and anticipation, *Nonlinear Studies*, Vol. 19, (2012), no. 1, 99-122.

[10] V. Lakshmikantham, M. Rama Mohana Rao, *Theory of integro-differential equations*, Gordon and Breach Science Publishers, S.A, 1995.

[11] V. Lakshmikantham, S. Leela and M. Rama Mohana Rao, D Integral and Integro-differential Inequalities, Vol. I, *Applicable Analysis*, 34 (1987) 157-164.

[12] V. Lakshmikantham, S. Leela, *Differential and Integral Inequalities*, Vol. I, Academic press, New York, 1969.

[13] J. Vasundhara Devi, Ch. V. Sreedhar, S. Nagamani, Monotone iterative technique for integro-differential equations with retardation and anticipation, *Communicaiton in Applied Analysis*, 14 (2010), No.4, 325-336.

[14] Wen-Li Wang, Jing-Feng Tian, Generalized monotone iterative method for nonlinear boundary value problems with casual operators, *Boundary Value Problems*, 2014, 1-12.

[15] A. Cabada, J. A. Cid, L. Sanchez, Positivity and lower and upper solutions for fourth order boundary value problems, *Nonlinear Analysis*, Vol. 67, 2007, 1599-1612.

In: Understanding Integro-Differential Equations ISBN: 979-8-89113-040-1
Editors: J. Vasundhara Devi et al. © 2023 Nova Science Publishers, Inc.

Chapter 2

A Brief Overview of the Stability Theory of Integro-Differential Equations

Z. Drici[1]
F. A. McRae[2]
and J. Vasundhara Devi[3] *
[1]Department of Mathematics, lllinois Wesleyan University, Bloomington, Illinois, USA
[2]Department of Mathematics, Catholic University of America, Washington DC, USA
[3]Department of Mathematics and GVP Prof. V. Lakshmikantham Institute for Advanced Studies Gayatri Vidya Parishad College of Engineering, Visakhapatnam, AP India

Abstract

In this chapter an overview of the theory of Lyapunov stability of integro-differential equations (InDEs) is presented. The chapter is divided into two parts. In the first part the stability theory of different types of linear InDEs is discussed and different approaches to study stability concepts are given. The second part covers the stability theory of nonlinear Volterra

*Corresponding Author's Email: jvdevi.lias@gvpce.ac.in.

InDEs using the comparison method. Various types of stability notions, such as, practical stability and stability in terms of two measures are some of the topics discussed.

Keywords: Lyapunov stability, comparison theorems, practical stability, stability in two measures

AMS Subject Classification: 34A08, 34A12, 34A34, 34A37, 34A45, 34D05, 34D20

1. Introduction

In this section we will briefly introduce the concepts leading to Lyapunov stability and then proceed to trace the developments in the stability theory in general. The word stability is derived from the Latin 'stare' and is used in the sense of staying firm or durable. This word was first used in mechanics in the definition of a type of equilibrium of a material particle or a system. As this definition was inadequate to study many other setups, new definitions were introduced. Foremost among these definitions is Lyapunov stability[1], as it pertains to a general differential equation and not to any specific particle, and it applies to a solution of a differential equation and not only to an equilibrium or a critical point.

In addition to the above features, the beauty of Lyapunov's ideas is that the approach and the theoretical concepts involved do not require knowledge of the solution of the problem under study[1]. Further, they can be adapted to study other important aspects of system behaviour. Both theorems introduced by Lyapunov involve a function $V(t,x)$ called Lyapunov function. The criteria satisfied by the Lyapunov function determine the types of stability possessed by the solution of the problem under consideration.

Since its introduction, the usefulness of the Lyapunov function has increased substantially. It is employed to study various qualitative and quantitative properties of dynamic systems. It is also used as a vehicle to reduce the study of a complicated system to the study of a relatively simpler system. The properties of the complicated system are then assessed using the properties of the simple system.

In many situations a system may be unstable in the sense of Lyapunov stability but for all practical purposes it is stable. An example of such a system is an aeroplane under turbulence. To accommodate such situations the notion of practical stability [2] was introduced. It must be noted that this concept is neither weaker nor stronger than Lyapunov stability.

The notions of Lyapunov stability enriched the stability theory by giving rise to many new concepts such as, eventual stability, partial stability, and relative stability. These concepts were also extended to practical stability and to boundedness of solutions, also known as Lagrange stability.

As generalization of ideas is core to development in mathematics, it is natural to ask whether all the above mentioned concepts could be unified. This resulted in the notion of stability in terms of two measures [3], which provides a unified structure to study several nonlinear problems.

The literature dealing with the stability theory of InDEs is vast [4] and varied in the sense that many types of InDEs were studied. InDEs are broadly classified as linear and nonlinear InDEs. The linear InDEs studied include both scalar InDEs and systems of InDEs (InDS). The development of the theory of linear InDSs and InDEs is similar to that of ordinary differential sytems and ordinary differential equations (ODEs). One approach is to consider the integral part of the InDE as a perturbation of the ODE and the InDE is studied using the theory of ODEs. The study of nonlinear InDEs is comparatively difficult and one way to study them is to consider the nonlinear term as a perturbation of the linear InDE.

In this chapter, an attempt is made to showcase several types of linear InDEs and present the stability results pertaining to them. With respect to nonlinear InDEs the stability results presented here deal with Lyapunov theorems in this setup and their generalizations. These results are developed using comparison theorems and by reducing the study of a complicated system to the study of a simpler system using a Lyapunov function or a Lyapunov-like function [4].

2. Part I - Linear Integro-Differential Equations

2.1. InDE of Convolution Type

A very simple type of integro-differential equations is the convolution type. Since the integral can be considered as a convolution, one can use Laplace transforms to obtain a solution of the InDE. We consider an InDS of convolution type given by,

$$y'(t) = Cy(t) + \int_0^t H(t-s)y(s)ds \tag{1}$$

where C is an nxn constant matrix and $H(t)$ is an nxn continuous matrix on $\mathbb{R}_+$. Suppose $y(t_0) = \phi$, where $\phi(t)$ exists for $0 \le t \le t_0$, $t_0 > 0$. System (1) can be written as

$$y(t) = Z(t-t_0)\phi(t_0) + \int_{t_0}^{t} Z(t-s)[\int_{0}^{t_0} H(s-\sigma)\phi(\sigma)d\sigma]ds \tag{2}$$

where $Z(t)$ is a solution of the equation,

$$Z'(t) = CZ(t) + \int_0^t H(t-s)Z(s)ds, \ \ Z(0) = I. \tag{3}$$

Define $p(t) = \int_0^\infty |\int_0^t Z(t-\sigma)H(\sigma+u)d\sigma|du$ and assume that it exists and is finite for all $t \ge 0$.

Before proceeding to define the stability concepts, we need the following notation. Let $t \ge 0$ be fixed. Set $B(t) = \{\phi : [0,t] \to \mathbb{R}, \phi$ is continuous and $|\phi|_t = max_{0\le s\le t}|\phi(s)|$ is bounded$\}$

We now present the definition of stability for InDS (1) and proceed to give some stability results from [4, 5]

Definition 1. *The trivial solution of InDS* (1) *is said to be stable if given* $\varepsilon > 0$ *and* $t_0 \in \mathbb{R}_+$, *there exists a* $\delta = \delta(t_0, \varepsilon) > 0$ *such that* $|\phi|_{t_0} = max_{0\le s\le t_0}|\phi(s)| < \delta$ *implies* $|y(t)| < \varepsilon$.

Definition 2. *The trivial solution of InDS* (1) *is said to be uniformly stable if the* δ *in the above Definition* 1 *depends only on* ε, $t \ge t_0$.

Definition 3. *The trivial solution of InDS* (1) *is said to be quasi-equi-asymptotically stable or attractive if given* $\varepsilon > 0$ *and* $t_0 \in \mathbb{R}_+$, *there exist a* $\delta = \delta(t_0)$ *and a* $T(t_0, \varepsilon)$ *such that* $|\phi|_{t_0} = max_{0\le s\le t_0}|\phi(s)| < \delta$ *implies* $|y(t)| < \varepsilon$, *for all* $t \ge t_0 + T$.

Definition 4. *The trivial solution of InDS* (1) *is said to be quasi-uniformly-asymptotically stable or uniformly attractive, if the* δ and T *in the Definition* 3 *depend only on* ε.

Definition 5. *The trivial solution of InDS* (1) *is asymptotically stable if it satisfies both Definitions* 1 *and* 3.

Definition 6. *The trivial solution of InDS* (1) *is uniformly asymptotically stable if it satisfies both Definitions* 2 *and* 4.

Remark 1. *The definitions of various types of stability for both the InDEs and InDSs are the same except that the norm for InDE is absolute value.*

We now state the following stability result from [5].

Theorem 1. *Suppose that $H \in L^1(\mathbb{R}_+)$. Then, the equilibrium solution $y \equiv 0$ of InDS (1) is*

1. *Uniformly stable if and only if the two functions $Z(t)$ and $p(t)$ are uniformly bounded on $\mathbb{R}_+$, and*

2. *Uniformly asymptotically stable if and only if it is uniformly stable and the functions $Z(t)$ and $p(t) \to 0$ as $t \to \infty$.*

2.2. InDE of Convolution Type with Infinite Memory

We make a change in the lower limit of the integral in InDS (1) and consider InDS of the type

$$y'(t) = Cy(t) + \int_{-\infty}^{t} H(t-s)y(s)ds, \tag{4}$$

for $t \geq t_0 \geq 0$ with $y(t) = \phi(t)$ on $-\infty \leq t \leq t_0$ for some $t_0 > 0$, where ϕ is a continuous function in $\mathbb{R}^n$ and the norm is suitably defined. Then, the various stability properties of the zero solution of InDS (4) can be defined in a similar fashion as that of InDS (1). We next present a result from [6] that establishes an equivalent criterion between InDS (1) and InDS (4).

Theorem 2. *Suppose that $H \in L^1(\mathbb{R}_+)$. Then the following statements are equivalent:*

1. *The zero solution of InDS (1) is uniformly stable;*

2. *The zero solution of InDS (4) is uniformly stable;*

3. *The function $Z(t)$ is bounded and for each $\phi \in C(\mathbb{R})$, the solution $y(t,0,\phi)$ of equation (4) is bounded on $\mathbb{R}_+$.*

2.3. Integral as Perturbation

An InDE can be considered as a perturbation of an ODE. We now present two results in that setup from [7]. Consider a linear InDS of the type

$$y'(t) = C(t)y(t) + \int_0^t H(t-s)y(s)ds \tag{5}$$

and its corresponding ordinary differential system

$$x'(t) = C(t)x(t). \tag{6}$$

Let $X(t)$ be the fundamental matrix solution of the InDS (6). We now state the following theorems that deal with the stability and asymptotic stability of InDS (5).

Theorem 3. *Assume that* $|X(t)X^{-1}(s)| \leq L$ *for* $t \geq s \geq 0$ *and* $\int_0^\infty \int_0^t |H(t,s)|ds\,dt \leq K$. *Then, the trivial solution of InDS* (5) *is uniformly stable.*

Theorem 4. *Assume that* $\int_0^t |X(t)X^{-1}(s)|ds \leq M$ *for* $t \geq 0$ *and* $\sup_{t\geq 0} \int_0^t |H(t,s)|ds \leq \frac{1}{M}$. *Further, let* $\lim_{s\to\infty} \int_0^t |H(s,u)|du = 0$ *for all* $t \geq 0$. *Then, the trivial solution of InDS* (5) *is uniformly asymptotically stable.*

2.4. Linear InDE with Nonlinear Perturbation

We now proceed to present a problem that is studied using an approach similar to the approach in the previous subsection, the difference being that the ODE is replaced by an InDE and the perturbation term is nonlinear. Consider the scalar linear Volterra InDE of the form

$$y'(t) = f(t)y(t) + \int_0^t H(at-s)y(s)ds \tag{7}$$

and its perturbation

$$z'(t) = f(t)z(t) + \int_0^t H(at-s)z(s)ds + g(t,z(t)), \tag{8}$$

where a is a constant with $a > 1$, $f(t)$ is continuous on $\mathbb{R}_+$ and $H : \mathbb{R}_+ \to \mathbb{R}_+$ is continuous in t and x, and g satisfies

$$|g(t,z(t))| \leq \lambda(t)|z(t)|, \tag{9}$$

where $\lambda(t)$ is continuous on $\mathbb{R}_+$.

It is assumed that for each $\phi \in B(t_0)$, $t_0 \geq 0$, there exists a unique solution $y(t,t_0,\phi)$ of InDE (8) on an interval $[t_0,\eta)$ with $y(s) = \phi$, $0 \leq t \leq t_0$. We now proceed to give stability results from [8]. The stability results were obtained by constructing the Lyapunov functional $V(t)$ and also using certain estimates given below.

1. Assume that for some constant $c < 0$, $G_c(t) = e^{-ct} \int_t^\infty H(u)e^{cu}du$ exists and $G_c(t) \in L^1[0,\infty)$;

2. Set $a_1(t) = \frac{f(t)}{a} + \frac{1}{a^2}G_c(at-t)$;

3. Let $a_2(t) = \lambda(t)[1+\frac{\lambda(t)}{2a^2}]$;

4. Let the Lyapunov functional be defined by
$V(t) = \frac{1}{2}[y(t) + \frac{1}{a}\int_0^t G_c(at-s)y(s)ds]^2 + p\int_0^t \int_{at-s}^\infty |G_c(u)|du\, y^2(s)ds$,
where p is a positive constant to be suitably defined.

Theorem 5. *1. Let $G_c(t) \in L^1[0,\infty)$ with $\frac{1}{a}\int_0^\infty |G_c(u)|du < 1$.*

2. Assume that $a a_1(t) + a_2(t) + \frac{(a_1(t)-c)^2}{2} + a_3 Q \leq -\beta$, for some $\beta \geq 0$,

3. and also assume that $Q = [\frac{1}{\sqrt{a}}\int_0^\infty |G_c(u)|du]^2$ and $a_3 = \frac{1}{2} + \frac{L^2}{2} + \frac{|c|}{a}$.
Then, the trivial solution of (8) is stable.

Theorem 6. *Suppose that the assumptions in Theorem 5 hold. Further assume that there exists a positive constant k such that $\int_{(a-1)t}^{at}\int_v^\infty |G_c(u)|dudv < k$ for all $t > 0$. Then, the zero solution of (8) is uniformly stable.*

Theorem 7. *Suppose that the assumptions in the Theorem 5 hold such that the second condition is satisfied for $\beta > 0$. Further, assume that $\int_v^\infty |G_c(u)|du \in L^1[0,\infty)$, then the zero solution of (8) is uniformly asymptotically stable.*

2.5. Quasilinear InDS

In [9] the authors considered a quasilinear system obtained by adding two terms to InDS (1), as follows.

$$y'(t) = Cy(t) + \int_0^t H(t-s)y(s)ds + f(t,y) + \int_0^t G(t,s,y(s)ds, \qquad (10)$$

where f and G are continuous for $0 \le s \le t < \infty, ||x|| \le r$ and $f(t,0) = G(t,s,0) \equiv 0$. Let $\hat{H}$ be the Laplace transform of H.

Theorem 8. *Let the matrix $zI - C - \hat{H}(z)$ be invertible for $Rez \ge 0$ and let $\sup_t ||f(t,x)|| = o(||x||), x \to 0$ and $\sup_t \int_0^t \sup_{||x|| \le r} ||G(t,s,x)||ds = o(\tau), \tau \to 0$. Then, the trivial solution of quasilinear system (10) is stable .*
If, in addition, $lim_{t\to\infty} \int_0^T \sup_{||x||\le r} ||G(t,s,x)||ds = 0$, for any $T > 0$ and $\tau \le r$, then, the trivial solution of system (10) is asymptotically stable.

Theorem 9. *Assume that*

1. *for certain $\alpha > 0$, $q_i(t), Q_i(t,s), i = 1,2$*

$$||f(t,x)|| \le q_1(t)o(||x||) + q_2(t)||x||^{1+\alpha}$$

$$||G(t,s,x)|| \le Q_1(t,s)o(||x||) + Q_2(t,s)||x||^{1+\alpha}$$

2. *$||C(t)|| \le Mexp(-\gamma t)$ for certain $\gamma > 0$, $M > 0$.*

Further, if the function $\frac{1}{t}\int_0^t [q_1(s) + \int_0^s exp(\gamma(s-\tau))Q_1(s,\tau)d\tau]ds$ is bounded on $[0,\infty)$ and the function $exp(-\beta t)[q_2(t) + \int_0^t Q_2(t,s)exp(\gamma(1+\alpha)(t-s))ds] \in L^1[0,\infty)$ for some $0 < \beta < \infty$, then the trivial solution of system (10) is asymptotically stable.

The following theorem gives criteria for the instability of system (10)

Theorem 10. *Let z_0 be a pole of the matrix $(zI - C - \hat{H}(z))^{-1}$ with the maximal real part $\alpha = Rez_0 > 0$, and let $(k+1)$ be the maximal order of poles located on the line $Rez = \alpha$. Let the assumptions (1) and (2) of the above theorem hold. Further assume that*

$$A(t) = q_1(t)(t+1)^k + \int_0^t Q_1(t,s,)exp(-\alpha(t-s))(s+1)^k ds \in L^1[0,\infty)$$

$$B(t) = q_2(t)(t+1)^{k(1+\alpha)} + \int_0^t Q_2(t,s,)exp(-\alpha(t-s)(1+\alpha))(s+1)^{k(1+\alpha)}ds$$

be bounded on $[0,\infty]$. *Then, the trivial solution of quasilinear InDS (10) is unstable.*

2.6. Linear Barbashin Type InDE

We discuss the stability and boundedness of a linear InDE of Barbashin type from [10]. Consider the following linear InDE of Barbashin type given by

$$\frac{\partial y(t,x)}{\partial t} = c(t,x)y(t,x) + \int_0^1 h(t,x,s)y(t,s)ds + f(t,x), t > 0, 0 \leq x \leq 1 \quad (11)$$

where $c(.,.) : [0,\infty) \times [0,1] \to \mathbb{R}$, $h(.,.,.) : [0,\infty) \times [0,1]^2 \to \mathbb{R}$ and $f(.,.) : [0,\infty) \times [0,1] \to \mathbb{R}$ are given functions and $y(.,.)$ is unkown.

This equation is transformed into an operator equation in a suitable space as follows.

Let $C(t)$ be such that $(C(t)\hat{y}(t))(x) = c(t,x)y(t,x)$ and $H(t)$ be such that $(H(t)\hat{y}(t))(x) = \int_0^1 h(t,x,s)y(t,s)ds$ and $B(t) = C(t) + H(t)$.

Then the operator equation

$$u'(t) = A(t)u(t), \ (t \geq 0),$$

is considered in a Hilbert space and apriori estimates are obtained on A and are then extended to $B(t)$. In order to do so, the following assumptions are made.

1. For almost all $x, s \in [0,1], c(t,x)$ and $h(t,x,s)$ have bounded measurable derivatives in t, $c_t'(t,x)$ and $h_t'(t,x,s)$.
2. The operators $B(t)$ and $B'(t)$ are $\in L^2$ and are defined as follows. For all $x \in [0,1]$ and $w \in L^2$,

$$B(t)w(x) = c(t,x)w(x) + \int_0^1 h(t,x,s)w(s)ds,$$

$$B'(t)w(x) = c_t'(t,x)w(x) + \int_0^1 h_t'(t,x,s)w(s)ds,$$

   are assumed to be bounded uniformly in $t, t \in [0,\infty)$.
3. The Frobenius norm or the Hilbert-Schmidt norm,

$$N_2(H_I(t)) := (\int\int_0^1 (h(t,x,s) - h(t,s,x))^2 dsdx)^{\frac{1}{2}} < \infty.$$

4. $\sup_{t\geq 0}\alpha(B(t)) < 0$, where $\alpha(B(t)) = \sup_{s\in\sigma(B(t))} Re\, s$, and $\sigma(B(t))$ is the spectrum of $B(t)$ and $H_I(t) = \frac{H(t)-H^*(t)}{2i}$ where $H^*(t)$ is the adjoint operator of $H(t)$.

The homogeneous equation corresponding to the linear Barbashin InDE (11) is given by,

$$\frac{\partial y(t,x)}{\partial t} = c(t,x)y(t,x) + \int_0^1 h(t,x,s)y(t,s)ds \quad (t>0,\ 0\leq x\leq 1), \tag{12}$$

with initial condition $y(0,x) = y_0(x),\ 0\leq x\leq 1$ and $y_0 \in L^2$ is given.

Definition 7. *The equation (12) is said to be exponentially stable if there exist positive constants M and β such that any solution $y(t,.)$ of (11) and (12) satisfies $||y(t,0)||_{L^2} \leq Me^{-\beta t}||y_0||_{L^2}$ $(t\geq 0)$.*

Theorem 11. *Let assumption (3) and the condition*

$$sup_{t\geq 0}\,\mu^2(t)||B'(t)||_{L^2} < 2$$

hold, where

$$\mu(t) = \sum_{j,k=1}^{\infty} \frac{N_2^{j+k}(H_I(t))(k+j)!}{2^{(j+k)/2}|\alpha(B(t))|^{j+k+1}(j!k!)^{3/2}}. \tag{13}$$

Then, equation (12) is exponentially stable. Further, if $||f(t,.)||_{L^2}$ is bounded on $[0,\infty)$, then any solution of (11) is bounded on $[0,\infty)$.

2.7. Impulsive InDS

Now we consider an impulsive InDE and proceed to present the stability results from [11]. The stability results are obtained by using the solution of the corresponding integral equation and the corresponding impulsive differential system.

$$\frac{dy(t)}{dt} = ay(t) + \int_0^{\infty} h(s)y(t-s)ds, t\neq \tau_i \tag{14}$$

$$y(s) = \phi(s), s\leq 0, \tag{15}$$

$$y/_{t=\tau_i} = p_i\phi(s),\ \ 0\leq p_i\leq p,\ \ i=1,2,3.... \tag{16}$$

$0<\tau_1<\tau_2<...<\tau_i<...,\ \tau_i\to\infty$ as $t\to\infty$, and $\tau_{i+1}-\tau_i\geq\theta>0,\ i=0,1,2,...$ where ϕ is bounded and continuous on $(-\infty,0]$ and a, b and $p=\max\{p_i\}, i=1,2,3...$ are real numbers. The kernel $h:[0,\infty)\to[0,\infty)$ is bounded and continuous on $[0,\infty)$ and is such that $\int_0^\infty h(s)ds<\infty$ and $\int_0^\infty sh(s)ds<\infty$.

The solution of system (14) - (16) can be obtained as the solution of the corresponding integral equation given by

$$y(t) = W(t,0)\phi(0) + b\int_0^t W(t,s)(\int_0^\infty h(u)y(s-u)du)ds, \tag{17}$$

where $W(t,s)$ is the fundamental solution of the associated unperturbed differential system, that is, system (14) (16) - without the integral term.

The following is a stability result in this set up.

Theorem 12. *Suppose the delay kernel h, the parameters a, b, θ and p are such that*

$$a - \frac{1}{\theta} ln(1+p) > |b|(1+p)\int_0^\infty h(s)ds,$$

then, the trivial solution of the impulsive InDS (14) -(16) is asymptotically stable.

3. Part II - Nonlinear Integro-Differential Equations

3.1. Introduction

In the second part of this chapter, stability properties of nonlinear integro-differential equations are presented. It is well known that Lyapunov's second method is one of the basic tools for the study of the qualitative behavior of the solutions of nonlinear differential equations. In extending this method to Volterra integro-differential equations, one can either use Lyapunov functionals or Lyapunov functions. The use of Lyapunov functionals requires prior knowledge of the solutions, whereas the use of Lyapunov functions requires choosing appropriate minimal subsets of functions in $C[\mathbb{R}, \mathbb{R}^n]$ along which the derivative of the Lyapunov function admits a convenient estimate [12 - 14].

However, the Lyapunov functionals used in applications are usually a combination of a Lyapunov function and a Lyapunov functional, constructed so that the corresponding derivative can be estimated suitably without requiring the minimal classes of functions or prior knowledge of solutions [4]. The development of the method of Lyapunov functions on product spaces is a consequence of this fact.

The stability and boundedness properties of solutions of InDEs were studied by many authors. Some of the early results developed using Lyapunov's second method are those of Miller [5], Grossman and Miller [15], Burton and Mahfoud [16], Mahfoud [17], Rama Mohana Rao and Srinivas [18], Lakshmikantham and Rama Mohana Rao [19], Hara, Yoneyama, and Itoh [7], Elaydi and Sivasundaram [20], Elaydi [21], Zhang [22], and Vanualailai and Nakagiri [23]. Numerous references of more recent results can be found in C.Tunc and O. Tunc [24, 25], Andreev and Sedova [13] and Sedova [26]. The work we present in this section uses the comparison method introduced in [27].

3.2. Lyapunov Functionals

3.2.1. Existence of Lyapunov Functionals

In this subsection we present results dealing with the existence of a Lyapunov functional for the following linear InDS

$$y'(t) = C(t)y(t) + \int_{t_0}^{t} H(t,s)y(s)ds,\ y(t_0) = y_0, \tag{18}$$

and the nonlinear InDS

$$y'(t) = \int_{t_0}^{t} H(t,s)G(s,y(s)ds,\ y(t_0) = y_0, \tag{19}$$

where $C(t), H(t,s)$ are $n \times n$ matrices defined on $t_0 \leq t < \infty$ and $t_0 \leq s \leq t < \infty$, $t_0 \geq 0$.

The following definition is necessary to introduce the Lyapunov function.

Definition 8. *A function $p \in C[\mathbb{R}_+, \mathbb{R}_+]$ is said to belong to the class $\mathbb{K}$ if $p(0) = 0$ and $p(t)$ is monotonically increasing in t.*

We now present the concept of exponential asymptotic stability

Definition 9. *The trivial solution of InDS (19) is generalized exponentially asymptotically stable (GEAS) if*

$$||y(t,t_0,y_0)|| \leq L(t)||y_0||\, exp[P(t_0) - P(t)]$$

where $L(t) > 0$ and is continuous for $t \in [t_0, \infty)$, $P \in \mathbb{K}$ and $P(t) \to \infty$ as $t \to \infty$.

Remark 2. *If $L(t) = L > 0$ and $P(t) = at$, with $a > 0$, then, the trivial solution of (18) is exponentially stable.*

We next present the following theorem from [28] dealing with the existence of a Lyapunov function.

Theorem 13. *Assume that the trivial solution of (19) is GEAS. Further suppose that there exists a function $P'(t)$ which is continuous on $J = [t_0, \infty)$ (say). Then, there exists a function $V(t,x)$ satisfying,*

1. $V \in C[J \times S_\rho, \mathbb{R}]$ *and* $|V(t,x) - V(t,y)| \leq L(t)||x-y||, t \in J, x, y \in S_\rho$,
2. $||x|| \leq V(t,x) \leq L(t)||x||,\ (t,x) \in J \times S_\rho$,
3. $D^-V(t,x) \leq -P'(t)V(t,x),\ (t,x) \in J \times S_\rho$.

Theorem 14. *Assume that the trivial solution of* (19) *is GEAS. Further suppose that,*

$$\sup_{t\geq 0}\{\int_0^t ||H(t,s)||ds\colon 0 \leq s \leq t < \infty)\} \leq M$$

and

$$||G(s,y)-G(s,x)|| \leq K(t,r)||y-x||, t\in J, x,y\in S_\rho,$$

where $K(t,r) > 0$ *such that* $\int_{t_0}^{\infty} K(s,r)ds = N$. *Then, there exists a function* $V(t,x)$ *satisfying:*

1. $V \in C[J\times S_\rho, \mathbb{R}]$ *and* $|V(t,x)-V(t,y)| \leq L(t)||x-y||, t\in J, x,y\in S_\rho,$
2. $||x|| \leq V(t,x) \leq L(t)||x||$, $(t,x)\in J\times S_\rho$
3. $D^- V(t,x) \leq -P'(t)V(t,x)$, $(t,x)\in J\times S_\rho$.

3.2.2. Lyapunov Functional for a Specific Scalar InDE

In this subsection, we will consider a scalar InDE corresponding to (18) and construct a Lyapunov functional following the work in [4]. Let

$$y'(t) = cy + \int_{t_0}^{t} h(t-s)y(s)ds, \tag{20}$$

where c is a constant and $h(t)$ is continuous for $0 \leq t < \infty$, with a strong sign condition.

Remark 3. *The definitions of stability concepts of InDS* (1) *are given in Section 2.1. Those definitions can be easily modified to define these concepts for scalar InDE* (20).

We now proceed to state the following results from [4, 16] and use a result involving a Lyapunov functional to prove stability results for (20).

Theorem 15. *Suppose that* $c < 0$, $h(t) > 0$ *and* $c+\int_0^{\infty} h(t)dt \neq 0$. *Then, the following statements are equivalent.*

1. *All the solutions of* (20) *tend to zero;*
2. $c+\int_0^{\infty} h(t)dt < 0$;
3. *Each solution of* (20) *is in* $L^1(\mathbb{R}_+)$;
4. *The zero solution of* (20) *is uniformly asymptotically stable;*
5. *The zero solution of* (20) *is asymptotically stable.*

Proof. We shall only prove 2 $\implies$ 3. Let $y(t,t_0,\phi)$ be any solution of the InDE (20) with the initial condition $\phi(t) = 2$ on $[0,t_0]$. We choose a Lyapunov functional in this set up as

$$V(t,y(.)) = |y| + \int_0^t \int_t^\infty h(\tau - s)d\tau|y(s)|ds.$$

Let $y(t)$ be a solution of the scalar InDE (20) and $y(t)$ not equal to the zero solution, then

$$\begin{aligned} V'(t,y(.)) &\leq c|y| + \int_0^t h(t-s)|y(s)|ds + \int_t^\infty h(\tau - t)d\tau|y| - \int_0^t h(t-s)|y(s)|ds \\ &= [c + \int_t^\infty h(\tau - t)d\tau]|y| \leq [c + \int_0^\infty h(\tau)d\tau]|y| = -\beta|y|, \end{aligned}$$

for some constant $\beta > 0$. This yields

$$0 \leq V(t,y(.)) \leq V(t_0,\phi(.)) - \beta\int_{t_0}^t |y(s)|ds,$$

which implies that $\int_{t_0}^\infty |y(s)|ds < \infty$, meaning that $y \in L^1(\mathbb{R}_+)$, completing the proof.

If we relax the condition on $h(t)$, we can modify the Lyapunov functional suitably to get asymptotic stability of the trivial solution of (20). We state this result. For a proof see [4, 16].

Theorem 16. *Suppose that in Theorem* 15, $c + \int_0^\infty |h(t)|dt < 0$. *Then, the trivial solution of scalar InDE* (20) *is uniformly asymptotically stable.*

To prove this theorem, choose the Lyapunov functional as

$$V(t,y(.)) = y^2 + \int_0^t \int_t^\infty |h(\tau - s)|d\tau|y(s)|ds,$$

and obtain an estimate for the Lyapunov functional as in the earlier theorem.

3.2.3. Lyapunov Functional for a General Scalar InDE

Next, we shall consider a more general scalar InDE [4, 16] and discuss the stability concepts using a Lyapunov functional. To that end, let

$$y'(t) = c(t)y + \int_{t_0}^t h(t,s)y(s)ds, \; t \in \mathbb{R}_+ \tag{21}$$

Theorem 17. *Assume that*

1. $c : \mathbb{R}_+ \to \mathbb{R}_+$ *and* h *are continuous for* $0 \leq s \leq t < \infty$;

2. *the integral* $\int_t^\infty |h(\tau,t)|d\tau$ *is defined and finite for all* $t \geq 0$, *and suppose there exists a* $\beta \in \mathbb{R}_+$ *such that*

$$\int_0^t |h(t,s)|ds + \int_t^\infty |h(\tau,t)|d\tau - 2|c(t)| \leq -\beta.$$

Then, the trivial solution of the InDE (21) is stable if and only if $c(t) < 0$.

Proof. Suppose that $c(t) < 0$ and consider the Lyapunov functional given by

$$V(t,y(.)) = y^2 + \int_0^t \int_t^\infty |h(\tau,s)|d\tau y^2 ds.$$

Then, we obtain that $V'(t,y(.)) \leq -\beta y^2$. Observing that V is positive definite and that $V'(t,y(.)) \leq 0$, we conclude that the solution of InDE (21) is stable. For the second part of the proof refer to [4, 16].

3.3. Method of Lyapunov Functions

In this section, we discuss the stability properties of a general class of InDEs using the method of Lyapunov functions. A Lyapunov function can be viewed as a transformation which reduces the study of a given complicated differential system to the study of a scalar differential equation. In the case of InDEs, this method relies on carefully selecting appropriate minimal subsets of functions in $C[\mathbb{R},\mathbb{R}^n]$, along which the derivative of the Lyapunov function admits a convenient estimate.

3.3.1. Stability Using Minimal Classes

Consider the InDS given by,

$$y'(t) = G(t,y(t),(Ty)(t)),\ t \geq t_0,\ y(t_0) = y_0, \tag{22}$$

where $(Ty)(t) = \int_{t_0}^t g(t,s,y(s))ds$, $G \in C[\mathbb{R}_+ \times S(\rho) \times \mathbb{R}^n, \mathbb{R}^n]$ and $g \in C[\mathbb{R}_+ \times \mathbb{R}_+ \times S(\rho), \mathbb{R}^n]$. Assume that $G(t,0,0) \equiv 0$ and $g(t,s,0) \equiv 0$, so that the InDS (22) admits the trivial solution.

As mentioned earlier, the estimate of the derivative of the Lyapunov function along continuous functions in the minimal class is used to construct the scalar differential equation mentioned above [4]. Then, the study of the behavior of solutions of the InDE reduces to the study of the behavior of the solutions of this scalar differential equation. For this purpose, let

$V \in C[\mathbb{R}_+ \times S(\rho), \mathbb{R}_+]$, and for $y \in C[\mathbb{R}_+, \mathbb{R}^n]$,

$$D_-V(t,y(t)) = \liminf_{h \to 0^-} \frac{1}{h}[V(t+h,y(t)+hG(t,y(t),(Ty)(t))) - V(t,y(t)).$$

Now we proceed to introduce three subsets of $C[\mathbb{R},\mathbb{R}^n]$ along which we estimate $D_-V(t,y(t))$ subject to the demands we impose on the solutions of the InDS (22).

1. $E_\alpha = \{y \in C[\mathbb{R}_+,\mathbb{R}^n] : V(s,y(s))\alpha(s) \leq V(t,y(t))\alpha(t),\ \ t_0 \leq s \leq t\}$;
2. $E_1 = \{y \in C[\mathbb{R}_+,\mathbb{R}^n] : V(s,y(s)) \leq V(t,y(t)),\ \ t_0 \leq s \leq t\}$;
3. $E_0 = \{y \in C[\mathbb{R}_+,\mathbb{R}^n] : V(s,y(s)) \leq f(V(t,y(t))),\ \ t_1 \leq s \leq t,\ t_1 \geq t_0\}$;

where $\alpha(t) > 0$ is a continuous function on R_+ and $f(r)$ is continuous on R_+, nondecreasing in r and $f(r) > r$ for $r > 0$.

We next present comparison theorems that are useful in establishing the results stated later [4, 19].

Theorem 18. *Let $V \in C[\mathbb{R}_+ \times S(\rho),\mathbb{R}_+]$ and $V(t,y)$ be Lipschitzian in y. Assume that for $t \geq t_0$ and $y \in E_1$,*

$$D_-V(t,y(t)) \leq p(t,V(t,y(t)), \tag{23}$$

where $p \in C[\mathbb{R}_+ \times \mathbb{R}_+,\mathbb{R}_+]$. Let $r(t) = r(t,t_0,u_0)$ be the maximal solution of the scalar ordinary differential equation:

$$u' = p(t,u),\ u(t_0) = u_0 \geq 0, \tag{24}$$

existing on $t_0 \leq t < \infty$. Let $y(t) = y(t,t_0,y_0)$ be any solution of the InDS (22) such that $y(t) \in S(\rho)$ for $t \in [t_0,t_1]$ and satisfying

$$V(t_0,y(t_0)) \leq u_0. \tag{25}$$

Then,

$$V(t,y(t)) \leq r(t) \text{ for all t} \in [\mathrm{t_0},\mathrm{t_1}]. \tag{26}$$

Proof. We briefly sketch the proof of the theorem. Let $y(t) = y(t,t_0,y_0)$ be any solution of the InDS (22) such that $y(t) \in S(\rho)$ for $t \in [t_0,t_1]$. Define $m(t) = V(t,y(t))$. For sufficiently small $\varepsilon > 0$, consider ε increment - scalar ODE,

$$u' = p(t,u)+\varepsilon,\ u(t_0) = u_0+\varepsilon, \tag{27}$$

whose solutions $u(t,\varepsilon)$ exist as far as $r(t)$ exists to the right of t_0.

Using the hypothesis on p, we note that the solutions of the scalar ODE (24) and the solutions, $u(t,\varepsilon)$, of the ε increment - scalar ODE (27), are monotonically increasing functions of t. We establish, using $m(t)$ and $u(t,\varepsilon)$, that $y(t) \in E_1$ for $t_0 \leq t \leq t_2$. Using the fact that $V(t,y)$ is Lipschitzian in y and the relation between V and p, we conclude that $V(t,y(t)) \leq r(t)$, completing the proof.

For a detailed proof of the theorem and the following results see [4, 29]. The following theorem deals with the solutions of InDS (22) in the set E_0.

Theorem 19. *Let $V \in C[\mathbb{R}_+ \times S(\rho), \mathbb{R}_+]$ and $V(t,y)$ be Lipschitzian in y. Assume that for $t \geq t_0$ and $y \in E_0$,*

$$D\_V(t,y(t)) \leq 0$$

Let $y(t) = y(t,t_0,y_0)$ be any solution of the InDS (22) such that $y(t) \in S(\rho)$ for $t \in [t_0,t_1]$. Then,

$$V(t,y(t)) \leq V(t_0,y(t_0)), \text{ for all } t \in [t_0,t_1]. \tag{28}$$

Proof. Proceeding as in Theorem 18 with $p \equiv 0$, we arrive at the inequality $V(s,y(s)) \leq V(t_2,y(t_2))$, $t_2 \in (t_0,t_1)$. Since $V(t_2,y(t_2)) \leq V(t_0,y(t_0)) + \varepsilon + \varepsilon(t_2 - t_0)$, from the assumption on $f(r)$ we get $V(s,y(s)) \leq f(V(t_2,y(t_2)))$, which means that $y(t) \in E_0$ for $t_0 \leq s,t \leq t_2$. The remaining proof follows the same pattern as in Theorem 18.

The next result is more general than Theorem 18 and is useful to establish the asymptotic stability of the trivial solution of InDS (22). The proof of the theorem is given below to draw attention to a subtle point in the proof.

Theorem 20. *Assume that the hypotheses of Theorem 18 hold except that the inequality (23) is replaced by*

$$\alpha(t)D\_V(t,y(t)) + V(t,y(t))D\_\alpha(t) \leq p(t,V(t,y(t)\alpha(t)), \tag{29}$$

for $t > 0$ and $y \in E_\alpha$, where $\alpha(t) > 0$, is continuous on $\mathbb{R}_+$ and

$$D\_\alpha(t) = \lim_{h \to 0^-} \inf \frac{1}{h}[\alpha(t+h) - \alpha(t)].$$

Then, $\alpha(t_0)V(t_0,y_0) \leq u_0$ implies that $\alpha(t)V(t,y(t)) \leq r(t)$, $t \geq t_0$.

Proof. Let $L(t,y(t)) = \alpha(t)V(t,y(t))$, then,
$L(t+h,y(t)+hG(t,y(t),(Ty)(t))) - L(t,y(t)) = V(t+h,y(t) +hG(t,y(t),(Ty)(t)))(\alpha(t+h) - \alpha(t)[V(t+h,y(t)+hG(t,y(t),(Ty)(t))) - V(t,y(t))]$.
Now using the relation (28), we obtain,

$$D\_L(t,y(t)) \leq p(t,L(t,y(t)), \text{for } t \in [t_0,t_1], y \in E_1. \tag{30}$$

Observe that L replaces V in E_1, and also that L is Lipschitzian in y. Hence, all the assumptions of Theorem 18 are satisfied and the conclusion of the theorem follows from the proof of Theorem 18.

The following results provide sufficient conditions for the stability and asymptotic stability of the trivial solution of InDS (22).

Theorem 21. *Assume that there exist functions $V(t,y)$ and $p(t,y)$ satisfying the following conditions:*

1. *$V \in C[\mathbb{R}_+ \times S(\rho), \mathbb{R}_+]$, and for $y \in C[\mathbb{R}_+, \mathbb{R}^n]$, $V(t,0) \equiv 0$, $V(t,y)$ is positive definite and locally Lipschitzian in y;*
2. *$p \in C[\mathbb{R}_+ \times \mathbb{R}_+, \mathbb{R}_+]$ and $p(t,0) \equiv 0$*
3. *$\alpha(t) > 0$ is continuous for $t \in \mathbb{R}_+$ and $\alpha(t) \to \infty$ as $t \to \infty$*
4. $$\alpha(t)D_-V(t,y(t)) + V(t,y(t))D_-\alpha(t) \leq p(t,V(t,y(t)\alpha(t))), \tag{31}$$
   for $t > t_0$ and $y \in E_\alpha$.

Then, stability of the trivial solution of scalar ODE (24) implies the asymptotic stability of the trivial solution of InDS (22).

The following theorem deals with the uniform asymptotic stability of the trivial solution of InDS (22)

Theorem 22. *Let $V \in C[\mathbb{R}_+ \times S(\rho), \mathbb{R}_+]$ and $V(t,y)$ be Lipschitzian in y. Assume that $t \geq t_0$, $y \in E_0$, and*

$$D_-V(t,y(t)) \leq -\phi(y(t)), \text{ where, } \phi \in \mathbb{K}. \tag{32}$$

Then, the trivial solution of InDS (22) is uniform asymptotically stable.

Remark 4. *All the results that were established in this section deal with InDS (22) with the initial condition as a value, that is, $y(t_0) = y_0$. However, the InDS (22) can be considered with the initial value replaced with a continuous function, representing the past history of the system. Thus, we have $y(t_0) = \phi$, for $t \geq 0$ and $\phi : [0,t_0] \to \mathbb{R}^n$ is continuous function. All the above theorems can also be proved in this case with suitable modifications.*

3.4. Lyapunov Functions on Product Spaces

As mentioned in the introduction, the Lyapunov functionals used in applications are usually a combination of a Lyapunov function and a Lyapunov functional, constructed so that the corresponding derivative can be estimated suitably without requiring the minimal class of functions or prior knowledge of solutions. The development of the method of Lyapunov functions on product spaces [4, 30] is a consequence of this fact.

Let $y_t(.) \in C[[0,t], \mathbb{R}^n]$. Suppose $y \in C[\mathbb{R}_+, \mathbb{R}^n] = C(\mathbb{R}_+)$, then for each $t \in \mathbb{R}_+$, $y_t(.)$ is the restriction of $y(s)$ given by $y_t(s) = y(s), 0 \leq s \leq t$, and $|y_t(.)|$ is the norm defined by

$$|y_t(.)| = \sup_{0 \leq s \leq t} |y(s)|.$$

Observe that $|y(t)| \leq |y_t(.)|$.

Consider a system of the form.

$$y'(t) = G(t, y(t), y_t(.)),\ t \in \mathbb{R}_+, \tag{33}$$

where $G \in C[\mathbb{R}_+ \times \mathbb{R}^n \times C(\mathbb{R}_+), \mathbb{R}^n]$. Assume that $G(t,0,0) \equiv 0$ and let $y(t) = y(t, t_0, y_{t_0}(.))$ be a solution of the system (32) with initial values $(t_0, y_{t_0}(.))$. A special case of the system (32) is the Volterra InDE given below:

$$y'(t) = c(t, y(t)) + \int_{t_0}^{t} h(t, s, y(s))ds,\ t \in \mathbb{R}_+, \tag{34}$$

where $c \in C[\mathbb{R}_+ \times \mathbb{R}^n, \mathbb{R}^n]$, $h \in C[\mathbb{R}_+ \times \mathbb{R}_+ \times C(\mathbb{R}_+), \mathbb{R}^n]$, $c(t,0) \equiv 0$ and $h(t,s,0) \equiv 0$.

We now proceed to employ Lyapunov functions on the product space $\mathbb{R}^n \times C(\mathbb{R}_+)$ and present some results pertaining to the stability of the system (32).

Consider $V \in C[\mathbb{R}_+ \times \mathbb{R}^n \times C(\mathbb{R}_+), \mathbb{R}^n]$, and suppose that $V(t,y)$ is positive definite and locally Lipschitzian in y. Let $y(t) = y(t, t_0, y_{t_0}(.))$ be a solution of the system (32). Now define,

$$D^+V(t, y, y_t(.)) = \lim_{h \to 0^+} \sup \frac{1}{h}[V(t+h, y+hG(t, y, y_t(.)), y_{t+h}(.)) - V(t, y, y_t(.))].$$

We shall now give sufficient conditions guaranteeing asymptotic stability of the trivial solution of the system (32).

Theorem 23. *Assume that there exists a function $V(t, y, y_t(.)) \in C[\mathbb{R}_+ \times S(\rho) \times C(\mathbb{R}_+), \mathbb{R}_+]$, such that*

1. $a(|y(t)|) \le V(t, y, y_t(.)) \le b(|y_t(.)|)$, *where* $a, b \in \mathbb{K}$;
2. $D^+V(t, y, y_t(.)) \le 0$

Then, the trivial solution of system (32) is uniformly stable.

Theorem 24. *Assume that there exists a function $V(t, y, y_t(.)) \in C[\mathbb{R}_+ \times S(\rho) \times C(\mathbb{R}_+), \mathbb{R}_+]$, such that*

1. $a(|y(t)|) \le V(t, y, y_t(.)) \le b(|y_t(.)|)$, *where* $a, b \in \mathbb{K}$;
2. $D^+V(t, y, y_t(.)) \le -c(|y_t(.)|)$, *where* $c \in \mathbb{K}$;

Then, the trivial solution of system (32) is uniformly asymptotically stable.

3.4.1. Stability in Two Measures

The theory of stability in terms of two measures [31, 32] unifies a variety of known stability concepts in a single set up. The next theorem gives a stability result using this approach. We begin with the following definitions.

Definition 10. *1.* $\Gamma = \{\hat{h} \in C[\mathbb{R}_+ \times \mathbb{R}^n, \mathbb{R}_+] : \inf_{y\in\mathbb{R}^n} \hat{h}(t,y) = 0 \textit{ for each } t \in \mathbb{R}_+\}$;

2. *For* $\hat{h}, h^0 \in \Gamma$ *and* $y_r(.) \in C(\mathbb{R}_+))$, *let* $h_0(t, y_t(.)) = \sup_{0\le s\le t} h^0(s, y(s))$ *and* $\tilde{h}(t, y_t(.)) = \sup_{0\ge s\ge t} \hat{h}(s, y(s))$.

It is clear that $h^0(s, y(s)) \le h_0(t, y_t(.))$ and $\tilde{h}(t, y_t(.)) \ge \hat{h}(s, y(s))$.

Definition 11. *Let* $\hat{h}, h^0 \in \Gamma$ *and let* $\tilde{h}, h_0$ *be as defined above. Then,* h_0 *is said to be finer than* $\tilde{h}$ *if there exist a* $\lambda > 0$ *and* $\phi \in \mathbb{K}$ *such that* $h_0(t, y_t(.)) \le \lambda$ *implies* $\tilde{h}(t, y_t(.)) \le \phi(h_0(t, y_t(.)))$.

Next, we define the criteria for V to be positive definite and decrescent in this setup.

Definition 12. *Consider* $V \in C[\mathbb{R}_+ \times \mathbb{R}^n \times C(\mathbb{R}_+), \mathbb{R}^n]$. *Then,* V *is said to be*

1. $\hat{h}$*-positive definite if there exist* $\rho > 0$ *and* $a \in \mathbb{K}$ *such that* $\hat{h}(t,y) < \rho$ *implies* $a(\hat{h}(t,y)) \le V(t, y, y_t(.))$;
2. h_0*-decrescent if there exist* $\rho_0 > 0$ *and* $b \in \mathbb{K}$ *such that* $h_0(t,y) < \rho_0$ *implies* $V(t, y, y_t(.)) \le b(h_0(t, y_t(.)))$.

We now proceed to define the stability concepts in the set up of two measures. We present the stability definition. All other concepts of stability and boundedness can be defined in a similar fashion.

Definition 13. *The trivial solution of system (32) is said to be*$(h_0 h - \hat{h})$ *uniformly stable, if given* $\varepsilon > 0$ *and* $t_0 \in \mathbb{R}_+$, *there exists a* $\delta = \delta(\varepsilon) > 0$ *such that* $t_0 \ge 0$ *and* $h_0(t_0, y_{t_0}(.)) < \delta$ *implies* $h(t, y(t)) < \varepsilon$, *for all* $t \ge t_0$.

To illustrate the generality of this concept, we give a couple of examples. For more details see [4,32].

1. If $h_0(t,y) = \hat{h}(t,y) = |y|$, which implies $h_0(t, y_t(.)) = |y_t(.)|$, where $|y|$ is the usual Euclidean norm and $|y_t(.)| = \sup_{0\le s\le t} |y(s)|$, then we obtain the well known uniform stability result for the trivial solution of system (32);
2. Let $z(t) = z(t, t_0, z_{t_0})$ be any prescribed solution of system (32). If $\hat{h}(t,y) = |y - z(t)|$, and $h_0(t, y_t(.)) = |y_t(.) - z_t(.)|$, where $|y|$ is the usual Euclidean norm and $|y_t(.)| = \sup_{0\le s\le t} |y(s)|$, then we obtain the stability of the prescribed solution of system (32).

As stability is a local concept, we describe the ρ-ball in this set up.

For any $h \in \Gamma$, define $S(\hat{h}, \rho) = \{(t,y) : \hat{h}(t,y) < \rho\}$ and for $h^* \in C[\mathbb{R}_+ \times C(\mathbb{R}_+), \mathbb{R}_+]$, let $S(h^*, \rho) = \{(t, y_t(.)) : h^*(t, y_t(.)) < \rho\}$. We state below a generalization of Lyapunov's second theorem in terms of two measures.

Theorem 25. *Assume that*

1. $\hat{h}, h^0 \in \Gamma$ *and* h_0 *is finer than* $\tilde{h}$;
2. *there exists a function* $V(t,y,y_t(.)) \in C[\mathbb{R}_+ \times S(\hat{h},\rho) \times C(\mathbb{R}_+), \mathbb{R}_+]$ *that is locally Lipschitzian in y,* $\hat{h}-$ *positive definite and* h_0- *decrescent;*
3. $D^+V(t,y,y_t(.)) \leq -\eta(h_0(t,y_t(.)))$, *for all* $(t,y,y_t(.) \in \mathbb{R}_+ \times S(\hat{h},\rho) \times S(h_0,\rho_0)$

Then, the trivial solution of the system (32) is $(h_o,\hat{h})-$ *uniformly asymptotically stable.*

3.4.2. Practical Stability for InDE

Consider the following initial value problem [33]

$$y'(t) = G(t,y(t),(Ty)(t)),\ t \geq t_0,\ y(t_0) = y_0, \tag{35}$$

where $(Ty)(t) = \int_{t_0}^{t} g(t,s,y(s))ds$, $G \in C[\mathbb{R}_+ \times \mathbb{R}^n \times \mathbb{R}^n, \mathbb{R}^n]$ and $g \in C[\mathbb{R}_+ \times \mathbb{R}_+ \times \mathbb{R}^n, \mathbb{R}^n]$.

In this subsection, we present a practical stability result using the comparison system [33]. In this case, the Lyapunov function V will serve as a vehicle to reduce the complex system to a simpler one. Then, the practical stability of the complex system is obtained through the practical stability of the corresponding scalar ODE. We assume the following hypotheses:

1. $V \in C[\mathbb{R}_+ \times \mathbb{R}^n, \mathbb{R}_+]$, $V(t,y)$ is locally Lipschitzian in y and

   $a(|y(t)|) \leq V(t,y) \leq b(|y(t)|)$, where $a,b \in \mathbb{K}$;

2. $f_0, f \in C[\mathbb{R}_+ \times \mathbb{R}_+, \mathbb{R}]$, $f_0(t,w) \leq f(t,w)$, $r(t,t_0,w_0)$ is the right maximal solution of

   $$w' = g(t,w), w(t_0) = w_0 \geq 0,$$

   existing on $[t_0,\infty)$ and $\mu(t,t^0,u^0)$ is the right maximal solution of

   $$u' = g_0(t,u), u(t^0) = u^0 \geq 0,\ t^0 > t_0, \tag{36}$$

   existing on $t_0 \leq t \leq t^0$;

3. $D\_V(t,y(t)) = \lim_{h\to 0^-} \inf \frac{1}{h}[V(t+h,y(t)+hG(t,y(t),(Ty)(t))) - V(t,y(t))$, such that,

   $$D^+V(t,y) \leq f(t,V(t,y)),\ (t,y) \in \Omega,$$

   where $\Omega = \{y \in C[\mathbb{R}_+,\mathbb{R}^n] : V(s,y(s)) \leq \mu(s,t,V(t,y(t)),t_0 \leq s \leq t\}$.

We now state a general comparison theorem used in establishing the practical stability results of the system (35).

Theorem 26. *Assume that all the forementioned hypotheses hold. Let $y(t,t_0,y_0)$ be any solution of system (35), such that $V(t_0,y_0) \leq w_0$. Then, $V(t,y(t,t_0,y_0)) \leq r(t,t_0,w_0)$, $t \geq t_0$.*

Having the comparison theorem at our disposal, we are in a position to state a very general result that establishes criteria for various types of practical + concepts.

Theorem 27. *Assume that all the forementioned hypotheses hold. Then, the practical stability properties of system (36) imply the corresponding practical stability properties of (35).*

Conclusion

This chapter dealt with the stability of various types of linear and nonlinear integro-differential equations. First, several approaches to the study the stability of various types of linear integro-differential equations were given. Next, stability results for nonlinear Volterra integro-differential equations, obtained using the comparison method, were presented. In the comparison method Lyapunov or Lyapunov-like functions and comparison theorems are used to reduce the study of a complicated system to the study of a simpler system. Different types of stability notions such as practical stability and stability in terms of two measures were also discussed.

References

[1] Rouche, N., Habets, P. and Laloy, M. *Stability Theory by Liapunov's Direct Method*, Springer-Verlag, New York, 1977.

[2] Lakshmikantham, V., Leela, S. and Martynyuk, A. A., *Practical Stability of Nonlinear Systems*, World Scientific, Singapore,1990.

[3] Lakshmikantham, V. and Liu, X., *Stability Analysis in Terms of Two Measures*, World Scientific, Singapore,1993.

[4] Lakshmikantham, V. and Rama Mohana Rao, M., *Theory of Integro-differential Equations*, Gordon and Breach Publishers, Switzerland, 1995.

[5] Miller R.K., Asymptotic stability properties of linear Volterra integro-differential equations, *Journal of Differential Equations*, 10, 485-506, 1971.

[6] Hara,T., Yoncyama,T. and Itoh,T., Characterization of stability concepts of Volterra integro-differential equations, *Journal of Mathematical Analysis and Applications* Vol. 142, 558-572, 1989.

[7] Hara,T., Yoneyama,T. and Itoh,T., Asymptotic stability criteria for nonlinear Volterra integro-differential equations, *Funkcial. Ekvac.* Vol. 33, 39-57, 1990.

[8] M. N. Islam and Y. N. Raffoul, Stability in linear Volterra integrodifferential equations with nonlinear perturbation, *Journal of Integral Equations and Applications* Vol. 17, No. 3, 259-276, 2005.

[9] O.Yu. Khvorost and Z.B.Tsalyuk, Stability and Unstability of quasilinear systems of integrodifferential equations, *Russian Mathematics*, Vol. 51, No.11, 76-79, 2007.

[10] Michael Gil, On stability of linear Barbashin type integrodiffferential equations, *Mathematical Problems in Engineering* 2015, Article ID 962565. http//dx.doi.org/10.1155/2015/962565.

[11] Sariyasa, Stability of integrodifferential Equations with Impulses, *J. of Indonesian Mathematical Society*, Vol. 13, No. 1,17-26, 2007.

[12] V. Lakshmikantham and Leela, S., *Differential and Integral Inequalities*, Vol. 2, Academic Press, New York, 1969.

[13] A. S. Andreev and N. O. Sedova, The method of Lyapunov-Razumikhin functions in stability analysis of systems with delay, *Automation and Remote control*, Vol. 80, No. 7, 1185-1229, 2019.

[14] V. S. Sergeev, Stability of solutions of Volterra integrodifferential equations, *Mathematical and Computer Modelling*, Vol. 45, 1376-1394, 2007.

[15] Grossman S.I. and Miller R.K. Perturbation theory for Volterra integro-differential equations, *J. Differential Equations* 8, 457-474, 1970

[16] Burton, T. A. and Mahfoud,W.E. Stability Criteria for Volterra Equations, *Transactions of American Mathemtical Society*, 279, 143-174,1983.

[17] Mahfoud,W.E. Boundedness properties in Volterra integro-differential systems. *Proceedings in American Mathematical Society*, 100, 37-45, 1987 .

[18] Rama Mohana Rao, M. and Srinivas, P. Asymptotic behavior of solutions of Volterra integro-differential equations, *Proceedings of the American Mathematical Society*, 94, 55-60,1985.

[19] V. Lakshmikantham and Rama Mohana Rao, M., Stability in variation for nonlinear integro-differential equations, *Applicable Analysis*, 24, 165-173,1987.

[20] Elaydi, S and Sivasundaram, S., A unified approach to stability in integro-differential equations via Lyapunov functions, *J. Math. Anal. Appl.* 144, 503 - 531, 1989.

[21] Elaydi, S., Stability of integrodifferential systems of nonconvolution type, *Math. Inequal. Appl.* 1, 423-430, 1998.

[22] Zhang, B., Construction of Liapunov functionals for linear Volterra integro-differential equations and stability of delay systems, Electron. *J. Qual. Theory Differ. Equ.* 30,1-17, 2000.

[23] Vanualailal, J. and Nakagiri, S., Stability of a system of Volterra integro-differential equations, *Journal of Mathematical Analysis and Applications*, 281,602-619, 2003.

[24] Cemil Tunc and Osman Tunc, New qualitative criteria for solutions of Volterra integro-differential equations, *Arab Journal of Basic and Applied Sciences*, 25:3, 158-165,DOI: 10.1080/25765299.2018.1509554, 2018.

[25] Cemil Tunc and Osman Tunc, A note on the qualitative analysis of Volterra integro-differential equations, *Journal of Talibah University for Science*, 13:1, 490-496, 2019

[26] Natalia Sedova, On uniform asymptotic stability for nonlinear integro-differential equations of Volterra type, *Cybernetics and Physics*, Vol. 8, No. 3, 161-166, 2019.

[27] Lakshmikantham, V. and Leela, S., *Differential and Integral Inequalities*, Vol. 1, Academic Press, New York,1969.

[28] Kandil, F. M., A note on stability of integrodifferential equations, *International Mathematical Forum*, Vol. 5, No. 67, pp 3343-3348, 2010.

[29] Lakshmikantham, V. and Rama Mohana Rao, M. Integro-differential equations and extensions of Lyapunovs method, *Journal of Mathematical Analysis*, 30, 435-447, 1970.

[30] Rama Mohana Rao, M. and Sanjay, K. S., Lyapunov functions on product spaces and stability of integro-differential equations, *Dynamical Systems Applications* 1, 93-102, 1992.

[31] Leela, S. and Rama Mohana Rao, M. (h_0, h, M_0)-stability for integro-differential equations, *Journal of Mathematical Analysis and Applications*, 130, 460-468, 1988.

[32] Lakshmikantham, V. and Liu, X. Z., *Stability Analysis in terms of two measures*, World Scientific Publishing Co. Pte. Ltd. NJ, USA 1993.

[33] Lakshmikantham, V., Leela, S. and Martynyuk, A.A., *Practical Stability of Nonlinear Systems*, World Scientific Publishing Co. Pte. Ltd. Teaneck, NJ, USA 1993.

In: Understanding Integro-Differential Equations ISBN: 979-8-89113-040-1
Editors: J. Vasundhara Devi et al. ©2023 Nova Science Publishers, Inc.

Chapter 3

Advances in the Qualitative Theory of Integro–Differential Equations

Osman Tunç[1,*]
Seenith Sivasundaram[2,†]
and Cemil Tunç[3,‡]
[1]Department of Computer Programing, Baskale Vocational School, Van Yuzuncu Yil University, Van, Turkey
[2]Department of Mathematics, College of Science Engineering and Mathematics, Daytona Beach, Florida, USA
[3]Department of Mathematics, Faculty of Sciences, Van Yuzuncu Yil University, Campus, Van, Turkey

Abstract

This work investigates the advances from the past until now in the qualitative properties of solutions of linear and nonlinear integro – differential equations (IDEs). Here, we present an extensive literature on the qualitative properties of solutions, including asymptotic stability, uniform stability, instability and global uniform asymptotic stability of the zero solution, as well as boundedness, square integrability and existence of solutions to various linear and non-linear Volterra IDEs, without delay and with delay. We also present some applications of

* Corresponding Author's E-mail: osmantunc89@gmail.com
† Corresponding Author's E-mail: seenithi@gmail.com
‡ Corresponding Author's E-mail: cemtunc@yahoo.com

such equations in sciences and engineering. Some examples are given to illustrate the results of this work and show their applications.

Keywords: qualitative theory, integro–differential equation, fractional integro-differential equation, Lyapunov function, Lyapunov–Krasovskii functional (LKF), Lyapunov-Razumikhin method, fixed point method, stabilty, Ulam stability, Ulam-Hyers stability, Ulam-Hyers-Rassias stability

202 k0 (MOS) Subject Classifications: 34D05, 34K06, 34K20, 34K37, 45J05

1. Introduction

Vito Volterra (1860 –1940) was an Italian mathematician and physicist, known for his contributions to mathematical biology and integral equations. He is one of the founders of functional analysis. Volterra's work on elasticity was the origin of his theory of IDEs: He found that for certain substances, the electric or magnetic polarization depends not only on the electromagnetic field at that moment, but also on the history of the electromagnetic state of the matter at all previous instants. These physical facts are modeled by IDEs. Indeed, from the qualitative theory of IDEs, it can be seen that IDEs occur very often as mathematical models in various disciplines, see, the sources [1-120]. The origins of the investigations on integral and integro-differential equations are based on the works of Abel, Lotka, Fredholm, Malthus, Verhulst and Volterra on problems in mathematical biology, elasticity, mechanics, electromagnetics, economics (see Burton [16], Lakshmikantham and Rama Mohana Rao [48], Volterra [96], Wazwaz [99] and the references therein). In addition, the work of Volterra [96] on the problems of competing species and electromagnetics is of fundamental importance in the development of mathematical modeling of various real-world problems. Also, Volterra assumed linear "heredity", i.e., that the strain is a linear functional of the stress. In this case, the fundamental equations are systems of linear IDEs. He proved that the strain in a definite interval of time can be determined, given the forces in the body and the stress and strain on its surface for this time-interval (see, for example, Hatamzadeh et al. [40], Kyselka [46], Volterra [96] and the references therein).

From those beginnings, the theory and applications of IDEs with bounded delays, unbounded delays and without delay have emerged as new areas of investigation. Over the past few decades, numerous research papers and books on the qualitative properties of IDEs and functional differential equations (FDEs) were published (see [1-39], [41-45], [47-110] and the references therein).

Indeed, mathematical models called IDEs, integral equations, integro-delay differential equations (IDDEs), ordinary differential equations, FDEs are very attractive and interesting equations in the literature due to their very effective roles in numerous scientific fields and applications. In the past and in recent years, an extensive literature such as the numerous books by Burton [16], Hale [37], Hale and Verduyn Lunel [38], Kress [47], Lakshmikantham and Rama Mohana Rao [48], Rahman [71], Reissig et al. [75] and the related papers in the references of this work have been devoted to discussions of stability, uniform stability, uniform asymptotically stability, exponential stability, instability, integrability, boundedness, existence of solutions, etc., of these kinds of mathematical models.

IDEs and numerous kind of FDEs are extensively used for modeling various problems from mechanics, power systems, control theory, physics, electricity, biology, population ecology, neural networks, medicine, economics, etc., and their fundamental properties were intensively investigated theoretically (Burton [16], Hale [37], Hale and Verduyn Lunel [38], Hatamzadeh et al.[40], Kyselka [46], Lakshmikantham and Rama Mohana Rao [48], Volterra [96]).

For example, as a specific example in physics, an RLC, standard closed electric circuit, can be mathematically modeled by the below linear IDE:

$$L\frac{d}{dt}I(t) + RI(t) + \frac{1}{C}\int_0^t I(s)ds = E(t),$$

where $I(t)$ is the electric current, R is the resistance, L is the inductance, $C$is the capacitance and the impressed voltage $E(t)$ (Bohner et al. [15]).

2. Basic Results

We consider the following system of delay differential equations (DDEs) of the form:

$$\frac{dx}{dt} = F(t, x_t), x_t(\theta) = x(t+\theta), -\tau \le \theta \le 0. \quad (1)$$

We suppose $F: (-\infty, \infty) \times C \to R^n$, where C is a set of continuous functions $\phi: [-\tau, 0] \to R^n$, $\tau > 0$. Here, F is continuous and takes closed bounded sets into bounded sets, and $F(t, 0) = 0$. Since $F(t, 0) = 0$, the DDE (1) includes the solution $x(t) \equiv 0$, with zero initial function $\phi \equiv 0$.

Recall that for $t \in R$,

$$C(t) = \{\phi: [-\tau, 0] \to R^n | \phi \text{ is continous}\}.$$

If $\phi \in C(t)$, then $\|\phi\|_{t_0} = \sup_{t_0 - \tau \le s \le t_0} |\phi(s)|$, where $|.|$ is a convenient norm on R^n (see, Burton [16]).

For any $\phi \in C([-\tau, 0], R^n)$, we refer the usual Euclidean norm $\|.\|$, which is defined by

$$\|\phi\| = \sup_{-\tau \le s \le 0} |\phi(s)|.$$

We mean a continuously differentiable function $x(s)$ on $0 \le s < T \le \infty$ such that (1) is satisfied on $0 \le t < T$ for $x_t(.)$ a segment of $x(s)$. If $T < \infty$, and there is no solution $y(s)$ of (1) on $0 \le s < T_1$ with $T_1 > T$ such that $x(s) = y(s)$ on $0 \le s < T$, we say that $x(s)$ is noncontinuable on $0 \le s < T$. If $T = \infty$, $x(s)$ is said to be noncontinuable i.e., defined for all $s \ge 0$. We assume that: If $x(s)$ is a noncontinuaable solution of (1) $0 \le s < T \le \infty$, then there exists a sequence $\{t_j\}$ such that $t_j \to T$ and $|x(t_j)| \to \infty$ as $j \to \infty$.

Let R denote the set of real numbers, and R^n the set of real n -tuples. If x and y are elements of $R^n$with $x = (x_1, \ldots, x_n), y = (y_1, \ldots y_n)$, then $(x, y) = \sum_{j=1}^{n} x_j y_j$ and $|x| = (x, x)^{1/2}$. $x_t(.)$denotes a function continuous on the interval $0 \le s \le t$ to R^n and by S_t the set $\{x_t(.)\}$of all such functions. If $x(s)$ is a function defined and continuous on $0 \le s < T \le \infty$ to R^n, then for each fixed t, $0 \le t < T$, this function defines a member $x_t(.)$of S_t given by $x(s)$, $0 \le s \le t$. The function $x_t(.)$ is called a segment of $x(s)$.

Definition 1 (Seifert [78]). The point $x = 0$ is stable for (1) if given $\varepsilon > 0$, there exists a $\delta(\varepsilon) > 0$ such that *if* $|x_0| < \delta(\varepsilon)$, then every solution $x(t, x_0)$ of (1) is defined for $t \ge 0$ and satisfies $|x(t, x_0)| < \varepsilon$ for $t \ge 0$.

Definition 2 (Seifert [78]). The point$x = 0$ is said to be asymptotically stable for (1) if it is stable and if there exists a $r_0 > 0$ such that $|x_0| < r_0$ implies $x(t, x_0) \to 0$ as $t \to +\infty$.

Definition 3 (Seifert [78]). The solutions of (1) are bounded if given x_0 in R^n, every solution $x(t, x_0)$ is defined for $t \geq 0$ and there exists a $B(x_0)$ such that $|x(t, x_0)| \leq B(x_0)$ for $t \geq 0$.

Theorem 1 (Burton [16]). Let $V(t, x_t)$ be a differentiable scalar functional defined when $x: [\alpha, t] \to R^n$ is continuous and bounded by some $D \leq \infty$. Assume the following conditions hold:

$(A1) V(t, 0) = 0,\ W_1(|x(t)|) \leq V(t, x_t)$, where $W_1(r)$ is a wedge;
$(A2)\ \dot{V}(t, x_t) \leq 0$.

Then, the zero solution of the system of DDEs (1) is stable.

Next, consider a system of DDEs of the form:

$$\dot{x}(t) = G(x_t), \tag{2}$$

which is a particular case of system of DDEs (1). We assume that $C = C([-\tau, 0], R^n)$ is the space of continuous functions from $[-\tau, 0]$ into R^n and $G: C \to G^n$ is continuous, and $G(0) = 0$, (see Yoshizawa [104]).

If H is a given positive constant, we use the notation C_H for the set $\{\phi \text{ in } C: \|\phi(t)\| < H\}$; that is, C_H is the open ball in C of radius H.

Lemma 1 (Sinha [81, Lemma 1]). Suppose $G(0) = 0$. Let V be a continuous functional defined on C_H with $V(0) = 0$ and let $u(s)$ be a function, non-negative and continuous for $0 \leq s < \infty$, $u(s) \to \infty$ as $s \to \infty$ with $u(0) = 0$. If for all φ in $C, \boldsymbol{u(\phi(0)) \leq V(\phi)}$, $V(\phi) \geq 0, V'(\varphi) \leq 0$, then the zero solution of (2) is stable.

Let $\mathbb{R} \subset C_H$ be a set of all functions $\phi \in C_H$ where $V'(\varphi) = 0$. If $\{0\}$ is the largest invariant set in $\mathbb{R}$, then the solution $x(t) = 0$ of (2) is asymptotically stable.

Let $V(t, x)$ denote a function continuous on $R \times R^n$ to R.

Throughout this work, x represents $x(t)$.

Lemma 2 (Hale and Verduyn Lunel [38, Theorem 4.2, pp.152])**.** The system of DDEs (1) is globally uniformly asymptotically stable if there exists a continuous function $V(t,x)$, positive definite functions $\boldsymbol{u}, \boldsymbol{v}, \boldsymbol{\omega}$ and a continuous non-decreasing function $q(s) > s$ for $s > 0$ such that the following conditions hold:

$$u(|x|) \le V(t,x) \le v(|x|)\ , \forall t \in R^+, \forall x \in \mathbb{R}^n,$$
$$\frac{d}{dt}V(t,x) \le -\omega(|x|)$$

if

$$V(t+s, x(t+s)) < q(V(t, x(t))), \forall s \in [-\tau, 0].$$

Theorem 2 (Seifert [78])**.** Let there exist functions $u(s), v(s)$ and $f(s)$ continuous for $s \ge 0$ and such that $u(0) = v(0) = 0$, f(s) is increasing, $f(s) > s$ for $s > 0$, and suppose $V(t,x)$ is a real-valued function continuous in (t,x) for $t \ge 0$ and x in D, D is an open subset of R^n containing the zero vector. Let V satisfy:

(H_2) $u(|x|) \le V(t,x) \le v(|x|)$ for $t \ge 0, x$ in$D$,
(H_3) $\dot{V}(t,x(t)) \le 0$ for any solution $x(t)$ of (1) for which $x(s)$ is in D

and

$f(V(t,x(t))) > V(s,x(s))$ for $0 \le s \le t$; here and henceforth
$\dot{V}(t,x(t)) = \lim_{h\to 0+} sup[V(t+h), x(t+h) - V(t,x(t))]/h.$

Then, the point $x = 0$ is stable for the system of DDEs (1).

Theorem 3. Let there exist a function V satisfying the hypotheses of Theorem 2 except that now $D = R^n$. Then, the solutions of the system of DDEs (1) are bounded, (see, Seifert [78]).

We now assume that the following conditions hold.

(a) There exist real-valued functions $u(s)$ and $v(s)$, continuous and increasing for $s \ge 0$ such that $u(0) = v(0) = 0$, and

$u(|x|) \le V(t,x) \le v(|x|)$ for $t \ge 0$, $x \in R^n$,

(b) Let $f(s)$ be a function continuous for $s \geq 0$ such that $f(s) > s$ for $s > 0$. Let $x(t)$ be a solution of (1) on $[0, T), T \leq \infty$. Then, there exists a number $r > 0$, and a function $w(s) > 0$ for $s > 0$, such that the condition

(i) $V(s, x(s)) < f(V(t, x(t)))$ for $s \in [t_0, t]$, $t > 0$, where $t_0 = max[0, t - r]$, implies.
(ii) $\dot{V}(t, x(t)) \leq -w(|x(t)|)$,

where $w(s)$ and r may depend on the solution $x(t)$, as well as on V and f, and

$$\dot{V}(t, x(t)) = \lim_{h \to 0} sup[V(t + h, x(t + h)) - V(t, x(t))]/h.$$

Let

$$V_1: R^+ \times C_H \to R^+, R^+ = [0, \infty),$$

be a continuous LKF with $V_1(t, 0) = 0$. Further, let $\frac{d}{dt} V_1(t, x)$ denote the derivative of $V_1(t, x)$ on the right along any solution of (1).

Theorem 4 (Burton [16, Theorem 4. 2.9]). Assume that

$(A1)$ The Lyapunov –Krasovskii functional (LKF) $V_1(t, x)$ satisfies the locally Lipschitz condition in x, i.e., for every compact $S \subset \mathbb{R}^n$ and $\gamma > t_0$, there is a $K_{\gamma S} \in R$ with $K_{\gamma S} > 0$ such that

$$|V_1(t, x_t) - V_1(t, y_t)| \leq K_{\gamma S} \|x - y\|_{[t_0 - \tau, t]}$$

for all $t \in [t_0, \gamma]$ and $x, y \in C([t_0 - \tau, t_0], S)$.

$(A2)$ Let $Z: R^+ \times C_H \to R^+$ be a continuous functional such that it satisfies the one –sided locally Lipschitz condition in t, i.e.,

$$Z(t_2, \phi) - Z(t_1, \phi \leq K(t_2 - t_1), 0 < t_1 < t_2 < \infty, K > 0, K \in R, \phi \in C_H.$$

$(A3)$ There are four strictly increasing functions ω, ω_1, ω_2, $\omega_3: R^+ \to R^+$ with value 0 at 0 such that

$$\omega(\|\phi(0)\|) + Z(t,\phi) \le V_1(t,\phi) \le \omega_1(\|\phi(0)\|) + Z(t,\phi),$$
$$Z(t,\phi) \le \omega_2(\|\phi\|_C),$$

and

$$\frac{d}{dt}V_1(t,x) \le -\omega_3(\|x(t)\|)$$

whenever $t \in \mathbb{R}^+$and $x \in C_H$. Then, the solution $x(t) = 0$ of (1) is uniformly asymptotically stable.

3. Lyapunov and Lyapunov –Krasovskii Qualitative Results of IDEs

The qualitative properties of solutions, such as stability, instability, and boundedness of solutions, are very important for IDEs and different kinds of FDEs. These concepts and various others have attracted the attention of many researchers during the last decades. Hence, numerous results related to the qualitative properties of certain IDEs and FDEs have been obtained in the literature (see, [1-39], [41-45], [47-110]).

Next, we summarize some of the work from the references cited in this chapter dealing with the qualitative proprieties of IDEs and systems of IDEs.

In the book by Burton [16], a reference book on integral equations and IDEs, stability, uniform stability, asymptotic stability, uniform asymptotic stability of the zero solution, as well as, integrability and boundedness of non-zero solutions when $F(t) \neq 0$, are discussed for the following systems of IDEs using Lyapunov's second method

$$x' = A(t)x + \int_0^t C(t,s)x(s)ds,$$

$$x' = Ax + \int_0^t C(t,s)x(s)ds + F(t),$$

$$x' = Ax + \int_0^t B(t-s)x(s)ds,$$

$$x' = Ax + \int_0^t D(t-s)x(s)ds + F(t),$$

$$x' = A(t)x + \int_0^t C_1(t,s)x(s)ds + \int_0^t C_2(t,s)x(s)ds,$$
$$x' = Ax + f(t,x) + \int_0^t C(t,s)x(s)ds.$$

In the book by Lakshmikantham and Rama Mohana Rao [48], a reference book on the theory of IDEs, various qualitative behaviors of solutions such as stability, uniform stability, asymptotic stability, uniform asymptotic stability of the zero solution, as well as, integrability and boundedness of non-zero solutions when $f(t,x) \neq 0$ and$g(t,y) \neq 0$, are investigated using Lyapunov's second method. Very interesting results are obtained for the following scalar and systems of IDEs :

$$u' = \alpha u + \int_0^t a(t-s)u(s)ds,$$
$$u' = \alpha(t)u + \int_0^t a(t,s)u(s)ds,$$
$$x' = A(t)x + \int_0^t K(t,s)x(s)ds,$$
$$x' = Ax + \int_0^t K(t,s)x(s)ds,$$
$$x' = A(t)x + \int_{-\infty}^t K(t,s)x(s)ds + f(t,x),$$
$$x' = Ax + \int_0^t C(t-s)x(s)ds,$$
$$y' = A(t)y + \int_0^t C(t-s)y(s)ds + g(t,y).$$

In [17], Burton first considered the following class of scalar IDEs:

$$x'(t) = A(t)f(x(t)) + \int_0^t B(t,s)g(x(s))ds,$$
$$x'(t) = -\int_0^t C(t,s)h(x(s))ds$$

and

$$x'(t) = A(t)f(x(t)) + \int_0^t [B(t,s) - C(t,s) - D(t,s)]f(x(s)ds.$$

He discussed the stability of the zero solution, the boundedness and convergence of the bounded solutions to the zero solution of the first equation by a Lyapunov–Krasovskii functional [17, Theorem 1]. He also discussed the stability of the zero solution, boundedness of solutions, the square integrability of $x'(t)$, the convergence of the solution $x(t)$ to the zero solution as $t \to \infty$ of the second equation via a Lyapunov –Krasovskii functional [17, Theorem 2], as well as, the stability of the zero solution, the boundedness of solutions and showed that $\int_0^\infty f^2(x(t))dt < \infty$ by a Lyapunov –Krasovskii functional [17, Theorem 4].

Later, Burton [17] considered the following system of IDEs and some of its modified versions:

$$x'(t) = A(t)f(x(t)) + \int_0^t B(t,s)E(x(s)x(s)ds.$$

In Burton [17], some sufficient conditions are established for this equation and its modified versions such that under those conditions the zero solution is stable or uniformly stable, the solutions are bounded, and all bounded solutions tend to zero for the considered equations [17, Theorems 5-8].

As for some recent works on the qualitative properties of solutions of IDEs, Matsunaga, Hashimoto [56] dealt with the problem of asymptotic stability for the following linear system of IDEs with distributed delay in the diagonal terms:

$$x' = Ax - a\int_{t-\tau}^t x(s)ds.$$

They also established some explicit conditions for the asymptotic stability of the zero solution of the above system. In particular, as the delay parameter increases monotonically under certain conditions, the zero solution switches in finite times from stability to instability to stability, and becomes eventually unstable.

Raffoul and Rai [70] used LKFs and discussed exponential stability of the following IDE with finite delay:

$$x' = Px + \int_{-\infty}^{t} C(t,s)g(x(s))ds.$$

In [70], the authors used modified versions of the Lyapunov –Krasovskii functional that were used previously to obtain a criterion for the stability of the zero solution of the infinite delay nonlinear Volterra IDE.

By using spectral properties of Metzler matrices and the comparison principle, Ngoc and Anh [59] established some new explicit criteria for uniform asymptotic stability and exponential stability of the following nonlinear IDE of the form:

$$x' = h(t,x) + \int_0^t q(t,s,x(s))ds.$$

In this paper, Ngoc and Anh presented a novel approach to problems of stability of Volterra IDEs. The approach of [59] is simple and relies upon the spectral properties of Metzler matrices and the comparison principle. Consequently, they obtained some explicit criteria for the uniform asymptotic stability and the exponential stability of the above nonlinear Volterra IDE. Some examples are given to illustrate the obtained results.

In Berezansky et al. [12], criteria for uniform exponential stability are derived for the linear IDEs with several point delays and several distributed delays of the form:

$$x' = \sum_{k=1}^{m} A_k(t)x(h_k(t)) + \sum_{k=1}^{k} \int_{g_k(t)}^{t} P_k(t,s)x(s))ds.$$

In Berezansky et al. [12], the main technique of the proofs is splitting the linear expressions in the equation (both with point and with distributed delays) into a "dominant" and a "remainder"" part, which can be done in a number of different ways, thus providing a number of different criteria. The next important ingredient is the use of a Bohl-Perron type result stating that a linear equation is exponentially stable if all of the solutions of the inhomogeneous counterpart of that equation are bounded.

Seifert [79] considered the following IDEs of Volterra type:

$$\dot{x}(t) = G(t,x(t)) + \int_0^t K(t,s,x(s))ds. \tag{3}$$

As an application, Seifert [79] considered the following particular case of IDE (3)

$$\dot{x}(t) = Ax(t) + \int_0^t K(t,s,x(s))ds, \tag{4}$$

where A is a real constant $n \times n$ matrix, the eigenvalues of which all have negative real parts, K is continuous in (t,s,x) for $0 \leq s \leq t < \infty$ and x in R^n, and is such that there exist constants μ and ρ for which

$$\left|\int_0^t K(t,s,x(s))ds\right| \leq \mu \sup_{0\leq s\leq t} |x(s)| \tag{5}$$

for any function $x(s)$ continuous on $0 \leq s \leq t$ with $|x(s)| \leq \rho$ on this interval.

It is well known from results in matrix theory that there exists a positive definite symmetric real matrix B such that $BA + A^T B = -I$, where I denotes the identity matrix and A^T the transpose of A. If λ is the least eigenvalue of B, then $\lambda > 0$ and $(x, Bx) \geq \lambda^2(x,x)$ for all x in R^n.

Choose a Lyapunov function as

$$V(t,x) = V(x) = (x, Bx).$$

Clearly (H_2) holds for the Lyapunov function V since we may take $v(s) = \lambda s^2$ and $u(s) = \Lambda s^2$, where λ and Λ are, respectively, the least and greatest eigenvalues of B.

The time derivative of the Lyapunov function V along (4) yields

$$\begin{aligned}\dot{V}(x(t)) &= (x(t), (A^T B + BA)x(t)) + 2\left(\int_0^t K(t,s,x(s))ds, Bx(t)\right)\\ &= -|x(t)|^2 + 2\left(\int_0^t K(t,s,x(s)), Bx(t)\right).\end{aligned} \tag{6}$$

Let $f(s) = q^2 s$ for $q > 1$, $t > 0$ and $x(s)$ be a solution of (4) such that $|x(s)| < \rho$ for $0 \leq s \leq t$. Suppose that

$$q^2(x(t), Bx(t)) > (x(s), Bx(s)) \text{ for } 0 \leq s \leq t.$$

This inequality implies

$$q^2(x(t), Bx(t)) > \lambda^2 |x_t|^2,$$

where we define $|x_t| = \sup_{0 \le s \le t} |x(s)|$.

If $B = (b_{ij})$ and $|B| = \sqrt{\Sigma_{i,j} b_{i,j}^2}$, using (5), it follows that

$$\begin{aligned}
(\textstyle\int_0^t K(t,s,x(s))ds, Bx(t)) &\le |B| \left|\textstyle\int_0^t K(t,s,x(s))ds\right| |x(t)| \\
&\le |B|\mu|x_t||x(t)| \\
&< |B|\mu q (x(t), Bx(t))^{1/2}|x(t)|\lambda \\
&\le |B|\mu q|x(t)|^2 \Lambda/\lambda.
\end{aligned}$$

If $2|B|\mu\Lambda/\lambda < 1$, then there obviously exists a $q > 1$ such that $2|B|\mu\Lambda/\lambda < 1$. For this choice of q, it follows from (6) that $\dot{V}(x(t)) \le -(1 - 2|B|\mu q\Lambda/\lambda)|x(t)|^2 \le 0$; thus, (H_3) holds for D the set of x in R^n such that $|x| < \rho$. As a consequence of the above result, we have, therefore, the following result.

Seifert [79] proved the following theorem .

Theorem 5. We assume that $\frac{2|B|\mu\Lambda}{\lambda} < 1$ and the conditions above hold. Then the trivial solution of IDE (4) is stable.

In [79], Seifert used Lyapunov functions of Razumkhim type to give conditions sufficient for the stability of the zero state of a system of ordinary differential equations involving an interval of delay which becomes unbounded as $t \to +\infty$. An example of such a system is an IDE of Volterra type:

$$\dot{x}(t) = G(t, x(t)) + \int_0^t K(t,s,x(s))ds,$$

where x, G and K denote functions with values in R^n. As an example, Seifert [79] also considered the following system of IDEs:

$$\dot{x}(t) = Ax(t) + h(t, x_t) + \int_0^t g(t,s,x(s))ds, \tag{7}$$

where A is a real stable $n \times n$ -matrix, h is a function continuous on $R \times C_r$, (C_r is the space (with the usual supremum norm) of continuous functions from $[-r, 0]$ to R^n), and satisfying

$$|h(t, \phi)| \leq \mu \|\phi\|,$$

where

$$\|\phi\| = \sup_{s \in [-r,0]} \geq |\phi(s)|,$$

and $g(t, s, x)$ is continuous on $R \times R \times R^n$, $t \geq s \geq 0$, and satisfies $|g(t, s, x)| \leq K(t, s)|x|$ with

$$\int_0^t K(t, s)ds \to 0 \text{ as} t \to +\infty.$$

In Seifert [79], it is shown that for μ sufficiently small, each bounded solution of (7) tends to zero as $t \to +\infty$. Since A is a stable matrix, by a well-known result, there exists a positive definite symmetric matrix B such that $BA + A^T B = -I$, where A^T denotes the transpose of A and I the identity matrix. Define the Lyapunov function $V = (Bx, x)$. Then, there exist positive numbers λ and Λ such that

$$\lambda^2 |x|^2 \leq Bx.x \leq \Lambda^2 |x|^2 \text{ for all } x \in R^n.$$

If

$$\mu < \lambda/(2|B|\Lambda),$$

where $|B| = \Sigma_{i,j=1}^n | b_{ii} |$, $B = (b_{ij})$. Then, there exists a $q > 1$ such that

$$\mu < \lambda/(2q|B|\Lambda).$$

Let $f(s) = q^2 s$. Then, for any positive integer k and any solution $x(t)$ of (7) such that

$$(Bx(s), x(s)) < q^2(Bx(t), x(t)) \text{ for } s \in [t - kr, t],\ t \geq kr,$$

it is clear that

$$q^2 \Lambda^2 |x(t)|^2 > \lambda^2 \|x_t\|^2,\ t \geq kr.$$

Thus,

$$|(Bx(t), h(t, x_t))| \leq |B|\ |x(t)|\mu\|x(t)\| \leq \mu|B|q\Lambda|x(t)|^2/\lambda,\ t \geq kr.$$

Fix $\mu_1 > 0$ such that

$$(2|B|q\Lambda/\lambda)(\mu + \mu_1) < 1.$$

Then, there exists a $k > 0$ such that for $t \geq kr$,

$$\int_0^t K(t,s)ds \leq \mu_1.$$

Since

$$\left|\int_{t-kr}^t g(t,s,x(s))ds\right| \leq \int_{t-kr}^t K(t,s)|x(s)|\,ds$$
$$\leq \sup_{\theta \in [t-kr,t]} |x(\theta)| \int_0^t K(t,s)ds,\ t \geq kr,$$

and $x(t)$ is a solution of (7) satisfying $(Bx(s), x(s)) < q^2(Bx(t), x(t))$, it follows that for such k, t and $x(t)$,

$$\left|2\left(Bx(t), \int_{t-kr}^t g(t,s,x(s))ds\right)\right| \leq 2q\Lambda\mu_1|B||x(t)|^2/\lambda.$$

If we define $\alpha = 1 - 2|B|q\Lambda(\mu + \mu_1)/\lambda$, and $w(s) = \alpha s^2$, then $(2|B|q\Lambda/\lambda)(\mu + \mu_1) < 1$, which implies that $\alpha > 0$, and

$$\left|\int_0^{t-kr} g(t,s,x)ds\right| \leq |x| \int_0^t K(t,s)ds \to 0 \text{ as } t \to +\infty,$$

Hence, it can be concluded that each bounded solution of (7) tends to zero as $t \to +\infty$, (see, Seifert [79, Theorem 2]).

In [95], Vanualailai and Nakagiri considered the IDE of the form:

$$\frac{d}{dt}[x(t)] = A(t)f(x(t)) + \int_0^t B(t,s)g(x(s))ds, \tag{8}$$

in which $A(t)$ is an $n \times n$ - matrix function continuous on $[0, \infty), B(t,s)$ is an $n \times n$ matrix continuous for $0 \leq s \leq t < \infty$, and f and g are $n \times 1$ vector functions continuous on $(-\infty, \infty)$.

In Vanualailai and Nakagiri [95], it is assumed that $A(t)$ is continuous for $0 \leq t < \infty$, $B(t,s)$ is continuous for $0 \leq s \leq t < \infty$, $\int_0^t |B(u,s)| du$ is defined and continuous for $0 \leq s \leq t < \infty$, $f(x)$ and $g(x)$ are continuous on $(-\infty, \infty), xf(x) > 0, \forall x \neq 0$, a$f(0) = g(0) = 0$.

Theorem 6 (Vanualailai and Nakagiri [95]). Let the conditions above hold and suppose there are constants $m > 0$ and $M > 0$ such that,

$$g^2(x) \leq m^2 f^2(x) \text{ if } |x| \leq M.$$

Define

$$\beta(t,k) = A(t) + k \int_t^{¥} |B(u,t)| du + \frac{1}{2} \int_0^t |B(t,s)| ds.$$

If there exists $k > 0$ with $m^2 < 2k$ and $\beta(t,k) \leq 0$ for $t \geq 0$, then the zero solution of IDE (8) is stable.

Theorem 7 (Vanualailai and Nakagiri [95]). Let the condition above hold, with $A(t) < 0$, and suppose there are constants $m > 0$ and $M > 0$ such that $g^2(x) \leq m^2 f^2(x)$ if $|x| \leq M$, $\alpha > 4$; and $N > 0$ such that $4x^2 \leq (\alpha - 4) f^2(x)$; if $|x| \leq N$, and $J \geq 1$ such that $-\frac{1}{4A(t)} \int_0^t |B(t,s)| ds < \frac{1}{J}$ for every $t \geq 0$. Suppose further there is some constant $k > 0$ such that $\frac{(1+\alpha) m^2}{J} < k$ and $A(t) + k \int_t^{\infty} |B(u,t)| du \leq 0$ for $t \geq 0$. Then, the zero solution of IDE (8) is stable.

Next, Vanualailai and Nakagiri [95] used the following LKF:

$$V_2(t, x(.)) = \frac{1}{2} x^2 + \sqrt{\alpha} \int_0^x \sqrt{u f(u)}\, du + \frac{1}{2} \alpha \int_0^x f(u)\, du$$
$$+ k \int_0^t \int_t^{\infty} |B(u,s)|\, du\, f^2(x(s)) ds$$

and its ~~a~~ modified version to prove the stability results therein.

Chang and Wang [22] dealt with the problem of the stability of the following perturbed n-dimensonial IDE:

$$\dot{x}(t) = A(t)x(t) + \int_{-\infty}^{t} C(t,s)x(s) + \int_{-\infty}^{t} D(t,s)\dot{x}(s)ds + b(t).$$

In [22], sufficient criteria for the stability and uniform stability of the zero solution of the above IDE were obtained.

Chang and Wang [22] also took into account the following nonlinear IDE:

$$\dot{x}(t) = A(t)x(t) + \int_{0}^{t} C(t,s)x(s)\,ds + f(t,x(t)), f(t,0) = 0, \tag{9}$$

where $A(t)$ is an $n \times n$ - matrix function continuous on $[0,\infty)$, $C(t,s)$ is an $n \times n$ matrix continuous for $0 \le s \le t < \infty$, and f is $n \times 1$ vector function continuous on $R^{+} \times R$. In this paper, the authors discussed the stability and uniform stability of the zero solution of IDE (9). The following theorem is proved in Chang and Wang [22].

Theorem 8 For IDE (9), if there exists a non-negative function $b(t)$ and a positive definite matrix B such that $\|f(t,x(t))\| \le b(t)\|x\|$ and

$$\frac{1}{2}\lambda_M^*(t)\lambda_m^{\frac{1}{2}}(B) + \lambda_M^{\frac{1}{2}}(t)(B)\left(\int_0^{+\infty}\|C(u,t)\|du + b(t)\right) \le 0,$$

where λ_M^* is the biggest eigenvalue of $BA(t) + A^T(t)B$. Then the zero solution of IDE (9) is stable.

Proof. We select $\|A\|_2$ as the matrix norm. Next, for IDE (9), Chang and Wang [22] chose the following LKF :

$$V(t,x(.)) = (x^T Bx)^{\frac{1}{2}} + \|B\|^{\frac{1}{2}} \int_0^t \int_t^{+\infty} \|C(u,s)\|du\,\|x(s)\|ds.$$

The time derivative of this LKF along the solution of IDE (9) is

$$\dot{V}_1(t,x(.)) = \frac{1}{2}(x^TBx)^{-\frac{1}{2}}(x^TBx)'$$
$$+ \|B\|^{\frac{1}{2}}\int_t^{+\infty}\|C(u,t)\|du\|x\|$$
$$- \|B\|^{\frac{1}{2}}\int_0^t\|C(t,s)\|.\|x(s)\|ds$$
$$= \frac{1}{2}(x^TBx)^{-\frac{1}{2}}\{x^T[A^T(t)B + BA(t)]x + x^T(t)Bf(t,x(t))$$
$$+ f^T(x,x(t))Bx(t)\}$$
$$+\frac{1}{2}(x^TBx)^{-\frac{1}{2}}\left\{\int_0^{t\int} x^T(s)C^T(t,s)ds\cdot Bx + x^TB\cdot\int_0^{t\int} C(t,s)x(s)ds\,\{\}\right\}$$
$$+\|B\|^{\frac{1}{2}}\int_0^{+\infty}\|C(u,t)\|du\cdot\|x\| - \|B\|^{\frac{1}{2}}\int_0^t\|C(t,s)\|\cdot\|x(s)\|ds$$
$$\le \frac{1}{2}(x^TBx)^{-\frac{1}{2}}\{x^T[A^T(t)B + BA(t)]x\} + \frac{1}{2}(x^TBx)^{-\frac{1}{2}}\cdot 2\|x(t)\|\|B\|$$
$$\times\|g(t,x(t))\| + \|B\|^{\frac{1}{2}}\int_0^{+\infty}\|C(u,t)\|du\cdot\|x\| - \|B\|^{-\frac{1}{2}}\|x\|^{-1}$$
$$\times\int_0^t\|x^T(s)\|\cdot\|C^T(t,s)\|ds\|B\|\cdot\|x\| - \|B\|^{\frac{1}{2}}\int_0^t\|C(t,s)\|\cdot\|x(s)\|ds$$
$$\le \frac{1}{2}(x^TBx)^{-\frac{1}{2}}\{x^T[A^T(t)B + BA(t)]x\} + \|B\|^{\frac{1}{2}}\int_0^{+\infty}\|C(u,t)\|du\cdot\|x\|$$
$$+ \|B\|^{\frac{1}{2}}b(t)\|x\|.$$

Since B is positive definite, $A^T(t)B + BA(t)$ is a symmetric matrix. Choosing a non-singular matrix $Q(t)$ and using $x = Q(t)y$ into the above formula (see, Chang and Wang [22]), it is obtained that

$$\frac{1}{2}(y^T\cdot I\cdot y)^{-\frac{1}{2}}\sum_{i=1}^n\lambda_i^*y_i^2 + \|B\|^{\frac{1}{2}}\|Q(t)y\|\int_0^{+\infty}\|C(u,t)\|du +$$
$$\|B\|^{\frac{1}{2}}b(t)\|Q(t)y\|$$
$$= \|y\|\left(\frac{1}{2}\frac{\sum_{i=1}^n\lambda_i^*(t)y_i^2}{\|y\|^2} + \|B\|^{\frac{1}{2}}\frac{\|Q(t)y\|}{\|y\|}\int_0^{+\infty}\|C(u,t)\|du + \|B\|^{\frac{1}{2}}b(t)\|Q(t)y\|\right)$$
$$\le \|y\|\left(\frac{1}{2}\lambda_M^*(t) + \|B\|^{\frac{1}{2}}\|Q(t)\|\int_0^{+\infty}\|C(u,t)\|du + \|B\|^{\frac{1}{2}}b(t)\|Q(t)\|\right).$$

It is also known that if $B = P \cdot P$, we have $\|Q(t)\| = \|P^{-1}\|$, but $P^{-1}$is also a positive definite matrix, $\|P^{-1}\| = \lambda_m^{-1}(P) = \lambda_m^{-\frac{1}{2}}(B) = \|B^{-1}\|^{\frac{1}{2}}\|B\|^{\frac{1}{2}} = \lambda_M^{\frac{1}{2}}(B)$. Bringing this into the above formula, we obtain

$$\dot{V}(t, x(.)) \leq \left(\frac{1}{2}\lambda_M^*(t) + \left(\frac{\lambda_M(B)}{\lambda_m(B)}\right)^{\frac{1}{2}} \int_0^{+\infty} \|C(u,t)\| du + \lambda_M(B)^{\frac{1}{2}} b(t)\|Q(t)\|\right) \|Q^{-1}(t)x\|$$

$$\leq \left[\frac{1}{2}\lambda_M^*(t)\lambda_m^{\frac{1}{2}}(B) + \lambda_M^{\frac{1}{2}}(B)\left(\int_0^{+\infty}\|C(u,t)\|du + b(t)\right)\right] \|x\| \, \|Q^{-1}(t)\| \, \|Q(t)\| \leq 0.$$

Since $V(t, x(\cdot))$is positive definite and $\dot{V}(t, x(.)) \leq 0$, then the zero solution of IDE (9) is stable (see, Chang and Wang [22]).

The next result of Chang and Wang [22] is given by the following theorem.

Theorem 9 If the conditions in Theorem 8 are satisfied and there exists a constant M such that

$$\int_0^t \int_t^{+\infty} \|C(u,s)\| duds \leq M, t \geq 0,$$

Then, the zero solution of IDE (9) is uniformly stable (for proof see, Chang and Wang [22]).

The final result of Chang and Wang [22] is given by the following theorem.

Theorem 10 [22]. For IDE (9), if there exists a non-negative matrix $b(t)$ such that

$$\|g(t, x(t))\| \leq b(t)\|x\|$$

and there exists $K > 1$ such that when $t \in R$ we have

$$\mu A(t) + K\left[\int_0^t \|C(t,s)\|ds + b(t)\right] \leq 0,.$$

Then, the zero solution of (9) is uniformly stable.

Example 1 [22]. Consider a functional differential equation such that

$$\lambda_M^*(t) = -4e^{-2t}\ , \|C(t,s)\| = e^{-t-s}, \|C(u,t)\| = e^{-u-s}.$$

Then,

$$\frac{1}{2}\lambda_M^*(t)\lambda_m^{\frac{1}{2}}(B) + \lambda_M^{\frac{1}{2}}(B)\int_0^{+\infty}\|C(u,t)\|du = \left(-2+\sqrt{2}\right)e^{-2t} < 0$$

and when $\int_0^t \int_t^{+\infty} C(u,s)duds = 1 - \frac{1}{2}e^{-2t}, t \geq 0$. This is obviously bounded, and then from Theorem 10 , the zero solution of the corresponding IDE is uniformly stable.

Example 2 [22]. Consider the integro-differential equation

$$\dot{x}(t) = \begin{pmatrix} -e^{2t} & 0 \\ 0 & -e^{3t} \end{pmatrix} x(t) + \int_0^t \begin{pmatrix} e^{-t+2s} & 0 \\ 0 & e^{-t+s} \end{pmatrix} x(s)ds.$$

Obviously, it follows that $\mu(A(t)) = e^{-2t}$, $\int_0^t \|C(t,s)\|ds = \frac{1}{2}(e^{-t} - e^{-t})$. Let $K = 2$. Then, we have $\mu(A(t)) + 2\left[\frac{1}{2}(e^t - e^{-t})\right] = -e^{2t} + e^t + e^{-t} \leq 0, t \in R$.

According to Theorem 10, the zero solution of this equation is uniformly stable, but because $\int_0^t \int_t^{+\infty}\|C(u,s)\|duds = e^t - 1$ when $t \geq 0$ is unbounded, from Theorem 9 we cannot determine its uniform stability.

In a recent and very interesting paper, Sedova [77] considered the following IDE:

$$\dot{x}(t) = G(t, x(t)) + \int_0^t H(t, s, x(s))ds, \tag{10}$$

where $x(t) \in R^n, t \in R^+, R^+ = [0, \infty), G \in C(R^+ \times R^n, R^n), H \in C(D \times R^n, R^n)$. Also, for a continuous function $x: (-\infty, A] \to R^n, 0 \leq A \leq +\infty$, for each $t \leq A$, the function $x_t : R^- \to R^n, R^- = (-\infty, 0]$, is defined as $x_t(s) = x(t + s)$. It is also assumed that $G(t, 0) = 0, H(t, s, 0) = 0$, so IDE (10) admits the zero solution. If the sequence $\{G(t + t_k, x)\}$ converges to $G * (t, x)$, and the sequence $\{H(t + t_k, t + t_k + s, x)\}$ converges to $H * (t, t + s, x)$ as $t_k \to +\infty$ uniformly on the corresponding compact subsets, then a limiting equation for IDE (10) has the form:

$$\dot{x}(t) = G * (t, x(t)) + \int_{-\infty}^{t} H * (t, s, x(s)) ds. \tag{11}$$

It is clear that equation (11) can be considered in the same phase space as the orginal one, and the limiting functions satisfy the same constraints as the original ones.

In [77], the specific applications of Razumikhin technique to the stability analysis of this IDE are considered and new sufficient conditions for uniform asymptotic stability of the zero solution are given using the phase space of a special construction and constraints on the right side of IDE (11).

Theorem 11 (Sedova [77])**.** Suppose that there exists a positive definite continuously differentiable function $V(x)$ such that one of the following conditions is satisfied:

1) the derivative of $V(x)$ by virtue of (10) is negative definite for $\varphi \in C_g$ satisfying $sup_{s \leq 0} V(\varphi(s)/g(s)) \leq V(\varphi(0))$ and for sufficiently large t.

2) the derivative $\frac{d}{dt} V(x(t))$ by virtue of every limiting equation (11) does not exceed a function $W(x(t))$ for $\varphi \in C_g, (\varphi \in C_g \Rightarrow |\varphi|_g \leq h, h > 0)$, satisfying $sup_{s \leq 0} V(\varphi(s)/g(s)) \leq V(\varphi(0))$, and $W(x)$ is negative definite. Then, the zero solution of IDE (10) is uniformly asymptotically stable.

Example 3 (Sedova [77])**.** Consider the following scalar IDE:

$$\dot{x}(t) = -h(t)x(t) - b(t)x^3(t) + \int_0^t c(at - s)x(s) ds, \tag{12}$$

where $c \in L_1[0, +\infty], a > 1, h(t) \geq 0$ and $b(t) \geq b_0 > 0$ are continuous functions. The following conditions for uniform asymptotic stability are obtained via an LKF:

$$2h(t) \geq [1 + (1/a)] \int_{(a-1)t}^{\infty} |c(v)| dv,$$

$$\int_{t}^{\infty} |c(u)| du \in L_1 0, +\infty).$$

We denote $M(t) = \int_{-\infty}^{0} |c((a-1)t - s)| ds = \int_{(a-1)t}^{\infty} |c(v)| dv$ (this function is bounded). Then, for an arbitrarily small $\delta > 0$ there exists an appropriate function g such that

$$\int_{-\infty}^{0} |c((a-1)t - s| g(s) ds < M(t) + \delta.$$

Using the Lyapunov function $V(x) = x^2$, we obtain that the zero solution of (12) is uniformly asymptotically stable if the functions $h(t)$ and $b(t)$ are bounded and satisfy the inequality

$$h(t) \geq \int_{(a-1)t}^{\infty} |c(v)| dv + \varepsilon$$

for some $\varepsilon > 0$.

Example 4 (Sedova [77]). Consider the following scalar IDE:

$$\dot{x}(t) = -p(t)x(t) + q(t, x(t)) + \int_{0}^{t} C(t-s) h(x(s)) ds,$$

Where

$$\begin{aligned} &h(0) = 0, |h(x)| \leq \gamma |x| \text{ for } x \neq 0, \\ &p(t) - \gamma \textstyle\int_{t}^{\infty} |C(u-t)| du > \delta > 0 \text{ for } t \in R^{+}, \\ &|q(t, x)| \leq r(t)|x| \text{ for some } r(t) \in L_1. \end{aligned} \tag{13}$$

By using an LKF, it is proved that under the given conditions the zero solution of IDE (13) is asymptotically stable. Next, the inequality $p(t) -$

$\gamma \int_t^{\infty} |C(u-t)|du > \delta > 0$ for $t \in R^+$ implies that $\int_{-\infty}^{0} |C(-s)|ds < \varepsilon = (inf_{t\in R^+} p(t) - \delta_1)/\gamma$ for some $\delta_1 \in (0, \delta)$. Then, there exists an appropriate function g such that $\int_{-\infty}^{0} |C(-s)|g(s)ds < \varepsilon$. Using the function $V(x) = x^2$, it can now be obtained that the time derivative $V'$is negative definite for $\varphi \in C_{,g}$ satisfying $sup_{s\le 0} V(\varphi(s)/g(s)) \le V(\varphi(0))$ and for sufficiently large t. Then, these results imply the uniform asymptotic stability of the zero solution of the above IDE.

Nieto and Tunç [64] considered the non-linear VIDE as follows:

$$\frac{dx}{dt} = -F(t,x)x + \int_0^t K(t,s,x(s))ds, \forall t > t_0 \ge 0, \tag{14}$$

where $t \in R^+$, $x \in R^n$, $F(t,x) \in C(R^+ \times R^n, R^n \times R^n), D = \{(u,v) \in R^2 : 0 \le v \le u < \infty\}$,and $K(u,v,x) \in C(D \times R^n, R^n)$ and $K(u,v,x) = 0 \Leftrightarrow x = 0$.

Initially, they investigated the asymptotic and uniform stability of the trivial solution of IDE (14) using the LKF approach, under the following conditions:

$(C1)$ $F(t,x) \in C(R^+ \times R^n, R^n \times R^n)$ and is positive definite such that

$$\sup_{(t,x)\in\Re^+\times\Re^n} \|F(t,x)\| < \infty,$$

and the eigenvalues of $F(t,x)$ satisfy,

$$f_{1i} \ge \lambda_i(F(t,x)) \ge f_{0i},$$
$$f_{0i} > 0$$
$$\forall t \in R^+, R^+ = [0,\infty), \forall x \in R^n, (i = 1,..,n).$$

$(C2)$ Let $\beta > 0$ be a positive constant such that

$$\|K(t,s,x(s))\| \le \|D(t,s)\| \, \|f(x(s))\|,$$
$$\|f(x(s))\| \le \beta\|x(s)\|,$$
$$\int_0^t \|D(t,s)\|ds \le \alpha_1(t), \int_t^{\infty} \|D(u,t)\|du \le \alpha_2(t),$$

where $\alpha_1, \alpha_2 \in C(R^+, R^+)$ and are bounded for $\forall t \in R$, and

$$\alpha(t) = f_0 - \frac{1}{2}\int_0^t \|D(t,s)\|\, ds - \frac{1}{2}\beta^2 \int_t^\infty \|D(u,t)\| du \geq c > 0, \forall t \in R,$$

where $c \in R$, $f_0 = min\{f_{01}, f_{02}, ..., f_{0n})$.

Theorem 12 (Nieto and Tunç [64]). The zero solution of IDE (14) is asymptotic stable if assumptions$(C1)$ and $(C2)$ hold.

Proof. To prove the theorem, we use the LKF method. Define the functional: $W = W(t, x_t) = \frac{1}{2}\langle x(t), x(t)\rangle + \sigma \int_0^t \int_t^\infty \|D(u,s)\|\ \|f(x(s))\|^2 duds$, where $\sigma > 0$, $\sigma \in R$, the constant σ is chosen later in the proof. For the first step, from the given functional, we derive $W(t,0) = 0$ and $W(t, x_t) \geq \frac{1}{2}\|x(t)\|^2$. Thus, clearly, we see that $W$is positive definite and has a lower bound. Next, differentiating W gives:

$$\frac{d}{dt}W(t,x_t) = \frac{1}{2}\langle x'(t), x(t)\rangle + \frac{1}{2}\langle x(t), x'(t)\rangle$$
$$+ \sigma \int_t^\infty \|D(u,t)\|\ \|f(x(t))\|^2 du$$
$$-\sigma \int_0^t \|D(t,s)\|\ \|f(x(s))\|^2 ds$$
$$= -\frac{1}{2}\langle F(t,x(t))x(t), x(t)\rangle - \frac{1}{2}\langle x(t), F(t,x(t))x(t)\rangle +$$
$$\frac{1}{2}\left\langle x(t), \int_0^t K(t,s,x(s))ds\right\rangle + \frac{1}{2}\left\langle \int_0^t K(t,s,x(s))ds, x(t)\right\rangle$$
$$+\sigma \int_t^\infty \|D(u,t)\|\ \|f(x(t))\|^2 du - \sigma \int_0^t \|D(t,s)\|\ \|f(x(s))\|^2 ds.$$

Since $f_{1i} \geq \lambda_i(F(t,x(t))) \geq f_{0i}$, $f_{0i} > 0$, we can assume that $f_0 = min\{f_{01}, f_{02}, ..., f_{0n})$.

Then, by assumptions $(C1)$, $(C2)$ and an elementary inequality, we have

$$\frac{d}{dt}W(t,x_t) \leq -f_0\|x(t)\|^2 + \int_0^t \|K(t,s,x(s))\| ds\, \|x(t)\|$$
$$+\sigma \int_t^\infty \|f(x(t))\|^2 \|D(u,t)\|\ du\ - \sigma \int_0^t \|f(x(s))\|^2 \|D(t,s)\|\ ds$$
$$\leq -f_0\|x(t)\|^2 + \int_0^t \|f(x(s))\| \|D(t,s)\|\ \|x(t)\| ds$$

$$+\sigma \int_t^{\infty}\|D(u,t)\|\ \|f(x(t))\|^2 du - \sigma \int_0^t \|D(t,s)\|\ \|f(x(s))\|^2 ds$$

$$\leq -f_0\|x(t)\|^2 + \frac{1}{2}\int_0^t \|D(t,s)\|\ [\|f(x(s))\|^2 + \|x(t)\|^2] ds$$

$$+\sigma \int_t^{\infty} \|f(x(t))\|^2 \|D(u,t)\| du - \sigma \int_0^t \|f(x(s))\|^2 \|D(t,s)\|\ ds$$

$$= -f_0\|x(t)\|^2 + \frac{1}{2}\int_0^t \|f(x(s))\|^2 \|D(t,s)\|\ ds + \frac{1}{2}\int_0^t \|D(t,s)\|\ \|x(t)\|^2 ds$$

$$+\sigma \int_t^{\infty} \|f(x(t))\|^2 \|D(u,t)\| du - \sigma \int_0^t \|f(x(s))\|^2 \|D(t,s)\| ds$$

$$\leq -f_0\|x(t)\|^2 + \frac{1}{2}\int_0^t \|f(x(s))\|^2 \|D(t,s)\| ds + \frac{1}{2}\int_0^t \|D(t,s)\|\ \|x(t)\|^2 ds$$

$$+\sigma\beta^2 \int_t^{\infty} \|D(u,t)\|\ \|x(t)\|^2 du - \sigma \int_0^t \|D(t,s)\|\ \|f(x(s))\|^2 ds.$$

Let $\sigma = \frac{1}{2}$. Then, we rearrange this inequality as follows:

$$\frac{d}{dt}W(t,x_t) \leq -f_0\|x(t)\|^2 + \frac{1}{2}\int_0^t \|D(t,s)\|\ \|x(t)\|^2 ds$$

$$+\frac{1}{2}\beta^2 \int_t^{\infty} \|D(u,t)\|\ \|x(t)\|^2 du$$

$$= -\ [f_0 - \frac{1}{2}\int_0^t \|D(t,s)\|\ ds - \frac{1}{2}\beta^2 \int_t^{\infty} \|D(u,t)\| du]\|x(t)\|^2 \leq c\|x(t)\|^2$$

$$\leq 0$$

by $(C2)$.

This result together with the discussion above imply that the trivial solution of IDE (14) is stable.

As for the last step of the proof, consider the set:

$$I \equiv \{(t,x(t)) : \frac{d}{dt}W(t,x_t) = 0\}.$$

We now observe that if $(t,x(t)) \in I$, then $\alpha(t) = 0$ or $\|x(t)\|^2 = 0$. Since $\alpha(t) \geq c > 0$, then $(t,x(t)) \in I$ implies that $\|x(t)\|^2 = 0$. Hence, we have $\|x(t)\|^2 = 0 \Leftrightarrow (x_1(t), x_2(t), \ldots, x_n(t)) = (0,0,\ldots,0)$. This equality and IDE (14), together, imply that $x(t) = 0$ since $K(t, s, x(t)) = 0 \Leftrightarrow x(t) = 0$. Thus, the largest invariant set contained in I is $(t, 0) \in I$. As a consequence, the trivial solution of IDE (14) is asymptotically stable (see the LaSalle's invariance principle (Reissig et al. [75] and Sinha [81, Lemma 1]). Hence, we complete the proof of Theorem 12.

Next, they discuss the uniform stability of the trivial solution of IDE (14). Before proceeding further, it is necessary to introduce additional assumptions given below.

$(C3)$ Let $\beta > 0$, $\Delta > 0$ and $\gamma > 0$ with $\beta, \Delta, \gamma \in R$ such that

$$\|K(t, s, x(s))\| \le \|D(t, s)\| \, \|f(x(s))\|,$$
$$\|f(x(s))\| \le \beta \|x(s)\|,$$
$$\int_0^t \|D(t, s)\| ds \le \alpha_1(t), \forall t \in R,$$
$$\int_0^{t_0} \int_{t_0}^{\infty} \|D(u, s)\| \, duds = \Delta < \infty,$$

where $\alpha_1 \in C(R^+, R^+)$, α_1 is also bounded for $\forall t \in R$, and

$$\alpha(t) = f_0 - \frac{1}{2} \int_0^t \|D(t, s)\| \, ds - \frac{1}{2} \beta^2 \int_t^{\infty} \|D(u, t)\| du \ge 0, \forall t \in R.$$

Theorem 13 (Nieto and Tunç [64]). The zero solution of IDE (14) is uniformly stable if assumptions $(C1)$ and $(C3)$ hold.

Proof. Let $x \in R^n$ and $|x|$ be any norm, C denote the Banach space of continuous functions $\phi: [t_0 - \tau, t_0] \to R^n$, $\tau > 0$, with $\|\phi\|_{t_0} = \sup_{t_0 - \tau \le s \le t_0} |\phi(s)|$.

In the proof of this theorem, the main tool is the LKF $W = W(t, x_t)$, which is used in the proof of Theorem 12. By the time derivative of the functional W and the conditions $(C1)$ and $(C3)$, it is easily derived that $\frac{d}{dt} W(t, x_t) \le 0$. Hence, it is clear that the functional $W(t, x_t)$ is decreasing. In view of this information, we can write that

$$\frac{1}{2} \|x(t)\|^2 \le W(t, x_t) \le W(t_0, \phi(t_0)), t \ge t_0.$$

From this point, we have

$$W(t_0,\phi(t_0)) = \frac{1}{2}\|\phi(t_0)\|^2 + \sigma\int_0^{t_0}\int_{t_0}^{\infty}\|D(u,s)\|\ \|f(\phi(s))\|^2 duds$$
$$\leq \frac{1}{2}\|\phi(t_0)\|^2 + \sigma\beta^2\int_0^{t_0}\int_{t_0}^{\infty}\|D(u,s)\|\,\|\phi(s)\|^2 duds$$
$$\leq \frac{1}{2}[1+2\sigma\beta^2]\|\phi\|_{t_0}^2\int_0^{t_0}\int_{t_0}^{\infty}\|D(u,s)\|\, duds$$
$$= \frac{1}{2}[1+2\sigma\beta^2\Delta]\|\phi\|_{t_0}^2.$$

Hence, we obtain that

$$\|x(t)\|^2 \leq [1+2\sigma\beta^2\Delta]\|\phi\|_{t_0}^2.$$

Now, by the definition of the stability, for each $\varepsilon > 0$, choose a constant $\delta = \left(\frac{1}{\sqrt{1+2\sigma\beta^2\Delta}}\right)\frac{\varepsilon}{2}$ such that if $\|\phi(t)\| < \delta$, $\forall t \in [-\tau, t_0]$, then $\|x(t,t_0,\phi)\| \leq \left(\sqrt{1+2\sigma\beta^2\Delta}\right)\delta < \varepsilon, \forall t \geq t_0$.

Then, since δdoes not depend on the constant $\boldsymbol{t_0}$, the solution $x(t) \equiv 0$ of IDE (14) is uniformly stable. This competes the proof.

Next, Nieto and Tunç [64] considered the following IDDE:

$$\frac{dx}{dt} = -F(t,x)x + \int_{t-\tau}^{t} K(t,s,x(s))ds, \tag{15}$$

with the initial condition $x(t) = \phi(t)$, $t \in [-\tau, 0]$, $\phi \in C([-\tau,0],R^n)$, where $t \in [-\tau,\infty)$, τ is a given positive constant, i.e., constant delay, $x \in R^n$, F is defined as in IDE (14) and $K(t,s,x) \in C(R\times R\times R^n, R^n)$ with $-\tau \leq s \leq t < \infty$ and $K(t,s,x) = 0$ iff $x = 0$.

Finally, for the last result of Nieto and Tunç [64], we require an additional assumption as follows.

$(C4)$ There exist positive constants β, f_0 and $\sigma_1$such that

$$\|K(t,s,x(s))\| \leq \|D(t,s)\|\ \|f(x(s))\|,$$
$$\|f(x(s))\| \leq \beta\|x(s)\|,$$
$$\int_{t-\tau}^{t}\|D(t,s)\|\ ds < +\infty,\ \int_0^t\|D(t,s)\|\ ds < +\infty,$$
$$\text{and } \vartheta(t) = f_0 - \beta\int_{t-\tau}^{t}\|D(t,s)\|\ ds \geq \sigma_1.$$

Theorem 14 (Nieto and Tunç [64]). The trivial solution of IDDE (15) is globally asymptotically stable if assumptions $(C1)$ and $(C4)$ hold.

Proof. The proof is by means of the Lyapunov- Razumikhin method (see, Hale ([37] and Zhou and Egorov [109]). For the first step, we choose a Lyapunov function $W_1(x(t)) = W_1(t, x(t))$ as follows:

$$W_1(t, x(t)) = \frac{1}{2}\langle x(t), x(t)\rangle = \frac{1}{2}\|x(t)\|^2.$$

Next, from the Lyapunov function $W_1(t, x(t))$, it follows that $W_1(t, 0) = 0$. Let $k_1 \in (0,1)$ and $k_2 \geq 1$, $k_1, k_2 \in R$. Then, it is clear that the Lyapunov function W_1 satisfies the inequality:

$$k_1\|x(t)\|^2 \leq W_1(t, x) \leq k_2\|x(t)\|^2.$$

We now consider an arbitrary initial data $(t_0, \phi) \in R^+ \times C([-\tau, 0], R)$ and a point $t > t_0$ such that the Razumikhin condition $W_1(t+\theta, x(t+s)) < W_1(t, x(t))$, $s \in [-\tau, 0]$, holds, i.e., $\frac{1}{2}\|x(t+s)\|^2 < \frac{1}{2}\|x(t)\|^2$holds for $s \in [-\tau, 0]$. Let $x(t) = x(t, t_0, \phi)$ denote the solution of IDDE (15) such that $x(t_0^+ + \theta) = \phi(\theta)$ for $\theta \in [-\tau, 0]$. Differentiating $W_1(t, x(t))$, we find:

$$\frac{d}{dt}W_1(t, x(t)) = -\frac{1}{2}\langle F(t, x(t))x(t), x(t)\rangle - \frac{1}{2}\langle x(t), F(t, x(t))x(t)\rangle$$
$$+\frac{1}{2}\left\langle x(t), \int_{t-\tau}^{t} K(t, s, x(s))ds\right\rangle + \frac{1}{2}\left\langle x(t), \int_{t-\tau}^{t} K(t, s, x(s))ds\right\rangle.$$

Next, assumptions $(C1)$, $(C4)$ and an elementary inequality give that

$$\frac{d}{dt}W_1(t, x(t)) \leq -f_0\|x(t)\|^2 + \|x(t)\| \int_{t-\tau}^{t}\|K(t, s, x(s))\|ds$$
$$\leq -f_0\|x(t)\|^2 + \|x(t)\| \int_{t-\tau}^{t} \|f(x(s))\| \, \|D(t, s)\|ds$$
$$\leq -f_0\|x(t)\|^2 + \beta\|x(t)\| \int_{t-\tau}^{t} \|x(s)\| \, \|D(t, s)\|ds$$
$$\leq -f_0\|x(t)\|^2 + \frac{\beta}{2}\int_{t-\tau}^{t} \|D(t, s)\| \, [\|x(s)\|^2 + \|x(t)\|^2]ds$$

$$= -f_0\|x(t)\|^2 + \frac{\beta}{2}\int_{t-\tau}^{t} \|x(s)\|^2\|D(t,s)\|\, ds$$

$$+\frac{\beta}{2}\int_{t-\tau}^{t} \|x(t)\|^2\|D(t,s)\|ds$$

$$\leq -f_0\|x(t)\|^2 + \frac{\beta}{2}\int_{t-\tau}^{t} \|x(s)\|^2\|D(t,s)\|\, ds$$

$$+\frac{\beta}{2}\|x(t)\|^2 \int_{t-\tau}^{t}\|D(t,s)\|ds. \tag{16}$$

We note that the term: $\frac{\beta}{2}\int_{t-\tau}^{t}\|x(s)\|^2\|D(t,s)\|\, ds$, is included in (16). We now apply the transformation $s - t = \xi$. Then, it follows that $ds = d\xi$. Hence, if $s = t - \tau$, then $\xi = -\tau$. Similarly, if $s = t$, then $\xi = 0$. From this point, we have

$$\frac{\beta}{2}\int_{t-\tau}^{t} \|x(s)\|^2\|D(t,s)\|\, ds = \frac{\beta}{2}\int_{-\tau}^{0} \|x(t+\xi)\|^2\|D(t,t+\xi)\|\, d\xi$$

$$\leq \frac{\beta}{2}\int_{-\tau}^{0} \|x(t)\|^2\|D(t,t+\xi)\|\, d\xi$$

$$= \frac{\beta}{2}\|x(t)\|^2 \int_{-\tau}^{0} \|D(t,t+\xi)\|\, d\xi$$

$$= \frac{\beta}{2}\|x(t)\|^2 \int_{t-\tau}^{t}\|D(t,s)\|\, ds. \tag{17}$$

Substituting the inequality (17) into the inequality (16) and using the given condition, we get

$$\frac{d}{dt}W_1(t,x(t)) \leq -f_0\|x(t)\|^2 + \frac{\beta}{2}\|x(t)\|^2 \int_{t-\tau}^{t} \|D(t,s)\|ds$$

$$+\frac{\beta}{2}\|x(t)\|^2 \int_{t-\tau}^{t} \|D(t,s)\|\, ds.$$

$$= -\left[f_0 - \beta\int_{t-\tau}^{t}\|D(t,s)\|\, ds\right]\|x(t)\|^2 \leq -\sigma_1\|x(t)\|^2 \leq 0.$$

Thus, as consequence, the zero solution of IDDE (15) is globally asymptotically stable. This inequality yields

$$\frac{d}{dt}W_1(t,x(t)) \leq 0.$$

From this point of view, since $W_1(t,x(t))$ is decreasing and positive definite, it is clear that

$$W_1(t,x(t)) \le W_1(0,x(0)) = L_0, L_0 > 0.$$

Next, by this discussion, we arrive at

$$k_1\|x(t)\|^2 \le W_1(t,x(t)) \le W_1(0,x(0)) = L_0,$$

and $\|x(t)\| \le \sqrt{k_1^{-1}L_0}, \forall t \ge t_0$. Finally, by this inequality, it is clear that the solutions of IDDE (15) are bounded as $t \to \infty$. This result completes the proof.

Example 5 (Nieto and Tunç [64]). We now have the IDE:

$$\begin{pmatrix} x_1' \\ x_2' \end{pmatrix} = -\begin{pmatrix} 9+\dfrac{1}{1+exp(t)+x_1^2+x_2^2} & 1 \\ 1 & 9+\dfrac{1}{1+exp(t)+x_1^2+x_2^2} \end{pmatrix}\begin{pmatrix} x_1 \\ x_2 \end{pmatrix} + \int_0^t \begin{pmatrix} exp(-2t+s)\dfrac{\sqrt{x_1^2(s)+x_2^2(s)}}{1+x_1^2(s)} \\ 2\,exp(-2t+s)\dfrac{sin\sqrt{x_1^2(s)+x_2^2(s)}}{1+x_2^2(s)} \end{pmatrix} ds, \quad (18)$$

where $t \ge 0$ and $\begin{pmatrix} x_1 \\ x_2 \end{pmatrix} = \begin{pmatrix} x_1(t) \\ x_2(t) \end{pmatrix} = x(t) = x \in \mathbb{R}^2$. Compare IDE (18) with IDE (14). Then, we have:

$$F(t,x_1,x_2) = \begin{pmatrix} 9+\dfrac{1}{1+exp(t)+x_1^2+x_2^2} & 1 \\ 1 & 9+\dfrac{1}{1+exp(t)+x_1^2+x_2^2} \end{pmatrix}$$

and

$$K(t,s,x(s)) = \begin{pmatrix} exp(-2t+s)\dfrac{\sqrt{x_1^2(s)+x_2^2(s)}}{1+x_1^2(s)} \\ 2\,exp(-2t+s)\dfrac{sin\sqrt{x_1^2(s)+x_2^2(s)}}{1+x_2^2(s)} \end{pmatrix}.$$

Now, some simple calculations give the following relations:

$$\|F(t,x_1,x_2)\| = \left\| \begin{pmatrix} 9+\dfrac{1}{1+exp(t)+x_1^2+x_2^2} & 1 \\ 1 & 9+\dfrac{1}{1+exp(t)+x_1^2+x_2^2} \end{pmatrix} \right\|,$$

$$\sup_{(t,x)\in\mathbb{R}^+\times\mathbb{R}^2} \|F(t,x_1,x_2)\| \le 22 < \infty,$$

$$\lambda_1(F(t,x_1,x_2)) = 8+\frac{1}{1+exp(t)+x_1^2+x_2^2},$$

$$\lambda_2(F(t,x_1,x_2)) = 10+\frac{1}{1+exp(t)+x_1^2+x_2^2},$$

$$f_{0i} = 8 \le 8+\frac{1}{1+exp(t)+x_1^2+x_2^2} \le \lambda_i(F(t,x_1,x_2)),$$

$$\lambda_i(F(t,x_1,x_2)) = 10+\frac{1}{1+exp(t)+x_1^2+x_2^2} \le 11 = f_{1i},$$

$$f_0 = 8,$$

$$8 \le f_{0i} \le \lambda_i(F(t,x)) \le f_{1i} \le 11, (i=1,2).$$

$$\|K(t,s,x(s))\| = \left\| \begin{pmatrix} exp(-2t+s)\dfrac{\sqrt{x_1^2(s)+x_2^2(s)}}{1+x_1^2(s)} \\ 2\,exp(-2t+s)\dfrac{sin\sqrt{x_1^2(s)+x_2^2(s)}}{1+x_2^2(s)} \end{pmatrix} \right\|$$

$$\le exp(-2t+s)\sqrt{x_1^2(s)+x_2^2(s)} + 2\,exp(-2t + s)\left|sin\sqrt{x_1^2(s)+x_2^2(s)}\right|$$

$$\le exp(-2t+s)\sqrt{x_1^2(s)+x_2^2(s)} + 2\,exp(-2t+s)\sqrt{x_1^2(s)+x_2^2(s)}$$

$$\le 3\,exp(-2t+s)\sqrt{x_1^2(s)+x_2^2(s)} = \|D(t,s)\|\,\|f(x(s))\|,$$

where

$$\|D(t,s)\| = 3\,exp(-2t+s), 0 \le s \le t,$$
$$\|D(u,t)\| = 3\,exp(-2u+t), 0 \le t \le u,$$
$$\|f(x_1(s),x_2(s))\| = \sqrt{x_1^2(s)+x_2^2(s)} = \|x(s)\|, \beta = 1.$$

Then,

$$\int_0^t \|D(t,s)\|ds = 3\int_0^t exp(-2t+s)ds = 3\,[exp(-t) - exp(-2t)]$$
$$= \alpha_1(t),$$

$$\int_t^\infty \|D(u,t)\|du = 3\int_t^\infty exp(-2u+t)du = \frac{3}{2}exp(-t) = \alpha_2(t).$$

After these estimates,

$$\alpha(t) = f_0 - \frac{1}{2}\int_0^t \|D(t,s)\|\,ds - \frac{1}{2}\beta^2\int_t^\infty \|D(u,t)\|du$$
$$= 8 - \frac{3}{2}exp(-t) + \frac{3}{2}exp(-2t) - \frac{3}{2}exp(-t)$$
$$\ge 8 - \frac{3}{2} - \frac{3}{2} = 5 = \rho > 0.$$

Thus, assumptions $(C1)$ and $(C2)$ hold, which leads that the trivial solution of IDE (18) is asymptotic stable. For the next step, we derive the following results:

$$\|D(u,s)\| = 3\,exp(-2u+s), 0 \le s \le u,$$
$$\int_0^{t_0}\int_{t_0}^{\infty}\|D(u,s)\|duds = 3\int_0^{t_0}\int_{t_0}^{\infty}\exp(-2u+s)duds = \frac{3}{2}[\exp(3t_0) - \exp(2t_0)] = \Delta < \infty,$$
$$t_0 \ge 0,$$
$$\omega(t) = f_0 - \beta^2\int_0^t \|D(t,s)\|\,ds$$
$$= 8 - 3\,\,exp(-t) + 3\,exp(-2t) \ge 8 - 3\,\,exp(-t) \ge 5 = \gamma.$$

Thus, assumption $(C3)$ of Theorem 13 is fulfilled. As a consequence of this fact, the trivial solution of IDE (18) is uniformly stable.

Example 6. We now consider the IDDE:

$$\begin{pmatrix} x_1' \\ x_2' \end{pmatrix} = -\begin{pmatrix} 9 + \dfrac{1}{1 + exp(t) + x_1^2 + x_2^2} & 1 \\ 1 & 9 + \dfrac{1}{1 + exp(t) + x_1^2 + x_2^2} \end{pmatrix} \begin{pmatrix} x_1 \\ x_2 \end{pmatrix} + \int_{t-1}^{t} \begin{pmatrix} exp(-2t+s)\dfrac{\sqrt{x_1^2(s)+x_2^2(s)}}{1+x_1^2(s)} \\ 2\,exp(-2t+s)\dfrac{sin\sqrt{x_1^2(s)+x_2^2(s)}}{1+x_2^2(s)} \end{pmatrix} ds, \tag{19}$$

where $t-1 \geq 0$, $\tau = 1$ is fixed constant delay and $\begin{pmatrix} x_1 \\ x_2 \end{pmatrix} = x(t) = x \in \mathbb{R}^2$.

Compare IDDE (19) and IDDE (15): $F(t,x)$ is the same as in Example 5 and satisfies assumption $(C1)$ of Theorem 14. Next, we have

$$\int_{t-\tau}^{t} K(t,s,x(s))ds = \int_{t-1}^{t} \begin{pmatrix} exp(-2t+s)\dfrac{\sqrt{x_1^2(s)+x_2^2(s)}}{1+x_1^2(s)} \\ 2\,exp(-2t+s)\dfrac{sin\sqrt{x_1^2(s)+x_2^2(s)}}{1+x_2^2(s)} \end{pmatrix} ds.$$

Then

$$K(t,s,x(s)) = \begin{pmatrix} exp(-2t+s)\dfrac{\sqrt{x_1^2(s)+x_2^2(s)}}{1+x_1^2(s)} \\ 2\,exp(-2t+s)\dfrac{sin\sqrt{x_1^2(s)+x_2^2(s)}}{1+x_2^2(s)} \end{pmatrix}.$$

Similarly, as in Example 4, we obtain the following:

$$\|D(t,s)\| = 3\,exp(-2t+s),\ 0 \leq s \leq t,$$

$$\int_{t-\tau}^{t} \|D(t,s)\| \; ds = 3\int_{t-1}^{t} exp(-2t+s) \; ds$$
$$= 3[exp(-t) - exp(-t+1)] < +\infty,$$
$$\|f(x_1(s), x_2(s))\| = \|x(s)\|, \beta = 1.$$
$$\vartheta(t) = f_0 - \frac{1}{2}(\beta^2+1)\int_{t-1}^{t} \|D(t,s)\| \; ds$$
$$= 8 - 3[exp(-t) - exp(-t+1)] \geq 5 = \sigma.$$

Subject to the previous discussion, condition $(C4)$ of Theorem 14 is fulfilled. Then, we conclude that the trivial solution of IDDE (19) is globally uniformly asymptotic stable. Furthermore, the solutions of IDDE (19) are square integrable and bounded as $t \to \infty$.

Example 7 (Bohner et al. [15]). We now consider the standard electric RLC circuit governed by the scalar IDDE with constant delay:

$$L\frac{d}{dt}I(t) + RI(t) + \frac{1}{C}\int_{t-1}^{t} I(s)ds = E(t), \tag{20}$$

where I, R, L, C and $E$are defined just as in the introduction. Equation (20) can be written as

$$\frac{dI}{dt} = -\frac{R}{L}I(t) - \frac{1}{CL}\int_{t-1}^{t} I(s)ds + \frac{1}{L}E(t).$$

Define a Lyapunov function by $v(t) = \frac{1}{2}I^2(t)$. Differentiating the Lyapunov function $\boldsymbol{v(t)}$ and considering IDDE (20), we have

$$v' = -\frac{R}{L}I^2(t) - \frac{1}{CL}I(t)\int_{t-1}^{t} I(s)ds + \frac{I(t)}{L}E(t)$$
$$\leq -\frac{R}{L}I^2(t) + \frac{1}{2CL}\int_{t-1}^{t} [I^2(t) + I^2(s)]ds + \frac{I(t)}{L}E(t)$$
$$\leq -\frac{R}{L}I^2(t) + \frac{1}{2CL}I^2(t) + \frac{1}{2CL}\int_{t-1}^{t} I^2(s)ds + \frac{I(t)}{L}E(t). \tag{21}$$

Note that the term: $\int_{t-1}^{t} I^2(s)\, ds$, is included in (21).

We now consider an arbitrary initial data $(t_0, \phi) \in \mathbb{R}^+ \times C([-1,0], R)$ and a point $t > t_0$ such that $I^2(t+s) < I^2(t)$, $s \in [-1,0]$. Then, we apply the transformation $s - t = \xi$. In this case, it follows that $ds = d\xi$. Hence, if $s = t - 1$, then $\xi = -1$. As before, if $s = t$, then $\xi = 0$. Thus,

$$\int_{t-1}^{t} I^2(s)\,ds = \int_{-1}^{0} I^2(t+\xi)\,d\xi \leq \int_{-1}^{0} I^2(t)\,d\xi = I^2(t).$$

Substitution of this inequality into (21) implies

$$v' \leq -\frac{R}{L} I^2(t) + \frac{1}{CL} I^2(t) + \frac{I(t)}{L} E(t).$$

Let $E(t) \leq \frac{1}{2C} I(t)$. Then,

$$v' \leq -\frac{R}{L} I^2(t) + \frac{1}{CL} I^2(t) + \frac{1}{2LC} I^2(t) = -\left[\frac{R}{L} - \frac{3}{2CL}\right] I^2(t).$$

We now have for a positive constant β that $v' \leq -\beta I^2(t)$ if $RC > \frac{3}{2}$. Thus, we proved that that the solutions of IDDE (20) are globally uniformly asymptotically stable.

Tunç [93] discussed the system of IIDEs of the form:

$$\frac{dx}{dt} = -f(t, x(t)) + g(t, x(t), x(t-\tau)) + \int_{t-\tau}^{t} h(t, s, x(s))ds + q(t, x(t), x(t - r(t))) \quad (22)$$

with $x(t) = \phi(t), t \in [t_0 - \tau, t_0]$, where ϕ is a continuous initial function, $x = (x_1, \ldots, x_n)^T \in R^n$, $t \in R$, $s \in [-\tau, \infty), \tau$ is a positive constant, $r(t)$, $r \in C([0,\infty),[0,\tau])$, is the variable delay, $f \in C(\mathbb{R} \times \mathbb{R}^n, \mathbb{R}^n), g \in C(R \times R^n \times C_H, R^n)$, $h \in C(R \times [-\tau, \infty) \times C_H, R^n)$ and $q \in C(R \times R^n \times C_H, R^n)$. we assume that $f(t, 0) = 0, g(t, 0,0) = 0$ and $h(t, s, 0) = 0$. Then, the system of IDDEs (22) includes the zero solution when $q(t, x(t), x(t - r(t))) = 0$.

Now, consider an unperturbed system of IDDEs, which is obtained from the system of IDDEs (22), when $q(.) \equiv 0$:

$$\frac{dx}{dt} = -f(t, x(t)) + g(t, x(t), x(t-\tau)) + \int_{t-\tau}^{t} h(t, s, x(s))ds. \quad (23)$$

Next, new sufficient conditions are given, which are needed in the proofs of Theorems 15 and 16.

We suppose that the following conditions hold:

$(H1) f(t,0) = 0, g(t,0,0) = 0, h(t,s,0) = 0, x_i(t) f_i(t,x(t)) > 0$ for all $x_i(t) \neq 0$, $x_i(t) \in R$;

$(H2) H(t,s,0) = 0, H(t,s,x) \equiv \int_t^{\infty} \|h(u,s,x(s))\| du \leq \tilde{L} < \infty$ for $t_0 \leq s \leq t, x \in R^n$,

where $\tilde{L} > 0, \tilde{L} \in R$;

$(H3)$ the function H satisfies the locally Lipschitz condition in x and

$$\|f(t,x(t))\| - \|g(t,x(t),x(t-\tau))\| - \int_t^{\infty} \|h(u,t,x(t))\| du \geq K\|x(t)\|$$

for $t_0 \leq s \leq t, x \in R^n$, where $K > 0, K \in R$.

Theorem 15 (Tunç [93]). The solution $x(t) = 0$ of IDDEs (23) is uniformly asymptotically stable if conditions$(H1)$, $(H2)$ and $(H3)$ hold.

Proof. We introduce an LKF $\Gamma := \Gamma(t,x)$ defined by

$$\begin{aligned}\Gamma(t,x) &:= \|x\| + \lambda \int_{t-\tau}^{t} \int_t^{\infty} \|h(u,s,x(s))\| \, duds \\ &= |x_1| + \ldots + |x_n| + \lambda \int_{t-\tau}^{t} \int_t^{\infty} \|h(u,s,x(s))\| \, duds,\end{aligned} \tag{24}$$

in which $\lambda > 0, \lambda \in R$, $\lambda$is an arbitrary constant. From this point of view, the LKF Γ satisfies that $\Gamma(t,0) = 0, \Gamma(t,x) \geq \|x\|$.

Next, by some simple calculations, we obtain that

$$\begin{aligned}|\Gamma(t,x) - \Gamma(t,y)| &\leq | \, \|x\| - \|y\| \, | \\ &+ \lambda \left| \int_{t-\tau}^{t} \int_t^{\infty} \|h(u,s,x(s))\| \, duds - \int_{t-\tau}^{t} \int_t^{\infty} \|h(u,s,y(s))\| \, duds \right| \\ &= | \, \|x\| - \|y\| \, | + \lambda \left| \int_{t-\tau}^{t} H(t,s,x(s)) \, ds - \int_{t-\tau}^{t} H(t,s,y(s)) \, ds \right| \\ &\leq \|x - y\| + \lambda \int_{t-\tau}^{t} |H(t,s,x(s)) - H(t,s,y(s))| \, ds \\ &\leq \|x - y\| + \lambda L_H \int_{t-\tau}^{t} \|x(s) - y(s)\| \, ds\end{aligned}$$

$$\leq \|x(t) - y(t)\| + \lambda\tau L_H \sup_{t-\tau\leq s\leq t} \|x(s) - y(s)\|$$
$$= K_0 \sup_{t-\tau\leq s\leq t} \|x(s)) - y(s)\|,$$

where $L_H > 0$is the Lipschitz constant and $K_0 := 1 + \lambda\tau L_H$. Hence, it is seen that the LKF Γ satisfies the locally Lipschitz condition in x. For the following step, let

$$Z(t,x) := \lambda \int_{t-\tau}^{t} \int_{t}^{\infty} \|h(u,s,x(s))\| \, duds.$$

Then, using condition $(H2)$, it follows that

$$Z(t,x) = \lambda \int_{t-\tau}^{t} \int_{t}^{\infty} \|h(u,s,x(s))\| \, duds = \lambda \int_{t-\tau}^{t} H(t,s,x(s)) \, ds$$
$$= \lambda \int_{t-\tau}^{t} |H(t,s,x(s)) - H(t,s,0)| \, ds$$
$$\leq \lambda L_H \int_{t-\tau}^{t} \|x(s) - 0\| \, ds$$
$$\leq \lambda\tau L_H \sup_{t-\tau\leq s\leq t} \|x(s)\| = \lambda\tau L_H \|x(s)\|_{[t-\tau,t]},$$

where $L_H > 0$ is the Lipschitz constant. From this point of view, for any two $t_1 > 0, t_2 > 0$ with $t_1 < t_2 < \infty$, adding and subtracting the term $\lambda \int_{t_1-\tau}^{t_2-\tau} H(t,s,x(s)) \, ds$, it follows that

$$Z(t_2,x) - Z(t_1,x) = \lambda \int_{t_2-\tau}^{t_2} H(t,s,x(s)) \, ds - \lambda \int_{t_1-\tau}^{t_1} H(t,s,x(s)) \, ds$$
$$= \lambda \int_{t_1}^{t_2} H(t,s,x(s)) \, ds - \lambda \int_{t_1-\tau}^{t_2-\tau} H(t,s,x(s)) \, ds$$
$$\leq \lambda \int_{t_1}^{t_2} H(t,s,x(s)) \, ds \leq \lambda \int_{t_1}^{t_2} \tilde{L} \, ds = \lambda\tilde{L}(t_2 - t_1).$$

Differentiating the LKF Γ in (24) along the solution of IDDEs (23) and using condition $(H1)$, we derive

$$\frac{d}{dt}\Gamma^+(t,x(t)) = \sum_{i=1}^{n} x_i'(t)\, sgn\, x_i(t+0) + \lambda \int_t^{\infty} \|h(u,t,x(t))\|du$$
$$-\lambda \int_t^{\infty} \|h(u,t-\tau,x(t-\tau))\|du - \lambda \int_{t-\tau}^{t} \|h(t,s,x(s))\|\, ds$$
$$\leq \sum_{i=1}^{n} [-|f_i(t,x(t))| + |g_i(t,x(t),x(t-\tau))| + \int_{t-\tau}^{t} |h_i(t,s,x(s))|ds]$$
$$+\lambda \int_t^{\infty} \|h(u,t,x(t))\|du - \lambda \int_t^{\infty} \|h(u,t-\tau,x(t-\tau))\|du$$
$$-\lambda \int_{t-\tau}^{t} \|h(t,s,x(s))\|\, ds$$
$$\leq -\|f(t,x(t))\| + \|g(t,x(t),x(t-\tau))\| + \int_{t-\tau}^{t} \|h(t,s,x(s))\|ds$$
$$+\lambda \int_t^{\infty} \|h(u,t,x(t))\|du - \lambda \int_{t-\tau}^{t} \|h(t,s,x(s))\|\, ds.$$

Choose $\lambda = 1$. Then, we have

$$\frac{d}{dt}\Gamma^+\big(t,x(t)\big) \leq -\big\|f\big(t,x(t)\big)\big\| + \big\|g\big(t,x(t),x(t-\tau)\big)\big\|$$
$$+ \int_t^{\infty} \|h(u,t,x(t))\|du. \tag{25}$$

Using condition $(H3)$, from (25) we conclude that

$$\frac{d}{dt}\Gamma^+(t,x(t)) \leq -\big[\big\|f\big(t,x(t)\big)\big\| - \big\|g\big(t,x(t),x(t-\tau)\big)\big\| - \int_t^{\infty} \|h(u,t,x(t))\|du\big]$$
$$\leq -K\|x(t)\|,$$

Thus,

$$\frac{d}{dt}\Gamma^+(t,x(t)) \leq -K\|x(t)\|. \tag{26}$$

Let $W = W_1 = \|x\|$, $W_2 = \lambda\tau L_H \|x(s)\|_{[t-\tau,t]}$ and $W_3 = K\|x\|$. Hence, the proof is completed (see, Burton [16, Theorem 4. 2.9]).

The following result is on the integrability of the unperturbed system of IDDEs (23).

Theorem 16 (Tunç [93]). The norms of solutions of the system of IDDEs (23) are integrable in the sense of Lebesgue on R^+, $R^+ = [t_0, \infty)$, if conditions $(H1)$, $(H2)$ and $(H3)$ hold.

Proof. We again consider the L.K.F. Γ, which is defined by (24). By conditions $(H1) - (H3)$, we have the inequality (26). Then, integrating the inequality (26), we obtain

$$\Gamma(t,x) - \Gamma(t_0, \phi(t_0)) \leq -K \int_{t_0}^{t} \|x(s)\|\, ds, t \geq t_0.$$

Since the LKF Γ is decreasing, it is clear that.

$$K \int_{t_0}^{t} \|x(s)\|\, d \leq \Gamma(t_0, \phi(t_0)) - \Gamma(t,x) \leq \Gamma(t_0, \phi(t_0)).$$

Following this inequality, it is therefore seen that

$$\int_{t_0}^{t} \|x(s)\|\, ds \leq K^{-1}[\Gamma(t, x(t) - \Gamma(t_0, \phi(t_0))] \leq K^{-1}\Gamma(t_0, \phi(t_0)).$$

Let $K^{-1}\Gamma(t_0, \phi(t_0)) = K_0, K_0 > 0, K_0 \in R$. Let $t \to +\infty$. Hence, we obtain that,

$$\int_{t_0}^{\infty} \|x(s)\|\, ds \leq K_0.$$

This result shows that the solutions of Volterra IDDEs (23) are integrable on R^+, $R^+ = [t_0, \infty)$.

We now consider the perturbed system of IDDEs (22). We give the following additional condition to obtain the next boundedness theorem.

$(H4)$ There is a continuous nonnegative function $q_0(t)$ such that

$$\|q(t, x, x(t - r(t)))\| \leq |q_0(t)|\, \|x(t)\| \text{ with} \int_{t_0}^{\infty} |q_0(s)| ds < \infty.$$

Theorem 17 (Tunç [93]). The solutions of the perturbed system IDDEs (22) are bounded as $t \to +\infty$ if conditions $(H1)$, $(H2)$, $(H3)$ and $(H4)$ hold.

Proof. Consider the LKF Γ, which is given by (24). Differentiating the LKF (24) and using conditions $(H1) - (H4)$, we obtain that,

$$\frac{d}{dt}\Gamma^+(t, x(t)) \leq \|q(t, x, x(t - r(t)))\| \leq |q_0(t)|\|x\| \leq |q_0(t)|\Gamma(t, x).$$

Integration of this inequality yields $\Gamma(t, x(t)) \leq \Gamma(t_0, \phi(t_0))\, exp(\int_{t_0}^{\infty}|q_0(s)|)ds$.

Hence, by the condition $\int_{t_0}^{\infty}|q_0(s)|ds < \infty$, we can obtain for a positive constant K_1 that $\|x(t)\| \leq K_1$. Calculating the limit of this inequality as $t \to \infty$, we have $\lim_{t\to\infty}\|x(t)\| \leq \lim_{t\to\infty} K_1 = K_1$.

Clearly, we obtain that the solution $x(t)$ is bounded at infinity. Hence, this completes the proof of Theorem 17.

We now give a result on the instability of the unperturbed system of Volterra IDDEs (23). We suppose that the following conditions hold.

$$(H5)\ h(t, s, 0) = 0, g(t, 0,0) = 0, f(t, 0) = 0,$$
$$x_i(t)f_i(t, x(t)) < 0 \text{ for all } x_i(t) \neq 0,\ x_i(t) \in R;$$
$$(H6)\ \|h(u,t,x(t))\| \leq h_0\|K(u,t)\|\ \|x(t)\|,\ \int_t^{\infty}\|K(u, t)\|du \leq K_0 < \infty$$
$$\text{for } t_0 \leq u \leq t, x \in \mathbb{R}^n, \text{ where } h_0,\ K_0 > 0,\ h_0, K_0 \in R;$$
$$(H7)\ \|f(t, x(t))\| - \|g(t, x(t), x(t - \tau))\| - h_0K_0\|x(t)\| \geq K_1\|x(t)\|$$
$$\text{for } t_0 \leq t, x \in \mathbb{R}^n, \text{ where } K_1 > 0, \boldsymbol{K_1} \in \mathbb{R}.$$

Theorem 18 (Tunç [93]). The solution $x(t) = 0$ of the system of Volterra IDDEs (23) is unstable if the following conditions $(H5) - (H7)$ and $\tau < \frac{1}{h_0K_0}$ hold.

Proof. Let us define a new LKF $\Gamma_1 := \Gamma_1(t, x)$ by

$$\Gamma_1(t, x) := \|x\| - \int_{t-\tau}^{t}\int_t^{\infty}\|h(u, s, x(s))\|\, duds. \tag{27}$$

Then, using (27) and $(H6)$,

$$\begin{aligned}
\Gamma_1(t,x) &= \|x\| - \int_{t-\tau}^{t}\int_{t}^{\infty} \|h(u,s,x(s))\|\, duds \\
&\geq \|x\| - h_0 \int_{t-\tau}^{t}\int_{t}^{\infty} \|K(u,s)\|\ \|x(s)\| duds \\
&\geq \|x\| - h_0 K_0 \int_{t-\tau}^{t} \|x(s)\|\, ds \\
&\geq \|x\| - h_0 K_0 \tau \inf_{t-\tau\leq s\leq t} \|x(s)\| \\
&\geq (1 - h_0 K_0 \tau) \inf_{t-\tau\leq s\leq t} \|x(s)\| > 0
\end{aligned}$$

$$\text{if } \tau < \frac{1}{h_0 K_0}.$$

Differentiating the LKF Γ_1 in (27) along the solutions of the system of IDDEs (23) and using condition ($H5$), we have

$$\begin{aligned}
\frac{d}{dt}\Gamma_1^{+}(t,x(t)) &= \sum_{i=1}^{n} x_i'(t)\, sgn\, x_i\,(t+0) - \int_{t}^{\infty} \|h(u,t,x(t))\| du \\
&\quad + \int_{t}^{\infty} \|h(u,t-\tau,x(t-\tau))\| du + \int_{t-\tau}^{t} \|h(t,s,x(s))\|\, ds \\
&= \sum_{i=1}^{n} [-f_i(t,x(t)) + g_i(t,x(t),x(t-\tau)) + \int_{t-\tau}^{t} h_i(t,s,x(s))ds]\, sgn\, x_i\,(t \\
&\quad + 0) \\
&\quad - \int_{t}^{\infty} \|h(u,t,x(t))\| du + \int_{t}^{\infty} \|h(u,t-\tau,x(t-\tau))\| du \\
&\quad + \int_{t-\tau}^{t} \|h(t,s,x(s))\|\, ds \\
&\geq \|f(t,x(t))\| - \|g(t,x(t),x(t-\tau))\| - \int_{t-\tau}^{t} \|h(t,s,x(s))\| ds \\
&\quad - \int_{t}^{\infty} \|h(u,t,x(t))\| du + \int_{t}^{\infty} \|h(u,t-\tau,x(t-\tau))\| du \\
&\quad + \int_{t-\tau}^{t} \|h(t,s,x(s))\|\, ds \\
&\geq \|f(t,x(t))\| - \|g(t,x(t),x(t-\tau))\| - \int_{t}^{\infty} \|h(u,t,x(t))\| du
\end{aligned}$$

$$\geq \|f(t,x(t))\| - \|g(t,x(t),x(t-\tau))\| - h_0 \int_t^{\infty} \|K(u,t)\| du \,\|x(t)\|$$
$$\geq \|f(t,x(t))\| - \|g(t,x(t),x(t-\tau))\| - h_0 K_0 \|x(t)\|$$
$$\geq K_1 \|x(t))\| \geq 0.$$

This result completes the proof of the instability of the system of IDDEs (23).

Recently, Tunç et al. [94] considered the following perturbed nonlinear system of IDDEs with state and distributed delays:

$$\dot{x}(t) = A(t)x(t) + A_d(t)F(x(t-h)) + \int_{t-h}^{t} A_D(s)G(x(s))ds$$
$$+H(t,x(t),x(t-h)), \tag{28}$$

where $x(t) \in R^n$, $t \in [-h,\infty)$, $x(t) = \phi(t)$ on $[-h,0], h \in R$, $h > 0$ is delay , $\phi \in C([-h,0],R^n)$ is an initial function, $A(t) = (a_{ij}(t))$, $A_d = (a_{d_{ij}}(t))$, $A(t), A_d(t) \in C[R^+, R^{n\times n}]$, $R^+ = [0,\infty)$, $A_D = (a_{Dij}(s)), i,j = 1,2,\ldots,n$, $-h \leq s \leq t < \infty$, $A_D \in C([-h,\infty), R^{n\times n})$, $F, G \in C[R^n, R^n]$, $F(0) = G(0) = 0$ and $H \in C(R^+ \times R^n \times R^n, R^n)$.

Consider now the following perturbed system

$$\dot{x}(t) = A(t)x(t) + A_d(t)F(x(t-h)) + \int_{t-h}^{t} A_D(s)G(x(s))ds, \tag{29}$$

which is obtained from system (28).

Tunç et al. [94] improved and generalized an asymptotic stability result which is available in the literature. They obtained an asymptotic stability result under less conservative conditions and gave additional new results for the system of IDDEs (29). The technique used in the proofs is based on the LKF approach.

The following conditions are needed in the next sections.

$(A1)$ The functions $F, G \in C[R^n, R^n]$ satisfy $\|F(x)\| \leq F_0\|x\|$ for all $x \in R^n$ with $F(0) = 0$, $\|G(x)\| \leq G_0\|x\|$ for all $x \in R^n$ with $G(0) = 0$ where F_0, G_0 and h are positive constants.

$(A2)$ There are constants F_0, G_0, h from $(A1)$ and $A_0 > 0$ such that

$$a_{ii}(t)+\sum_{j=1,j\neq i}^{n}\left|a_{ji}(t)\right|\leq -A_0 \text{ for all } t\in R^+$$
$$\text{and } A_0-hG_0\|A_D(t)\|-F_0\|A_d(t+h)\|\geq 0 \text{ for all } t\in[-h,\infty),$$

$(A3)$ There are constants $A_0, G_0, F_0, h>0$ from $(A2)$ and$K_0>0$ such that

$$a_{ii}(t)+\sum_{j=1,j\neq i}^{n}\left|a_{ji}(t)\right|\leq -A_0 \text{ for all } t\in R^+$$

and

$$A_0-hG_0\|A_D(t)\|-F_0\|A_d(t+h)\|\geq K_0 \text{ for all } t\in[-h,\infty).$$

The next theorem is the first main result in [94].

Theorem 19 (Tunç et al. [94]). We assume that conditions $(A1)$ and $(A2)$ are satisfied. Then, the solution$x(t)\equiv 0$ of system (29) is uniformly stable.

Proof. Consider a new LKF $\Pi := \Pi(t,x_t)$ given by

$$\Pi(t,x_t) := \|x(t)\|+\int_{t-h}^{t}\|A_d(s+h)\|\,\|F(x(s))\|ds+\int_{-h}^{0}\int_{t+\eta}^{t}\|A_D(s)\|\,\|G(x(s))\|dsd\eta, \tag{30}$$

where $-h\leq\eta\leq 0$. Next, note that the KLF Π satisfies

$$\Pi(t,0)=0,\ \Pi(t,x_t)\geq\|x\|.$$

Differentiating the LKFΠ in (30) along the solution of system (29), we find

$$\frac{d}{dt}\Pi(t,x_t)=\sum_{i=1}^{n}x_i'(t)\,sgn\,x_i(t+0)$$
$$+\|A_d(t+h)\|\,\|F(x(t))\|-\|A_d(t)\|\,\|F(x(t-h))\|$$
$$+\int_{-h}^{0}[\|A_D(t)\|\,\|G(x(t))\|-\|A_D(t+\eta)\|\,\|G(x(t+\eta))\|]d\eta$$
$$=\sum_{i=1}^{n}x_i'(t)\,sgn\,x_i(t+0)$$
$$+h\|A_D(t)\|\,\|G(x(t))\|-\int_{t-h}^{t}\|A_D(s)\|\,\|G(x(s))\|ds$$
$$+\|A_d(t+h)\|\,\|F(x(t))\|-\|A_d(t)\|\,\|F(x(t-h))\|$$
$$\leq\sum_{i=1}^{n}x_i'(t)\,sgn\,x_i(t+0)+hG_0\|A_D(t)\|\,\|x(t)\|$$

$$-\int_{t-h}^{t}\|A_D(s)\|\,\|G(x(s))\|ds$$
$$+\|A_d(t+h)\|\,\|F(x(t))\|-\|A_d(t)\|\,\|F(x(t-h))\|. \quad (31)$$

Using conditions$(A1)$ and$(A2)$, we obtain that

$$\sum_{i=1}^{n} sgn\,x_i\,(t+0)x_i'(t)\le\sum_{i=1}^{n}a_{ii}(t)\,|x_i(t)|+\sum_{i=1}^{n}\sum_{j=1,j\neq i}^{n}|a_{ji}(t)|\,|x_i(t)|$$
$$+\sum_{i=1}^{n}\sum_{j=1}^{n}|a_{dij}(t)|\,|F_j(x(t-h))|+\int_{t-h}^{t}\|A_D(s)\|\,\|G(x(s))\|ds$$
$$\le\sum_{i=1}^{n}\left(a_{ii}(t)+\sum_{j=1,j\neq i}^{n}|a_{ji}(t)|\right)|x_i(t)|$$
$$+\|A_d(t)\|\,\|F(x(t-h))\|+\int_{t-h}^{t}\|A_D(s)\|\,\|G(x(s))\|ds$$
$$\le -A_0\|x(t)\|+\|A_d(t)\|\,\|F(x(t-h))\|+\int_{t-h}^{t}\|A_D(s)\|\,\|G(x(s))\|ds$$

Combining this inequality and inequality (31) yields

$$\frac{d}{dt}\Pi(t,x_t)\le -A_0\|x(t)\|+\|A_d(t)\|\,\|F(x(t-h))\|$$
$$+\int_{t-h}^{t}\|A_D(s)\|\,\|G(x(s))\|ds$$
$$+hG_0\|A_D(t)\|\,\|x(t)\|-\int_{t-h}^{t}\|A_D(s)\|\,\|G(x(s))\|ds$$
$$+\|A_d(t+h)\|\,\|F(x(t))\|-\|A_d(t)\|\,\|F(x(t-h))\|$$
$$\le -A_0\|x(t)\|+hG_0\|A_D(t)\|\,\|x(t)\|+F_0\|A_d(t+h)\|\,\|x(t)\|.$$

Hence, we obtain

$$\frac{d}{dt}\Pi(t,x_t)\le -\left[A_0-hG_0\|A_D(t)\|-F_0\|A_d(t+h)\|\right]\|x(t)\|. \quad (32)$$

Using condition $(A2)$, from (32) we obtain $\frac{d}{dt}\Pi(t,x_t)\le 0$. Thus, it is proved that the zero solution of system (29) is uniformly stable.

The next theorem, Theorem 20, is the second main result of [94].

Theorem 20 (Tunç et al. [94]). We assume that conditions $(A1)$ and $(A3)$ are satisfied. Then, the solution $x(t) \equiv 0$ of system (29) is uniformly asymptotically stable.

Proof. Using condition$(A1)$ and revising the inequality (8) according condition $(A3)$, we have,

$$\begin{aligned}
\Pi(t, x_t) := \|x(t)\| &+ \int_{t-h}^{t} \|A_d(s+h)\| \, \|F(x(s))\| ds \\
&+ \int_{-h}^{0}\int_{t+\eta}^{t} \|A_D(s)\| \, \|G(x(s))\| ds d\eta \\
\leq \|x(t)\| + F_0 &\int_{t-h}^{t} \|A_d(s+h)\| \, \|x(s)\| ds \\
&+ G_0 \int_{-h}^{0}\int_{t+\eta}^{t} \|A_D(s)\| \, \|x(s)\| ds d\eta \\
\leq \|x(t)\| &+ hF_0 \sup_{t-h\leq s\leq t} \|A_d(s+h)\| \, \|x(s)\| - \\
h\eta G_0 \sup_{t+\eta\leq s\leq t} &\|A_D(s)\| \, \|x(s)\|, -h \leq \eta \leq 0, \Pi(t, x_t) \geq \|x(t)\|,
\end{aligned}$$

$$\begin{aligned}
\text{and } \frac{d}{dt}\Pi(t, x_t) &\leq -\left[A_0 - hG_0 A_D(t) - F_0\|A_d(t+h)\|\right]\|x(t)\| \\
&\leq -K_0 \|x(t)\| < 0, \\
&\|x(t)\| \neq 0.
\end{aligned}$$

Hence, conditions $(A1)$ and $(A3)$ are satisfied. Thus, the solution $x(t) \equiv 0$ of (29) is uniformly asymptotically stable. This is the end of the proof.

The integrability of solutions is given in the following theorem.

Theorem 21 (Tunç et al. [94]). We assume that conditions $(A1)$ and $(A3)$ are satisfied. Then, the norms of the solutions of system (29) are integrable on $\mathbb{R}^+$ in sense of Lebesgue.

Proof. Using the LKF $\Pi := \Pi(t, x_t)$, it follows that (see the proof of Theorem 2)

$$\frac{d}{dt}\Pi(t, x_t) \leq -K_0\|x(t)\|.$$

From this inequality, an integration gives that $\Pi(t,x_t)-\Pi(t_0,\phi(t_0))\le -K_0\int_{t_0}^{t}\|x(s)\|\,ds$. Hence, as the next step, we obtain $\int_{t_0}^{\infty}\|x(s)\|\,ds\le K_0^{-1}\Pi(t_0,\phi(t_0))<+\infty$. The proof is complete.

Example 8 (Tunç et al. [92]). We consider the nonlinear delay system

$$\begin{pmatrix} x_1'(t) \\ x_2'(t) \end{pmatrix}=\begin{pmatrix} -51+\cos t & 1 \\ 1 & -51+\cos t \end{pmatrix}\begin{pmatrix} x_1(t) \\ x_2(t) \end{pmatrix}$$
$$+\begin{pmatrix} \exp(-t) & 1 \\ 1 & \exp(-t) \end{pmatrix}\begin{pmatrix} \dfrac{x_1(t)}{1+x_1^2(t-5^{-1})} \\ \dfrac{x_2(t)}{1+x_2^2(t-5^{-1})} \end{pmatrix}$$
$$+\int_{t-\frac{1}{5}}^{t}\begin{pmatrix} \frac{1}{1+\exp(s)} & 1 \\ 2 & \frac{1}{1+\exp(s)} \end{pmatrix}\begin{pmatrix} \frac{x_1(s)}{1+x_1^2(s)} \\ \frac{x_2(s)}{1+x_2^2(s)} \end{pmatrix}ds, \tag{33}$$

where $t\ge 5^{-1}$. Comparing systems (33) and (29), we obtain the following relations:

$$A(t)=\begin{pmatrix} -51+\cos t & 1 \\ 1 & -51+\cos t \end{pmatrix}=\begin{pmatrix} a_{11}(t) & a_{12}(t) \\ a_{21}(t) & a_{22}(t) \end{pmatrix},$$
$$A_d(t)=\begin{pmatrix} \exp(-t) & 1 \\ 1 & \exp(-t) \end{pmatrix}=\begin{pmatrix} a_{d11}(t) & a_{d12}(t) \\ a_{d21}(t) & a_{d22}(t) \end{pmatrix},$$
$$F(x(t-h))=F(x_1(t-5^{-1}),x_2(t-5^{-1}))=\begin{pmatrix} \dfrac{x_1(t)}{1+x_1^2(t-5^{-1})} \\ \dfrac{x_2(t)}{1+x_2^2(t-5^{-1})} \end{pmatrix},$$
$$\mathrm{G}(x)=G(x_1,x_2)=\begin{pmatrix} \frac{x_1}{1+x_1^2} \\ \frac{x_2}{1+x_2^2} \end{pmatrix}, \text{ where } x=\begin{pmatrix} x_1 \\ x_2 \end{pmatrix}, A_D(t)=$$
$$\begin{pmatrix} \frac{1}{1+\exp(t)} & 1 \\ 2 & \frac{1}{1+\exp(t)} \end{pmatrix}.$$

Considering the above relations, we have the following estimates:

$$F(0)=0,\ \|F(x)\|=\left\|\begin{pmatrix}\frac{x_1}{1+x_1^2(t-5^{-1})}\\ \frac{x_2}{1+x_2^2(t-5^{-1})}\end{pmatrix}\right\|$$

$$=\frac{|x_1|}{1+x_1^2(t-5^{-1})}+\frac{|x_2|}{1+x_2^2(t-5^{-1})}\le F_0\|x\|,\ F_0=1,$$

$$G(0)=0,\ \|G(x)\|=\left\|\begin{pmatrix}\frac{x_1}{1+x_1^2}\\ \frac{x_2}{1+x_2^2}\end{pmatrix}\right\|=\frac{|x_1|}{1+x_1^2}+\frac{|x_2|}{1+x_2^2}\le G_0\|x\|,G_0=1,$$

$$a_{ii}(t)+\textstyle\sum_{j=1,j\neq i}^{n}\left|a_{ji}(t)\right|\le -49=-A_0$$

Since $a_{11}(t)+|a_{21}(t)|=-51+cos\,t+1\le -49=-A_0$ and

$$a_{22}(t)+|a_{12}(t)|=-51+cos\,t+1\le -49=-A_0,$$

$$\|A_d(t+h)\|=\left\|\begin{pmatrix}exp(-t-h) & 1\\ 1 & exp(-t+h)\end{pmatrix}\right\|\max_{1\le j\le 2}\sum_{i=1}^{2}\left|a_{d_{ij}}(t+h)\right|$$

$$=max\{exp(-t-h)+1,1+exp(-t-h)\}$$

$$=1+exp(-t-h)\le 2=d_0,h=\tfrac{1}{5},\ t\ge 5^{-1},$$

$$\|A_D(t)\|=\left\|\begin{pmatrix}\frac{1}{1+exp(t)} & 1\\ 2 & \frac{1}{1+exp(t)}\end{pmatrix}\right\|=\max_{1\le j\le 2}\sum_{i=1}^{2}\left|a_{D_{ij}}(t)\right|$$

$$=max\left\{\frac{1}{1+exp(t)}+2,\frac{1}{1+exp(t)}+1\right\}=\frac{1}{1+exp(t)}+2\le 3=D_0.$$

As for the last step, from these estimates, we arrive at

$$A_0-hG_0\|A_D(t)\|-F_0\|A_d(t+h)\|$$
$$\ge 49-5^{-1}\times 3-1-exp(-t-5^{-1})>46=K_0.$$

Thus, conditions $(A1)-(A3)$ hold, so the solution $x(t)\equiv 0$ of system (33) is uniformly stable, uniformly asymptotically stable and the norms of the solutions of the system are integrable.

For the next boundedness result of this paper, the following conditions are needed.

$(A6)$ The function $H \in C(R^+ \times R^n \times R^n, R^n)$ satisfies,

$\|H(t, x(t), x(t-h))\| \leq |H_0(t)| \, \|x(t)\|$ for all $t \in R^+$ and for all $x, x(t-h) \in R^n$, where $H_0 \in C(R^+, R)$, the function H_0 is bounded and there are constants F_0,
$G_0, A_0 > 0$ and $h > 0$ from $(A2)$ such that

$$a_{ii}(t) + \sum_{j=1, j \neq i}^{n} |a_{ji}(t)| \leq -A_0 \text{ for all } t \in R^+,$$

and $A_0 - hG_0\|A_D(t)\| - F_0\|A_d(t+h)\| - |H_0(t)| \geq 0$ for all $t \in [-h, \infty)$,

The following theorem, Theorem 22, is the boundedness result of [94].

Theorem 22 (Tunç et al. [94])**.** The solutions of perturbed system (28) are bounded at infinity if conditions$(A1)$ and$(A6)$ are satisfied.

Proof. In this proof, we use again the LKF given by (30). Clearly, using conditions $(A1)$ and $(A6)$, one can find that,

$$\begin{aligned}\frac{d}{dt}\Pi(t, x_t) &\leq -[A_0 - F_0\|A_d(t+h)\| - hG_0\|A_D(t)\| \,]\|x(t)\| \\ &\quad + \|H(t, x(t), x(t-h))\| \\ &\leq -[A_0 - F_0\|A_d(t+h)\| - hG_0\|A_D(t)\| - |H_0(t)| \,]\|x(t)\| \leq 0.\end{aligned}$$

As a result of this inequality, we find

$$\Pi(t, x_t) \leq \Pi(t_0, \phi(t_0)), t \geq t_0.$$

Hence, it is notable from the definition of LKF (30) that $\|x(t)\| \leq \Pi(t_0, \phi(t_0))$. If $t \to +\infty$, then $\lim_{t \to +\infty} \||x(t)|\| \leq \lim_{t \to +\infty} \Pi(t_0, \phi(t_0)) = \Pi(t_0, \phi(t_0)) \equiv$ a positive constant. Thus, the solutions of system (28) are bounded as $t \to +\infty$. The proof of Theorem 22 is complete.

Example 9 (Tunç et al. [94])**.** We now take into consideration the following nonlinear delay system:

$$\begin{pmatrix} x_1'(t) \\ x_2'(t) \end{pmatrix} = \begin{pmatrix} -51+\cos t & 1 \\ 1 & -51+\cos t \end{pmatrix} \begin{pmatrix} x_1(t) \\ x_2(t) \end{pmatrix} + \begin{pmatrix} \exp(-t) & 1 \\ 1 & \exp(-t) \end{pmatrix} \begin{pmatrix} \dfrac{x_1(t)}{1+x_1^2(t-5^{-1})} \\ \dfrac{x_2(t)}{1+x_2^2(t-5^{-1})} \end{pmatrix}$$

$$+\int_{t-\frac{1}{5}}^{t} \begin{pmatrix} \frac{1}{1+\exp(s)} & 1 \\ 2 & \frac{1}{1+\exp(s)} \end{pmatrix} \begin{pmatrix} \frac{x_1(s)}{1+x_1^2(s)} \\ \frac{x_2(s)}{1+x_2^2(s)} \end{pmatrix} ds,$$

$$+\begin{pmatrix} \frac{\sin x_1(t)}{1+\exp(t)+\exp(|x_1(t-5^{-1})|)} \\ \frac{\sin x_2(t)}{1+\exp(t)+\exp(|x_2(t-5^{-1})|)} \end{pmatrix}. \tag{34}$$

Comparing systems (34) and (28), it follows that

$$H(t,x_1,x_2,x_1(t-5^{-1}),x_2(t-5^{-1})) = \begin{pmatrix} \dfrac{\sin x_1}{1+\exp(t)+\exp(|x_1(t-5^{-1})|)} \\ \dfrac{\sin x_2}{1+\exp(t)+\exp(|x_2(t-5^{-1})|)} \end{pmatrix}.$$

$$\|H(t,x_1,x_2,x_1(t-5^{-1}),x_2(t-5^{-1}))\| \le \frac{|\sin x_1|}{1+\exp(t)+\exp(|x_1(t-5^{-1})|)} + \frac{|\sin x_2|}{1+\exp(t)+\exp(|x_2(t-5^{-1})|)} \le \frac{|\sin x_1|+|\sin x_2|}{1+\exp(t)} \le \frac{|x_1|+|x_2|}{1+\exp(t)} = \frac{1}{1+\exp(t)}\|x\|,$$

where $|H_0(t) =| \frac{1}{1+\exp(t)}$.

Next, the matrices $A(t), A_d(t), A_D(t)$, the functions $G(x_1, x_2)$ and $F(x_1(t-5^{-1}), x_2(t-5^{-1}))$ are the same as those in Example 1. Hence, it is seen that,

$$A_0 - hG_0\|A_D(t)\| - F_0\|A_d(t+h)\| - |H_0(t)| = 49 - 5^{-1}\times 3 - 2 - \frac{1}{1+\exp(t)} > 0.$$

Then, it is obvious that $\begin{pmatrix} x_1(t) \\ x_2(t) \end{pmatrix}$ of system (34) are bounded.

Recently, Ngoc and Hieu [61] considered the vector IDE of the form:

$$\dot{x}(t) = F(t, x(t), \int_0^t G(t, s, x(s))ds), t \geq t_0, \tag{35}$$

$$x(t) = \varphi(t), t \in [0, t_0], \tag{36}$$

where $F: R_+ \times R^n \times R^n \to R^n, R_+ = [0, \infty)$, and $G: \Delta \times R^n \to R^n$ $(\Delta := \{(t, s) \in R^2 : t \geq s \geq 0\})$ are given matrix-valued continuous functions. In [61], using a novel approach, the authors presented some new scalar criteria for the uniform asymptotic stability of general nonlinear the vector IDE above. Discussion and illustrative examples of the obtained results are also given.

Vector IDE (35) encompasses a large class of Volterra integro-differential equations. In particular, it includes nonlinear IDEs of the form

$$\dot{x}(t) = f(t, x(t)) + \int_0^t g(t, s, x(s))ds$$

and $\dot{x}(t) = A(t)f(x(t)) + \int_0^t B(t, s)g(x(s))ds.$

Let R^n be endowed with the Euclidean norm $\|.\|$. For given $\delta > 0$, let $B_\delta(R^n) := \{x \in R^n : \|x\| \leq \delta\}$. Denote by $C_{t_0} := C([0, t_0], R^n)$, the Banach space of all continuous functions on $[0, t_0]$, endowed with the norm $\|\vartheta\|_{t_0} := sup_{s\in[0,t_0]}\|\vartheta(s)\|$, for $\vartheta \in C_{t_0}$.

Consider the nonlinear IDE (35). For fixed $t_0 \in R^+$ and given $\vartheta \in C_{t_0}$, an initial value condition to IDE (35) is defined by (36). To ensure existence and uniqueness of solutions of the initial-value problem (35)–(36), it is assumed throughout that the function F is (locally) Lipschitz continuous with respect to second and third arguments on each compact subset of $R^+ \times R^n \times R^n$ and the function G is (locally) Lipschitz continuous with respect to the third argument on each compact subset of $\Delta \times R^n$.

Definition 4 (Ngoc and Hieu [61]).

(i) The zero solution of IDE (35) is said to be uniformly stable if for any there exists $r_1 = r_1(\varepsilon) > 0$ such that $t_0 \geq 0, \vartheta \in C_{t_0}, \|\vartheta\|_{t_0} \leq r_1 \Rightarrow \|x(t, t_0, \vartheta)\| < \varepsilon, \forall t \geq t_0$.

(ii) The zero solution of IDE (35) is said to be uniformly asymptotically stable if it is uniformly stable and there exists $r_2 > 0$ with the property that, for each $\varepsilon > 0$ there is a number $T = T(\varepsilon) > 0$ such that $t \geq t_0, \vartheta \in C_{t_0}$, $\|\vartheta\|_{t_0} \leq r_2 \Rightarrow \|x(t, t_0, \vartheta)\| < \varepsilon, \forall t \geq t_0 + T$.

We are now in the position to state the main result in this setup.

Theorem 23 (Ngoc and Hieu [61]).

Suppose there exists $\delta > 0$ and continuous functions $\alpha, \beta: R^+ \to R^+$ and $\gamma: \Delta \to R^+$such that

$$\alpha(t) + \beta(t) \int_0^t \gamma(t, \theta) d\theta \leq 0,\ t \in R^+,$$
$$\sup_{t \in R^+} \int_0^t \gamma(t, s) ds < \infty,$$
$$x^T F(t, x, y) \leq -\alpha(t)\|x\|^2 + \beta(t)\|x\|\|y\|, t \geq 0, x, y \in B_\delta(R^n),$$

and

$$\|G(t, s, z)\| \leq \gamma(t, s)\|z\|,\ (t, s) \in \Delta, z \in B_\delta(R^n).$$

If the the following linear IDE

$$\dot{y}(t) = -\alpha(t) y(t) + \int_0^t \beta(t, s) y(s) ds,\ t \geq t_0,$$

is uniformly asymptotically stable, then the zero solution of IDE (35) is uniformly asymptotically stable.

Therorem 24 (Ngoc and Hieu [61]).

Suppose there exist positive numbers δ, α, β and a continuous function $\gamma: R^+ \to R^+$ such that,

$$x^T F(t, x, y) \leq -\alpha\|x\|^2 + \beta\|x\|\ \|y\|$$

and

$$\|G(t,s,z)\| \le \gamma(t-s)\|z\|,$$

for any $(t,s) \in \Delta$ and for any $x,y,z \in B_\delta(R^n)$. If $-\alpha + \beta \int_0^\infty \gamma(\theta)d\theta < 0$, then the zero solution of IDE (35) is uniformly asymptotically stable.

4. Ulam Stability, Ulam- Hyers Stability, Ulam -Hyers-Rassias Stability of IDEs

Kucche and Shikhare [118] investigated the existence and uniqueness of a solution and Ulam- type stabilities for Volterra delay integro-differential equations on a finite interval. The analysis of [118] is based on the Pachpatte's inequality and Picard operator theory. Examples are provided to illustrate the stability results obtained in the case of a finite interval. Also, an example is given to illustrate that the Volterra delay integro-differential equations are not Ulam–Hyers stable on the infinite interval.

Indeed, in [118], the authors consider the following nonlinear Volterra delay integro-differential equation:

$$x'(t) = f\left(t, x(t), x(g(t)), \int_0^t h\left(t,s,x(s),x(g(s))\right)ds\right), t \in I, \quad (37)$$
$$x(t) = \phi(t), t \in [-r,0],$$

where $\phi \in C([-r,0],R)$.

For a given $\varepsilon > 0$ and a positive nondecreasing continuous function $\psi \in C([-r,b],R_+)$, consider the following inequalities:

$$\left|y'(t) - f\left(t, y(t), y(g(t)), \int_0^t h\left(t,s,y(s),y(g(s))\right)ds\right)\right| \le \varepsilon, t \in I, \quad (38)$$

$$\left|y'(t) - f\left(t, y(t), y(g(t)), \int_0^t h\left(t,s,y(s),y(g(s))\right)ds\right)\right| \le \psi(t), t \in I, \quad (39)$$

$$\left|y'(t) - f\left(t, y(t), y(g(t)), \int_0^t h\left(t,s,y(s),y(g(s))\right)ds\right)\right| \le \varepsilon\psi(t), t \in I. \quad (40)$$

Definition 5 (Kucche and Shikhare [118]). Equation (37) is said to be Ulam-Hyers stable if there exists a real number $C > 0$such that for each $\varepsilon > 0$ and for each solution $y \in C^1([-r,b],R)$ of (38) there exists a solution $x \in C^1([-r,b],R)$ of (37) with $|y(t) - x(t)| \leq C\varepsilon$ for $t \in [-r,b]$.

Definition 6 (Kucche and Shikhare [118]). Equation (37) is said to be generalized Ulam-Hyers stable if there exists $\theta_f \in C(R_+,R_+), \theta_f(0) = 0$ such that for each solution $y \in C'([-r,b],R)$ of (39) there exists a solution $x \in C^1([-r,b],R)$ of (37) with $|y(t) - x(t)| \leq \theta_f(\varepsilon)$ for $t \in [-r,b]$.

Definition 7 (Kucche and Shikhare [118]). Equation (37) is said to be Ulam-Hyers-Rassias stable with respect to the positive nondecreasing continuous function $\psi: [-r,b] \to R_+$ if there exists $C_\psi > 0$ such that for each $\varepsilon > 0$ and for each solution $y \in C^1([-r,b],R)$ of (40) there exists a solution $x \in C^1([-r,b],\mathbb{R})$ of (37) with $|y(t) - x(t)| \leq C_\psi \varepsilon \psi(t)$ for $t \in [-r,b]$.

Definition 8 (Kucche and Shikhare [118]). Equation (37) is said to be generalized Ulam-Hyers-Rassias stable with respect to the positive nondecreasing continuous function $\psi: [-r,b] \to \mathbb{R}_+$ if there exists $C_\psi > 0$ such that for each solution $y \in C'([-r,b],\mathbb{R})$ of (40) there exists a solution $x \in C'([-r,b],R)$ of (37) with $|y(t) - x(t)| \leq C_\psi \psi(t)$ for $t \in [-r,b]$.

The following assumptions are needed to state and prove the main results of Kucche and Shikhare [118].

(H1) (i) Let $f \in C([0,b] \times R^3, R)$, $h \in C([0,b] \times [0,b] \times R^2, R)$ and $g \in C([0,b],[-r,b])$ be such that $g(t) \leq t$.

(ii) There exist constants $L_f, L_h > 0$ such that

$$|f(t,u_1,u_2,u_3) - f(t,v_1,v_2,v_3)| \leq L_f(|u_1 - v_1| + |u_2 - v_2| + |u_3 - v_3|);$$
$$|h(t,s,u_1,u_2,) - h(t,s,v_1,v_2,)| \leq L_h(|u_1 - v_1| + |u_2 - v_2|)$$

for all $t, s \in I, u_i, v_i \in R, (i = 1,2,3)$.

(H2) The function $\psi: [-r, b] \to R_{+}$ is positive, nondecreasing and continuous and there exists $\lambda > 0$ such that $\int_0^t \psi(s)\, ds \leq \lambda\psi(t), t \in [0, b]$.

Theorem 25 (Kucche and Shikhare [118]). Let the functions f and h in (37) satisfy (H1) and assume that (H2) holds. If $bL_f|2 + L_h b| < 1$, then the following assertions hold:

The initial value problem of (37) has a unique solution,

$$x \in C([-r, b], R) \cap C^1([0, b], R);$$

The equation (37) is Ulam-Hyers-Rassias stable with respect to the function ψ.

Corollary 1 Let the functions f and h in (37) satisfy (H1) and assume that (H2) holds. If $bL_f|2 + L_h b| < 1$, then the initial-value problem of (37) has a unique solution and the equation (37) is generalized Ulam-Hyers-Rassias stable with respect to the function ψ.

Proof. By taking $\varepsilon = 1$ in the proof of Theorem 24, we obtain

$$|y(t) - x(t)| \leq C_\psi \psi(t)\,,\ t\ \in [-r, b],$$

showing that (37) is generalized Ulam-Hyers-Rassias stable with respect to the function ψ.

Using arguments similar to those applied in the proof of Theorem 24, one can prove Ulam-Hyers stability of equation (37).

Observing that for $\psi(t) = 1,\ \forall t\ \in [-r, b]$, the assumption (H2) holds, we can state the following corollary of Theorem 24.

Corollary 2 Let the functions f and h in (37) satisfy the hypothesis (H1). If $bL_f|2 + L_h b| < 1$, then the initial-value problem of (37) has a unique solution and the equation (37) is Ulam-Hyers stable.

Proof. By taking $\psi(t) = 1,\ \forall t\ \in [-r, b]$ in the proof of Theorem 24, we obtain

$|y(t) - x(t)| \leq C_\psi \varepsilon\,,\ t\ \in [-r, b]$, and the result follows.

Corollary 3 Let the functions f and h in (37) satisfy the hypothesis (H1). If $bL_f|2 + L_h b| < 1$, then the intial-value problem of (37) has a unique solution and the equation (37) is generalized Ulam-Hyers stable.

Proof. The result follows from Corollary 2, by taking $\theta_f(\varepsilon) = C\varepsilon$.

Applications (Kucche and Shikhare [118]). Consider some important special cases of the initial-value problem (37). Fix any $r > 0$, and define $g_1(t) = t - r,\ t \in [0, b]$. Then, we get the following special case of the initial value problem of (37):

$$x'(t) = f_1\left(t, x(t), x(t-r), \int_0^t h_1\left(t, s, x(s), x(s-r)\right)ds\right), t \in [0, b], \tag{41}$$

$$x(t) = \phi(t), t \in [-r, 0], \tag{42}$$

which is an initial-value problem for a nonlinear Volterra integro-differential difference equation.

Consider the following inequality:

$$\left|y'(t) - f_1\left(t, y(t), y(t-r), \int_0^t h_1\left(t, s, y(s), y(s-r)\right)ds\right)\right| \leq \varepsilon\psi(t),$$
$$t \in [0, b],$$

where ε, ψ and ϕ are as specified before.

As an application, we have the following theorem for the problem (41), (42).

Theorem 24 (Kucche and Shikhare [118]). Suppose that the following assumptions are fulfilled:

(A1) (i) $f_1 \in C([0, b] \times R^3, R), h_1 \in C([0, b] \times [0, b] \times R^2, R)$ and $g_1 \in C([0, b], [-r, b])$ be such that $g_1(t) \leq t$;

(ii) there exist constants $L_{f_1}, L_{h_1} > 0$ such that

$$|f_1(t, u_1, u_2, u_3) - f_1(t, v_1, v_2, v_3)| \leq L_{f_1}(|u_1 - v_1| + |u_2 - v_2| + |u_3 - v_3|);$$

$$|h_1(t, s, u_1, u_2,) - h_1(t, s, v_1, v_2,)| \leq L_{h_1}(|u_1 - v_1| + |u_2 - v_2|);$$

for all $t, s \in [0, b], u_i, v_i \in \mathbb{R}, (i = 1,2,3)$;

(A2) the function $\psi: [-r, b] \to R_+$ is positive, nondecreasing and continuous, and there exists$\lambda > 0$such that $\int_0^t \psi(s)ds \leq \lambda\psi(t)$, $t \in [0, b]$;

(A3) $bL_{f_1}|2 + L_{h_2}b| < 1$.

Then, the problem (40), (41) has a unique solution$x \in C([-r, b], R) \cap C^1([0, b], R)$, and the equation (40) is Ulam-Hyers-Rassias stable with respect to the function ψ.

Theorem 25 (Kucche and Shikhare [118]). Suppose that the following assumptions are fulfilled:

(B1) (i) $f_2 \in C([0,1] \times R^3, R), h_2 \in C([0,1] \times [0,1] \times R^2, R)$ and $g_2 \in C([0,1], [-r, 1])$

be such that $g_2(t) \leq t$;

(ii) there exists constants $L_{f_2}, L_{h_2} > 0$such that

$$|f_2(t, u_1, u_2, u_3) - f_2(t, v_1, v_2, v_3)| \leq L_{f_2}(|u_1 - v_1| + |u_2 - v_2| + |u_3 - v_3|);$$

$$|h_2(t, s, u_1, u_2,) - h_2(t, s, v_1, v_2,)| \leq L_{h_2}(|u_1 - v_1| + |u_2 - v_2|);$$

for all $t, s \in [0,1], u_i, v_i \in R, (i = 1,2,3)$;

(B2) the function $\psi: [-r, 1] \to R_+$ is positive, nondecreasing and continuous, and there exists$\lambda > 0$such that $\int_0^t \psi(s)ds \leq \lambda\psi(t), t \in [0,1]$;

(B3) $bL_{f_2}|2 + L_{h_2}| < 1$.

Then, the problem (41), (42) has a unique solution $x \in C([-r, 1], R) \cap C^1([0,1], R)$, and the equation (41) is Ulam-Hyers-Rassias stable with respect to the function ψ.

Example 10 (Kucche and Shikhare [118]). Consider the following nonlinear delay Volterra integro-differential equation:

$$x'(t) = 1 + \frac{t\,cos(x(t))}{140} - \frac{3x(t)}{140} + \frac{t\,cos(x(g(t)))}{70}$$
$$+\frac{1}{20}\int_0^t \frac{t}{70}\left\{sin(x(s) - sin\left(x(g(s))\right)\right\}ds, t \in [0,5], \tag{43}$$

$$x(t) = 0, t \in [-1,0], \tag{44}$$

where $g(t) = \frac{t}{2}, t \in [0,5]$. Clearly, we have $g(t) \leq t, t \in [0,5]$.

(i) Define h: $[0,5] \times [0,5] \times R \times R \to R$ by

$$h(t,s,x(s),x(g(s))) = \frac{t}{70}[sin(x(s)) - sin(x(g(s)))], t,s \in [0,5].$$

Then, for any $t, s \in [0,5]$ and $x_1, x_2, y_1, y_2 \in R$, we have

$$|h(t,s,x_1,x_2) - h(t,s,y_1,y_2)| \leq \frac{t}{70}\{|\sin x_1 - \sin y_1| + |\sin x_2 - \sin y_2|\}$$
$$\leq \frac{5}{70}\{|x_1 - y_1| + |x_2 - y_2|\}.$$

(ii) Define f: $[0,5] \times R \times R \times R \to R$ by

$$f\left(t, x(t), x(g(t)), \int_0^t h\left(t,s,x(s),x(g(s))\right)ds\right)$$
$$= 1 + \frac{t\,cos(x(t))}{140} - \frac{3x(t)}{140} + \frac{t\,cos(x(g(t)))}{70}$$
$$+\frac{1}{20}\int_0^t \frac{t}{70}\left[sin(x(s) - sin\left(x(g(s))\right)\right]ds$$
$$= 1 + \frac{t\,cos(x(t))}{140} - \frac{3x(t)}{140} + \frac{t\,cos(x(g(t)))}{70}$$
$$+\frac{1}{20}\int_0^t h(t,s,x(s),x(g(s)))ds, t \in [0,5].$$

Then, for any $t \in [0,5]$ and $x_1, x_2, x_3, y_1, y_2, y_3 \in R$, we have

$$|f(t,x_1,x_2,x_3) - f(t,y_1,y_2,y_3)|$$

$$\leq \left\{\frac{t}{140}|\cos x_1 - \cos y_1| + \frac{3}{140}|x_1 - y_1|\right\} + \frac{t}{70}|\cos x_2 - \cos y_2|$$
$$+\frac{1}{20}|x_3 - y_3|.$$

Next, for any $x, y \in R$ with $x < y$, by the Mean Value Theorem, there exists $p, x < p < y$ such that

$$\frac{\cos x - \cos y}{x-y} = -\sin p \Rightarrow |\cos x - \cos y| \leq |x - y|.$$

Therefore, we have

$$|f(t,x_1,x_2,x_3) - f(t,y_1,y_2,y_3)| \leq \left\{\frac{5}{140}|x_1 - y_1| + \frac{3}{140}|x_1 - y_1|\right\} +$$
$$\frac{5}{70}|x_2 - y_2| + \frac{1}{20}|x_3 - y_3|$$
$$\leq \frac{5}{70}\{|x_1 - y_1| + |x_2 - y_2| + |x_3 - y_3|\}.$$

Hence, the above defined functions f and h verify the assumptions (H1) and (H2) with

$$L_f = \frac{5}{70}, L_h = \frac{5}{70}, b = 5.$$

Further, we see that $bL_f(2 + bL_h) = 5\frac{5}{70}\left[2 + \frac{5}{70}5\right] = 0.84183673 < 1.0.84183673 < 1$. Therefore, by Corollary 2, the problem (43), (44) has a unique solution on $[-1,5]$ and the equation (43) is Ulam-Hyers stable on $[0,5]$.

Castro et al. [111] studied the Hyers-Ulam stability and the Hyers-Ulam-Rassias stability for the following class of Volterra integro-differential equation

$$y'(x) = f(x, y(x), \int_a^x k(x,\tau,y(\tau),y(\alpha(\tau)))d\tau), y(a) = c \in R, \quad (45)$$

with $y \in C^1([a,b])$, for $x \in [a,b]$ where, a and b are fixed real numbers, $f: [a,b] \times \mathbb{C} \times \mathbb{C} \to \mathbb{C}$ and $k: [a,b] \times [a,b] \times \mathbb{C} \times \mathbb{C} \to \mathbb{C}$ are continuous functions, and $\alpha: [a,b] \to [a,b]$ is a continuous delay function (i.e., fulfilling $\alpha(\tau) \leq \tau$ for all $\tau \in [a,b]$).

The formal definitions of the abovementioned stabilities are now introduced for our integro-differential equation (45).

If for each function y satisfying

$$\left|y'(x) - f(x, y(x), \int_a^x k(x,\tau,y(\tau),y(\alpha(\tau)))d\tau)\right| \leq \theta, x \in [a,b],$$

where $\theta \geq 0$, there is a solution y_0 of the integro-differential equation and a constant $C > 0$ independent of y and y_0 such that$|y(x) - y_0(x)| \leq C\theta$, for all $x \in [a,b]$, then we say thay the integro-differential equation (45) has the Hyers-Ulam stability.

If for each function y satisfying

$$\left|y'(x) - f(x, y(x), \int_a^x k(x,\tau,y(\tau),y(\alpha(\tau)))d\tau)\right| \leq \sigma(x), x \in [a,b],$$

where σ is a non-negative function, there is a solution y_0 of the integro-differential equation and a constant $C > 0$ independent of y and y_0 such that$|y(x) - y_0(x)| \leq C\sigma(x)$, for all $x \in [a,b]$, then we say they the integro-differential equation (45) has the Hyers-Ulam Rassias stability.

Theorem 26 Castro et al. [111]**.** Let $\alpha: [a,b] \to [a,b]$ be a continuous delay function with $\alpha(t) \leq t$ for all $t \in [a,b]$ and $\sigma: [a,b] \to (0,\infty)$ a non-decreasing continuous function. In addition, suppose that there is$\beta \in [0,1)$ such that $\int_a^x \sigma(\tau)d\tau \leq \beta\sigma(x)$,for all $x \in [a,b]$. Moreover, suppose that $f: [a,b] \times \mathbb{C} \times \mathbb{C} \to \mathbb{C}$ is a continuous function satisfying the Lipchitz condition

$$|f(x,u(x),g(x)) - f(x,v(x),h(x)| \leq M(|u(x) - v(x)| + |g(x) - h(x)|)$$

with $M > 0$ and the kernel $k: [a,b] \times [a,b] \times \mathbb{C} \times \mathbb{C} \to \mathbb{C}$ is a continuous function satisfying the Lipschitz condition

$$|k(x,t,u(t),u(\alpha(t))) - k(x,t,v(t),v(\alpha(t)))| \leq L|u(\alpha(t)) - v(\alpha(t))|$$

with $L > 0$. If $y \in C^1([a,b])$ is such that

$$\left|y'(x) - f(x, y(x), \int_a^x k(x,\tau,y(\tau),y(\alpha(\tau)))d\tau)\right| \leq \sigma(x), x \in [a,b],$$

and $M(\beta + L\beta^2) < 1$, then there is a unique function $y_0 \in C^1([a,b])$ such that

$$y_0'(x) = f(x, y_0(x), \int_a^x k(x, \tau, y_0(\tau), y_0(\alpha(\tau)))d\tau$$

and $|y(x) - y_0(x)| \leq \frac{\beta}{1-M(\beta+L\beta^2)}\sigma(x)$ for all $x \in [a,b]$. This means that under the above conditions, the integro-differential equation (1.1) has the Hyers-Ulam-Rassias stability.

Theorem 27 (Castro et al. [111]). Let $\alpha: [a,b] \to [a,b]$ be a continuous delay function with $\alpha(t) \leq t$ for all $t \in [a,b]$ and $\sigma: [a,b] \to (0,\infty)$ a non-decreasing continuous function. In addition, suppose that there is $\beta \in [0,1)$ such that $\int_a^x \sigma(\tau)d\tau \leq \beta\sigma(x)$, for all $x \in [a,b]$. Moreover, suppose that $f: [a,b] \times \mathbb{C} \times \mathbb{C} \to \mathbb{C}$ is a continuous function satisfying the Lipschitz condition

$$|f(x, u(x), g(x)) - f(x, v(x), h(x)| \leq M(|u(x) - v(x)| + |g(x) - h(x)|)$$

with $M > 0$ and $k: [a,b] \times [a,b] \times \mathbb{C} \times \mathbb{C} \to \mathbb{C}$ is a continuous kernel function satisfying the Lipschitz condition

$$|k(x, t, u(t), u(\alpha(t))) - k(x, t, v(t), v(\alpha(t)))| \leq L|u(\alpha(t)) - v(\alpha(t))|$$

with $L > 0$. If $y \in C^1([a,b])$ is such that

$$\left|y'(x) - f(x, y(x), \int_a^x k(x, \tau, y(\tau), y(\alpha(\tau)))d\tau)\right| \leq \theta, x \in [a,b],$$

where $\theta > 0$ and $M(\beta + L\beta^2) < 1$, then there is a unique function $y_0 \in C^1([a,b])$ such that

$$y_0'(x) = f(x, y_0(x), \int_a^x k(x, \tau, y_0(t), y_0(\alpha(t)))dt)$$

and $|y(x) - y_0(x)| \leq \frac{(b-a)\sigma(b)}{[1-M(\beta+L\beta^2)]\sigma(a)}\theta$ for all $x \in [a,b]$.

Finally, some interesting results on Ulam stability, Ulam- Hyers stability, Ulam -Hyers-Rassias stability of IDEs can be outlined as follows:

Castro and Guerra [112] address the problem of Hyers-Ulam-Rassias stability for nonlinear Volterra integral equations with delay and obtain conditions using the Banach fixed point theorem in a suitable complete metric space. Some illustrative examples are also given in [112].

Castro and Simões [113] discuss different types of stability of general nonlinear integro-differential equations. They introduce a new type of stability, in between the two notions of Hyers-Ulam-Rassias stability and Hyers-Ulam stability, namely the so-called semi-Hyers-Ulam-Rassias stability, in the framework of an appropriate metric space. At the end they introduce some examples to illustrate their results.

Castro and Simões [114] analyse different types of Hyers-Ulam-Rassias stabilities for a class of integro-differential equations and Hyers-Ulam stabilities of certain integral equations with point and distributed delays of Fredholm and Volterra types. Finite and infinite intervals are considered as integration domains. A fixed point argument within the framework of the Bielecki metric is used, and the results are illustrated by concrete examples.

Castro and Simões [115] obtain results regarding the Hyers-Ulam and the Hyers-Ulam-Rassias stabilities for the following class of Volterra integrodifferential equations:

$$y'(t) = f\left(t, y(x), \int_a^x k(t,\tau, y(s\tau), x(\alpha(\tau)))ds\right), y(a) = c \in R,$$

with $y \in C[a,b]$.

de Oliveira et al. [116] consider the problems–generated by some fractional integro-differential equations. Using the Banach fixed point theorem, the authors introduce some stability theorems of Ulam Hyers, Ulam Hyers-Rassias and semi-Ulam-Hyers-Rassias types.

In Kucche [117], Pachpatte's inequality is employed to discuss the Ulam-Hyers stabilities for Volterra integro-differential equations and Volterra delay integro-differential equations in Banach spaces on both finite and infinite intervals. Examples are given to show the applicability of our obtained results.

In Vu and Hoa [119], the Ulam-Hyers stability and the Ulam-Hyers-Rassias stability for the nonlinear Volterra integro-differential equations are established by employing the method of successive approximations. Some simple examples are given to illustrate the main results.

In Shah et al. [120], the stability in terms of Bielecki-Ulam-Hyers and stability in terms of Bielecki-Ulam-Hyers-Rassias of non-linear impulsive

Hammerstein integro-dynamic system with delay and non-linear impulsive mixed integro-dynamic system on time scales are achieved by utilizing a fixed point approach along with a Lipschitz condition and Gronwall's inequality. Examples are also provided for the illustration of the results.

Conclusion

This chapter provides numerous qualitative results with examples in relation to the scalar and vector linear and nonlinear integro –differential equations with and without delays. The given qualitative results can found in the related literature. The qualitative results of these chapter are called stability, instability, uniformly stability asymptotically stability, integrability and boundedness of solutions, etc., of IDEs, IDDEs, and Ulam stability, Ulam-Hyers stability, Ulam -Hyers-Rassias stability, etc., of IDEs, IDDEs. Here, it can be seen that there is an extensive literature on the mentioned qualitative concepts. According to the given results, it can be seen that most of the qualitative results have been obtained using Lyapunov-Krasovskii functional method. Meanwhile a few of these qualitative results have been proved depending upon Lyapunov-Razumikhin technique. On the other hand, the qualitative results of Ulam stability, Ulam- Hyers stability, Ulam -Hyers-Rassias stability, etc., have been proved using the fixed point method. Today, researchers are working effectively on the mentioned all the concepts. By this chapter, our aim is to provides researchers a bundle of scientific works on qualitative theory of IDEs, IDDEs, etc., which may be useful for researchers.

References

[1] Appleby, John A. D., Reynolds, David W., On necessary and sufficient conditions for exponential stability in linear Volterra integro-differential equations. *J. Integral Equations Appl.* 16 (2004), no. 3, 221–240.

[2] Adeyanju, A. A., Ademola, A. T., Ogundare, B. S., On stability, boundedness and integrability of solutions of certain second order integro-differential equations with delay. *Sarajevo J. Math.* 17(30) (2021), no. 1, 61–77.

[3] Adıvar, M., Raffoul, Y. N., Inequalities and exponential stability and instability in finite delay Volterra integro-differential equations. *Rend. Circ. Mat. Palermo* (2) 61 (2012), no. 3, 321–330.

[4] Alahmadi, F., Raffoul, Y. N., Alharbi, S., Boundedness and stability of solutions of nonlinear Volterra integro-differential equations. *Adv. Dyn. Syst. Appl.* 13 (2018), no.1,19–31.

[5] Alquran, M., Jaradat, I., Sivasundaram, S., Elegant scheme for solving Caputo-time-fractional integro-differential equations. *Nonlinear Stud.* 25 (2018), no. 2, 385–393.

[6] Barbu, V., Grossman, S. I., Asymptotic behavior of linear integro-differential systems. *Trans. Amer. Math. Soc.* 173 (1972), 277–288.

[7] Becker, L. C., Function bounds for solutions of Volterra equations and exponential asymptotic stability. *Nonlinear Anal.* 67 (2007), no. 2, 382–397.

[8] Becker, L. C., Uniformly continuous L^2 solutions of Volterra equations and global asymptotic stability. *Cubo* 11 (2009), no. 3, 1–24.

[9] Berezansky, L., Braverman, E., Stability conditions for scalar delay differential equations with a non-delay term. *Appl. Math. Comput.* 250 (2015), 157–164.

[10] Berezansky, L., Domoshnitsky, A., On stability of a second order integro-differential equation. *Nonlinear Dyn. Syst. Theory* 19 (2019), no. 1-SI, 117–123.

[11] Berezansky, L., Domoshnitsky, A., On stability of a second order integro-differential equation. *Nonlinear Dyn. Syst. Theory* 19 (2019), no. 1-SI, 117–123.

[12] Berezansky, L., Diblík, J., Svoboda, Z., Šmarda, Z., Uniform exponential stability of linear delayed integro-differential vector equations. *J. Differential Equations* 270 (2021), 573–595.

[13] Berezansky, L., Domoshnitsky, A., Koplatadze, R., *Oscillation, nonoscillation, stability and asymptotic properties for second and higher order functional differential equations.* CRC Press, Boca Raton, FL, 2020.

[14] Bohner, M., Tunç, O., Qualitative analysis of integro-differential equations with variable retardation. *Discrete & Continuous Dynamical Systems* – B, 2022, 1-19.

[15] Bohner, M., Tunç, O., Tunç, C., Qualitative analysis of Caputo fractional integro-differential equations with constant delays. *Comput. Appl. Math.* 40 (2021), no. 6, Paper No. 214, 17 pp.

[16] Burton, T. A., Volterra integral and differential equations. Second edition. *Mathematics in Science and Engineering*, 202. Elsevier B. V., Amsterdam, 2005.

[17] Burton, T. A., Construction of Liapunov functionals for Volterra equations. *J. Math. Anal. Appl.* 85 (1982), no. 1, 90–105.

[18] Burton, T. A., Mahfoud, W. E., Stability criteria for Volterra equations. *Trans. Amer. Math. Soc.* 279 (1983), no. 1, 143–174.

[19] Burton, T. A., Purnaras, I. K., Singular integro-differential equations with small kernels. *J. Integral Equations Appl.* 25 (2013), no. 1, 1–20.

[20] Cao, Ai Min, An, Wei, Stability of solutions to a nonlinear Volterra integro-differential equation. (Chinese) *Qufu Shifan Daxue Xuebao Ziran Kexue Ban* 32 (2006), no. 4, 57–60.

[21] Castro, L.P., Simões, A.M., Hyers-Ulam and Hyers-Ulam-Rassias Stability for a Class of Integro-Differential Equations. In: Taş, K., Baleanu, D., Machado, J. (eds) *Mathematical Methods in Engineering. Nonlinear Systems and Complexity,* (2019), vol 23. Springer, Cham.

[22] Chang, Xiao, Wang, Ruiqi, Stability of perturbed n-dimensional Volterra differential equations. *Nonlinear Anal.* 74 (2011), no. 5, 1672–1675.

[23] Crisci, M. R., Kolmanovskii, V. B., Russo, E., Vecchio, A., Stability of continuous and discrete Volterra integro-differential equations by Liapunov approach. *J. Integral Equations Appl.* 7 (1995), no. 4, 393–411.

[24] Du, X. T., Stability of Volterra integro-differential equations with respect to part of the variables. (Chinese) *Hunan Ann. Math.* 12 (1992), no. 1-2, 110–115.

[25] Du, X. T., Some kinds of Liapunov functional in stability theory of RFDE. *Acta Math. Appl. Sinica* (English Ser.) 11 (1995), no. 2, 214–224.

[26] Dung, Nguyen Tien, On exponential stability of linear Levin-Nohel integro-differential equations. *J. Math. Phys.* 56 (2015), no. 2, 022702, 10 pp.

[27] El-Borhamy, M., Ahmed, A., Stability analysis of delayed fractional integro-differential equations with applications of RLC circuits. *J. Indones. Math. Soc.* 26 (2020), no. 1, 74–100.

[28] Eloe, P., Islam, M., Zhang, B., Uniform asymptotic stability in linear Volterra integrodifferential equations with application to delay systems. *Dynam. Systems Appl.* 9 (2000), no. 3, 331–344.

[29] Funakubo, M., Hara, T., Sakata, S., On the uniform asymptotic stability for a linear integro-differential equation of Volterra type. *J. Math. Anal. Appl.* 324 (2006), no. 2, 1036–104.

[30] Gözen, M., Tunç, C., Stability in functional integro-differential equations of second order with variable delay. *J. Math. Fundam. Sci.* 49 (2017), no. 1, 66–89.

[31] Graef, J. R., Tunç, C., Continuability and boundedness of multi-delay functional integro-differential equations of the second order. *Rev. R. Acad. Cienc. Exactas Fís. Nat. Ser. A Mat. RACSAM* 109 (2015), no. 1, 169–173.

[32] Graef, J. R., Tunç, C., Şevgin, Sç, Behavior of solutions of nonlinear functional Volterra integro-differential equations with multiple delays. *Dynam. Systems Appl.* 25 (2016), no. 1-2, 39–46.

[33] Graef, J. R.,Tunç, C., Tunç, O., Stability of time-delay systems via the Razumikhin method. *Bol. Soc. Mat. Mex.* (3) 28 (2022), no. 2, Paper No. 26, 13 pp.

[34] Graef, J. R., Tunç, O., Asymptotic behavior of solutions of Volterra integro-differential equations with and without retardation. *J. Integral Equations Appl.* 33 (2021), no. 3, 289 .

[35] Grossman, S. I., Existence and stability of a class of nonlinear Volterra integral equations. *Trans. Amer. Math. Soc.* 150 (1970), 541–556.

[36] Grossman, S. I., Miller, R. K., Perturbation theory for Volterra integrodifferential systems. *J. Differential Equations* 8 (1970), 457–474.

[37] Hale, J., Theory of functional differential equations. Second edition. *Applied Mathematical Sciences,* Vol. 3. Springer-Verlag, New York-Heidelberg, 1977.

[38] Hale, J. K., Verduyn Lunel, S. M., Introduction to functional-differential equations. *Applied Mathematical Sciences,* 99. Springer-Verlag, New York, 1993.

[39] Hammami, M. A., Hnia, N., On the stability of perturbed Volterra integro-differential equations. *J. Integral Equations Appl.* 32 (2020), no. 3, 325–339.

[40] Hatamzadeh, S., Naser-Moghadasi, M., Babolian, E., Masouri, Z., Numerical approach to survey the problem of electromagnetic scattering from resistive strips based on using a set of orthogonal basis functions. *Progress In Electromagnetics Research, PIER* 81, 393–412, 2008.

[41] Hino, Y., Murakami, S., Stability properties of linear Volterra integro-differential equations in a Banach space. *Funkcial. Ekvac.* 48 (2005), no. 3, 367–392.

[42] Hristova, S., Tunc, C., Stability of nonlinear Volterra integro-differential equations with Caputo fractional derivative and bounded delays. *Electron. J. Differential Equations* 2019, Paper No. 30, 11 pp.

[43] Iskandarov, S., Komartsova, E. A., Specific asymptotic stability of solutions to a linear homogeneous Volterra integro-differential equation of the fourth order. Translation of Vestnik Moskov. Univ. Ser. I Mat. Mekh. 2021, no. 1, 22–28. Moscow Univ. *Math. Bull.* 76 (2021), no. 1, 22–28.

[44] Jakovleva, G. F., The stability of the solutions of integro-differential equations with retarded argument. (Russian) *Differencial'Uravnenija 5* (1969) 1218–1224.

[45] Jin, C., Luo, J., Stability of an integro-differential equation. *Comput. Math. Appl.* 57 (2009), no. 7, 1080–1088.

[46] Kyselka, A., Properties of systems of integro-differential equations in the statistics of polymer chains. *Polymer Science* U.S.S.R. 19 (1977), no.11, 2852-2858.

[47] Kress, R., Linear integral equations. Third edition. *Applied Mathematical Sciences,* 82. Springer, New York, 2014.

[48] Lakshmikantham, V., Rama Mohana Rao, M., Theory of integro-differential equations. *Stability and Control: Theory, Methods and Applications,* 1. Gordon and Breach Science Publishers, Lausanne, 1995.

[49] Liberman, L. H., On the stability of solutions of integro-differential equations. (Russian) *Izv. Vysš. Učebn. Zaved. Matematika* 1958 1958 no. 3 (4), 142–151.

[50] Liu, James H., Uniform asymptotic stability via Liapunov-Razumikhin technique. *Proc. Amer. Math. Soc.* 123 (1995), no. 8, 2465–2471.

[51] Maurice A. Jaswon, George T. Symm, Integral equation methods in potential theory and elastostatics. *Computational Mathematics and Applications.* Academic Press [Harcourt Brace Jovanovich, Publishers], London-New York, 1977.

[52] Mesmouli, M. B., Ardjouni, A., Djoudi, A., Stability in nonlinear Levin-Nohel integro-differential equations. *Nonlinear Stud*. 22 (2015), no. 4, 705–718.

[53] Miller, R. K., Asymptotic stability properties of linear Volterra integro-differential equations. *J. Differential Equations* 10 (1971), 485–506.

[54] Miller, R. K., Preservation of uniform asymptotic stability under perturbations. *Proc. Amer. Math. Soc.* 51 (1975), 155–158.

[55] Mahfoud, W. E., Boundedness properties in Volterra integro-differential systems. *Proc. Amer. Math. Soc.* 100 (1987), no. 1, 37–45.

[56] Matsunaga, H., Hashimoto, H., Asymptotic stability and stability switches in a linear integro-differential system. *Differ. Equ. Appl.* 3 (2011), no. 1, 43–55.

[57] Mohana Rao, M. Rama, Srinivas, P., Asymptotic behavior of solutions of Volterra integro-differential equations. *Proc. Amer. Math. Soc.* 94 (1985), no. 1, 55–60.

[58] Nápoles Valdés, Juan E., A note on the boundedness of an integro-differential equation. *Quaest. Math.* 24 (2001), no. 2, 213–216.

[59] Ngoc, Pham Huu Anh, Anh, Tran, The New stability criteria for nonlinear Volterra integro-differential equations. *Acta Math. Vietnam.* 43 (2018), no. 3, 485–501.

[60] Ngoc, Pham Huu Anh, Hieu, Le Trung, Exponential stability of integro-differential equations and applications. *Appl. Math. Lett.* 117 (2021), Paper No. 107127, 8 pp.

[61] Ngoc, Pham Huu Anh, Hieu, Le Trung, On uniform asymptotic stability of nonlinear Volterra integro-differential equations. Internat. *J. Control* 95 (2022), no. 3, 729–735.

[62] Ngoc, Pham Huu Anh, On stability of a class of integro-differential equations. *Taiwanese J. Math.* 17 (2013), no. 2, 407–425.

[63] Nieto, J. J., Tunç, O., An application of Lyapunov-Razumikhin method to behaviors of Volterra integro-differential equations. *Rev. R. Acad. Cienc. Exactas Fís. Nat. Ser. A Mat. RACSAM* 115 (2021), no. 4, Paper No. 197, 21 pp.

[64] Pi, Dingheng, Stability conditions of second order integro-differential equations with variable delay. *Abstr. Appl. Anal.* 2014, Art. ID 371639, 11 pp.

[65] Pi, Dingheng, Study the stability of solutions of functional differential equations via fixed points. *Nonlinear Anal.* 74 (2011), no. 2, 639–651.

[66] Grimmer, Ronald, Seifert, George, Stability properties of Volterra integrodifferential equations. *J. Differential Equations* 19 (1975), no. 1, 142–166.

[67] Raffoul, Y. N., Construction of Lyapunov functionals in functional differential equations with applications to exponential stability in Volterra integro-differential equations. *Aust. J. Math. Anal. Appl.* 4 (2007), no. 2, Art. 9, 13 pp

[68] Raffoul, Youssef, Exponential stability and instability in finite delay nonlinear Volterra integro-differential equations. *Dyn. Contin. Discrete Impuls. Syst. Ser. A Math. Anal.* 20 (2013), no. 1, 95–106.

[69] Raffoul, Youssef N. Necessary and sufficient conditions for stability of Volterra integro-dynamic equation on time scales. *Arch. Math. (Brno)* 52 (2016), no. 1, 21–33.

[70] Raffoul, Youssef, Rai, Habib, Uniform stability in nonlinear infinite delay Volterra integro-differential equations using Lyapunov functionals. *Nonauton. Dyn. Syst.* 3 (2016), no. 1, 14–23.

[71] Rahman, M., *Integral equations and their applications.* WIT Press, Southampton, 2007.

[72] Rautian, N. A., Exponential stability of semigroups generated by Volterra integro-differential equations. *Ufa Math. J.* 13 (2021), no. 4, 63–79.

[73] Razumihin, B. S., On stability of systems with retardation. (Russian) P*rikl. Mat. Meh.* 20 (1956), 500–512.

[74] Razumihin, B. S., The application of Lyapunov's method to problems in the stability of systems with delay. *Avtomat. i Telemeh.* 21 740–748, translated as *Automat. Remote Control* 21 (1960) 515–520.

[75] Reissig, R., Sansone, G., Conti, R., *Non-linear differential equations of higher order.* Translated from the German. Noordhoff International Publishing, Leyden, 1974.

[76] Samedova, S. A., Asymptotic stability of the solution of an integro-differential equation of a certain class. (Russian) Akad. Nauk Azerbaĭdžan. *SSR Dokl.* 14 1958 419–423.

[77] Sedova, N., On uniform asymptotic stability for nonlinear integro-differential equations of Volterra type. *Cybernetics and Physics* 8 (2019), no. 3, 2019, 161–166.

[78] Seifert, G., Liapunov-Razumikhin conditions for stability and boundedness of functional differential equations of Volterra type. *J. Differential Equations* 14 (1973), 424–430.

[79] Seifert, G., Liapunov-Razumikhin conditions for asymptotic stability in functional differential equations of Volterra type. *J. Differential Equations* 16 (1974), 289–297.

[80] Seifert, G., On certain bounded solutions of a Volterra integral equation. Volterra equations (*Proc. Helsinki Sympos. Integral Equations,* Otaniemi, 1978), pp. 267–270, Lecture Notes in Math., 737, Springer, Berlin, 1979.

[81] Sinha, A. S. C., On stability of solutions of some third and fourth order delay-differential equations. *Information and Control* 23 (1973), 165–172.

[82] Sivasundaram, S., Uvah, J., Controllability of impulsive hybrid integro-differential systems. *Nonlinear Anal. Hybrid Syst.* 2 (2008), no. 4, 1003–1009.

[83] Sivasundaram, S., Controllability and observability of nonlinear integro-differential systems. *Proceedings of Dynamic Systems and Applications,* Vol. 2 (Atlanta, GA, 1995), 507–516, Dynamic, Atlanta, GA, 1996.

[84] Tran The Anh, Ngoc, Pham Huu Anh, New stability criteria for linear Volterra time-varying integro-differential equations. *Taiwanese J. Math.* 21 (2017), no. 4, 841–863.

[85] Tunç, C., New stability and boundedness results to Volterra integro-differential equations with delay. *J. Egyptian Math. Soc.* 24 (2016), no. 2, 210–213.

[86] Tunç, C, Asymptotic stability and boundedness criteria for nonlinear retarded Volterra integro-differential equations. *J. King Saud Univ. Sci.* 30(2016), no.4, 3531–3536.

[87] Tunç, C., Stability and boundedness in Volterra integro-differential equations with delay. *Dynam. Systems Appl.* 26 (2017), no. 1, 121–130.

[88] Tunç, C., Akbulut, I., Stability of a linear integro-differential equation of first order with variable delays. *Bull. Math. Anal. Appl.* 10 (2018), no. 2, 19–30.

[89] Tunç, C., Tunc, O., On the exponential study of solutions of Volterra integro-differential equations with time lag. *Electron. J. Math. Anal. Appl.* 6 (2018), no. 1, 253–265.

[90] Tunç, C., Tunç, O., New results on the stability, integrability and boundedness in Volterra integro-differential equations. *Bull. Comput. Appl. Math.* 6 (2018), no. 1, 41–58.

[91] Tunç, C., Tunç, O., On the stability, integrability and boundedness analyses of systems of integro-differential equations with time-delay retardation. Rev. R.

Acad. Cienc. *Exactas Fís. Nat. Ser. A Mat. RACSAM* 115 (2021), no. 3, Paper No. 115, 17 pp.

[92] Tunç, C., Tunç, O., New results on the qualitative analysis of integro-differential equations with constant time-delay. *J. Nonlinear Convex Anal.* 23 (2022), no. 3, 435–448.

[93] Tunç, O., Stability, instability, boundedness and integrability of solutions of a class of integro-delay differential equations. *J. Nonlinear Convex Anal.* 23 (2022), no. 4, 801–819.

[94] Tunç, C., Wang, Y., Tunç, O., Yao, J. -C., New and improved criteria on fundamental properties of solutions of integro–delay differential equations with constant delay. *Mathematics.* 2021, 9(24):3317.

[95] Vanualailai, Jito, Nakagiri, Shin-ichi, Stability of a system of Volterra integro-differential equations. *J. Math. Anal. Appl.* 281 (2003), no. 2, 602–619.

[96] Volterra, V., *Theory of functionals and of integral and integro-differential equations.* Translated from the English and with an appendix by M. K. Kerimov. Edited and with a preface by P. I. Kuznetsov. With a biographical sketch by Edmund Whittaker. "Nauka"", Moscow, 1982.

[97] Ved', Ju. A., Ražapov, G., The stability of solutions of nonlinear integro-differential equations. (Russian) *1967 Studies in Integro-differential Equations in Kirghizia,* No. 4 (Russian) pp. 196–201 Izdat. "Ilim ", Frunze.

[98] Vivek, D., Kanagarajan, K., Sivasundaram, S., On the behavior of solutions of fractional differential equations on time scale via Hilfer fractional derivatives. *Fract. Calc. Appl. Anal.* 21 (2018), no. 4, 1120–1138.

[99] Wazwaz, A.M., *Linear and nonlinear integral equations. Methods and applications.* Higher Education Press, Beijing; Springer, Heidelberg, 2011.

[100] An, Wei , Jin, Zhi Ming, Stability of Volterra integro-differential equations. *Acta Math. Sci.* (Chinese) 16 (1996), no. 2, 214–219.

[101] Yang, Z. C., Stability of impulsive Volterra integro-differential equations. (Chinese) *Sichuan Daxue Xuebao* 40 (2003), no. 1, 16–19.

[102] Yankson, E.,Stability results for neutral integro-differential equations with multiple functional delays. *Khayyam J. Math.* 3 (2017), no. 1, 1–11.

[103] Yassine, H., Stability of global bounded solutions to a nonautonomous nonlinear second order integro-differential equation. *Z. Anal. Anwend.* 37 (2018), no. 1, 83–99.

[104] Yoshizawa, T., Stability Theory by Liapunov's Second Method. Publications of the Mathematical Society of Japan, no. 9., *The Mathematical Society of Japan,* Tokyo 1966.

[105] Zhang, Bo, Necessary and sufficient conditions for stability in Volterra equations of nonconvolution type. *Dynam. Systems Appl.* 14 (2005), no. 3-4, 525–549.

[106] Zhang, Bo, Contraction mapping and stability in an integro-differential equation of nonconvolution type. *Dynam. Systems Appl.* 23 (2014), no. 4, 665–675.

[107] Zhao, Jing, Meng, Fanwei, Stability analysis of solutions for a kind of integro-differential equations with a delay. *Math. Probl. Eng.* 2018, Art. ID 9519020, 6 pp.

[108] Zhang, Z. D., Asymptotic stability of Volterra integro-differential equations. (Chinese) *J. Harbin Inst. Tech.* 1990, no. 4, 11–19.

[109] Zhou, B., Egorov, A. V., Razumikhin and Krasovskiĭ stability theorems for time-varying time-delay systems. *Automatica J. IFAC* 71 (2016), 281–291.

[110] Zhuang, W., Existence and uniqueness of solutions of nonlinear integro-differential equations of Volterra type in a Banach space. *Appl. Anal.* 22 (1986), no. 2, 157–166.

[111] Castro, L. P., Simões, A. M., Hyers-Ulam and Hyers-Ulam-Rassias stability for a class of integro-differential equations. Mathematical methods in engineering, 81–94, *Nonlinear Syst. Complex.*, 23, Springer, Cham, 2019.

[112] Castro, L. P.,Guerra, R. C., Hyers-Ulam-Rassias stability of Volterra integral equations within weighted spaces. *Lib. Math.* (N.S.) 33 (2013), no. 2, 21–35.

[113] Castro, L. P., Simões, A. M., Different types of Hyers-Ulam-Rassias stabilities for a class of integro-differential equations. *Filomat* 31 (2017), no. 17, 5379–5390.

[114] Castro, L. P., Simões, A. M., Hyers-Ulam-Rassias stability of nonlinear integral equations through the Bielecki metric. *Math. Methods Appl. Sci.* 41 (2018), no. 17, 7367–7383.

[115] Castro, L. P., Simões, A. M., Hyers-Ulam and Hyers-Ulam-Rassias stability for a class of integro-differential equations. *Mathematical methods in engineering,* 81–94, Nonlinear Syst. Complex., 23, Springer, Cham, 2019.

[116] de Oliveira, E. Capelas, Sousa, J. Vanterler da C. Ulam-Hyers-Rassias stability for a class of fractional integro-differential equations. *Results Math.* 73 (2018), no. 3, Paper No. 111, 16 pp.

[117] Kucche, K. D., Shikhare, Pallavi U., Ulam-Hyers stability of integrodifferential equations in Banach spaces via Pachpatte's inequality. *Asian-Eur. J. Math.* 11 (2018), no. 4, 1850062, 19 pp.

[118] Kucche, K. D., Shikhare, P. U., Ulam stabilities for nonlinear Volterra delay integro-differential equations. Izv. Nats. Akad. Nauk Armenii Mat. 54 (2019), no. 5, 27–43; reprinted in *J. Contemp. Math. Anal.* 54 (2019), no. 5, 276–287.

[119] Vu, Ho, Hoa, N. V., Ulam-Hyers stability for a nonlinear Volterra integro-differential equation. *Hacet. J. Math. Stat.* 49 (2020), no. 4, 1261–1269.

[120] Shah, Syed Omar, Tunç, Cemil, Rizwan, Rizwan, Zada, Akbar, Khan, Qayyum Ullah, Ullah, Iftikhar, Ullah, I., Bielecki-Ulam's types stability analysis of Hammerstein and mixed integro-dynamic systems of non-linear form with instantaneous impulses on time scales. *Qual. Theory Dyn. Syst.* 21 (2022), no. 4, Paper No. 107, 21 pp.

In: Understanding Integro-Differential Equations ISBN: 979-8-89113-040-1
Editors: J. Vasundhara Devi et al. © 2023 Nova Science Publishers, Inc.

Chapter 4

MATHEMATICAL METHODS FOR INTEGRO-DIFFERENTIAL EQUATIONS AND THEIR APPLICATIONS

Gunvant A. Birajdar[1,*]
and N. Giribabu[2]
[1]School of Rural Development,
Tata Institute of Social Sciences Tuljapur,
Osmanabad, India
[2]Department of Mathematics,
Gayatri Vidya Parishad College of Engineering,
Visakhapatnam, India

Abstract

In this chapter, we have concentrated on powerful methods for solving intero-differential equations. Methods like the differential transform method, Laplace transform method, and Adomian decomposition method to obtain the solution of initial value problems of integro-differential equations.

*Corresponding Author's Email: gabirajdar11@gmail.com.

Keywords: differential transform method, adomian decomposition method, Laplace transform method, integro-differential equations

AMS Subject Classification: 53D, 37C, 65P

1. Introduction

Several scientific and engineering applications are usually described by integral equations or integro differential equations, standard or singular. We have seen a large class of initial and boundary value problems that can be converted to Volterra or Fredholm integral equations. Integral equations and integro-differential equations arise in the potential theory more than any other field. Integro-differential equations arise also in diffraction problems, conformal mapping, water waves, scattering in quantum mechanics, and Volterra's population growth model. The electrostatic, electro magnetic scattering problems and propagation of acoustical and elastical waves are scientific fields where integral and integro-differential equations appear. We have presented in this text a variety of traditional methods and some methods to handle integro-differential equations, Fredholm or Volterra type. Our concern in this text is the determination of the exact solutions in an easy computable fashion. Moreover, our aim is to present these selected methods to facilitate the computational work, and we avoided the abstract theorems that can be found in many other texts.

In this chapter we focus on some powerful method, like Differential transform method (DTM), Laplace transform method (LTM) and, Adomian decomposition method (ADM) for initial value problems for integro-differential equations. Their has been lot of work has done in the area of numerical techniques integro-differential equations parallel to ordinary differential equations have been established for integro-differential equations also. Avudainayagam and Vani [1] developed the numerical technique as Wavelet-Galerkin method (WGM) for integro-differential equations. Rashed [2] employed the Lagrange interpolation method to obtain solutions differential integral and integro-differential equations. We observe that Hosseini and Shahmorad [3] obtained the solution of Fredholm integro-differential equations using arbitrary polynomial bases through Tau method. The Taylor series method is used by Kanwal & Liu [4] and Maleknejad & Mahmoudi [5] for obtaining the solution of integro-differential equations as well as higher order nonlinear integro-differential equations respectively. Maleknejad et. al. [6] come up with rationalized Haar functions method to solve the system of integro-differential equations. The references ([7–11]) also presented the use differential transform methods for various examples. In the second part we have discussed about the Adomian decomposition method. In 1980, Adomian proposed decomposition method popularly known as Adomian decomposition method (ADM) [12–14] has been intensively applied in research to obtain solution of varieties of initial boundary value problems of integral and integro-differential equation. Wazwaz used Adomian decomposition method very extensively in his research [15–17]. The Taylor's method discussed by the researchers in [10, 18, 19]

2. Differential Transform Method

The Differential Transform Method (DTM) is a numerical method for solving differential as well as integral equations. The concept of the differential transform method was first proposed by Zho [20].This method construct, for differential equations, an analytical solution in the form of a polynomial.

The transformation of the kth derivative of a function in one variable is as follows:

$$P(k) = \frac{1}{k!}\left[\frac{d^k p(x)}{dx^k}\right]_{x=x_0} \tag{1}$$

and the inverse transformation is defined by

$$p(x) = \sum_{k=0}^{\infty} P(k)(x-x_0)^k \tag{2}$$

The following theorems that can deducted from (1) and (2) are given as:

Theorem 1. **1.** *If* $u(x) = v(x) \pm w(x), then \quad U(k) = V(k) \pm W(k).$

2. *If* $u(x) = cv(x)$, *then* $U(k) = cV(k)$, *where* c *is a constant.*

3. *If* $u(x) = \dfrac{d^n v(x)}{dx^n}$ *then* $U(k) = \dfrac{(k+n)!}{k!} V(k+n)$.

4. *If* $u(x) = v(x)w(x)$, *then* $U(k) = \sum_{k_1=0}^{k} V(k_1)W(k-k_1)$.

5. *If* $u(x) = x^n$ *then* $U(k) = \delta(k-n)$ *where,* $\delta(k-n) = \begin{cases} 1, & k=n; \\ 0, & k \neq n. \end{cases}$

6. *If* $u(x) = \int_{x_0}^{x} v(t)dt$ *then,* $U(k) = \dfrac{V(k-1)}{k}$, *where* $k \geq 1$.

7. *If* $u(x) = v(x)\int_{x_0}^{x} h(t)dt$ *then* $U(k) = \sum_{k_1=1}^{k} \dfrac{1}{k_1} V(k-k_1)H(k_1-1)$, *where* $k \geq 1$.

Example 1. *The Volterra integral equation*

$$u^{(iv)}(x) = x(1+e^x) + 3e^x + u(x) - \int_0^x u(t)dt \tag{3}$$

with boundary conditions $u(0) = 1, \quad u'(0) = 1, \quad u(1) = 1+e, \quad u'(1) = 2e.$ (4)

Solution We can obtained the another boundary condition (BCs) by using the equation (4) by putting $x = 0$, in equation (3), we get $u^{(iv)}(0) = 4$
We know that the differential transform for the exponential function is e^x is $\frac{1}{k!}$. Also, using the results in the theorem, we obtain

$$U(k+4) = \frac{\delta(k-1) + \sum_{k_1=0}^{k} \delta(k_1-1)\frac{1}{(k-k_1)!} + \frac{3}{k!} + U(k) - \frac{1}{k}U(k-1)}{(k+1)(k+2)(k+3)(k+4)} \tag{5}$$

After simplifying equation (5), we get

$$U(k+1) = \frac{1}{(k+4)!}\left[k!\delta(k-1) + 3 + k + k!U(k) - (k-1)!U(k-1)\right] \tag{6}$$

By using equations (1), (4) and $u^{iv}(0) = 4$ the following transformed BCs at $x = 0$ can be obtained

$$U(0) = 1, \quad U(1) = 1, \quad U(2) = a, \quad U(3) = b, \quad U(4) = \frac{1}{6} \tag{7}$$

where, according to equation (1) , $a = \frac{u''(0)}{2!}$ and $b = \frac{u'''(0)}{3!}$. Applying the recurrence relation in equation (5) and transformed the boundary conditions in equation (6), U(k) for $k \geq 5$ are easily obtained and the using equation (2), we get the terms up to N = 10:

$$\begin{aligned} u(x) &= 1 + x + ax^2 + bx^3 + \frac{x^4}{6} + \frac{x^5}{24} + \left(\frac{1}{180} + \frac{a}{360}\right)x^6 \\ &+ \left(\frac{1}{840} - \frac{a}{2520} + \frac{b}{840}\right)x^7 + \left(\frac{11}{40320} - \frac{b}{6720}\right)x^8 + \frac{x^9}{40320} \\ &+ \left(\frac{1}{453600} + \frac{a}{1814400}\right)x^{10} + O(x^{11}) \end{aligned} \tag{8}$$

The constants a and b are evaluated from boundary conditions given in equation (3), for $x = 1$, by taking $N = 10$, as follows:

$$a = 0.999998, \quad b = 0.500003.$$

For N = 20, these values are obtained as

$$a = 1.000000000000001, \quad b = 0.4999999999999991.$$

It is clear that in the limit case $N \longrightarrow \infty$, a converges to 1 and b converges to $\frac{1}{2}$. With these values of a and b, equation (8) becomes:

$$u(x) = 1 + x + x^2 + \frac{x^3}{2!} + \frac{x^4}{3!} + \frac{x^5}{4!} + \frac{x^6}{5!} + \frac{x^7}{6!} + \frac{x^8}{7!} + \frac{x^9}{8!} + \frac{x^{10}}{9!} + \ldots \tag{9}$$

which can be written in the closed from as follows:

$$u(x) = 1 + xe^x.$$

The Fig. 1 shows the graphical representation and its comparison with exact solution.

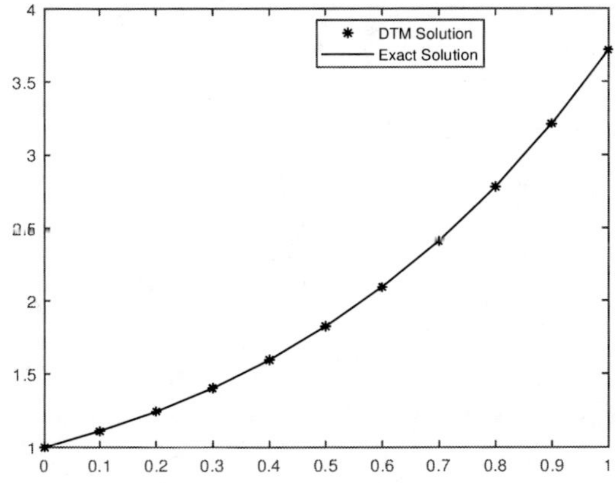

Figure 1. Graphical solution of boundary value problem (3).

Example 2. *In this example we will consider the following nonlinear integro-differential equation of initial value type:*

$$u'(x) = -1 + \int_0^x u^2(t)dt \tag{10}$$

with the initial boundary condition $u(0) = 0.$ (11)

The given initial value problem is a nonlinear in nature. At the point $x = 0$, equations (10) and (11) gives out the following boundary condition

$$u'(0) = -1. \tag{12}$$

Applying theorem and equation (10), we get the following recurrence relation;

$$U(k+1) = \frac{1}{k+1}\left(-\delta(k) + \frac{1}{k}\sum_{k_1=0}^{k-1} U(k_1)U(k-k_1-1)\right). \tag{13}$$

As $k \geq 1$, $\delta(k) = 0$ and equation (13) can be simplified to following form:

$$U(k+1) = \frac{1}{k(k+1)}\left(\sum_{k_1=0}^{k-1} U(k_1)U(k-k_1-1)\right). \tag{14}$$

The differential transformation of the initial boundary conditions in equations (11) and (12) are

$$U(0) = 0 \quad U(1) = -1. \tag{15}$$

By using equations (14) and (15) and (2), the following series solution is obtained:

$$u(x) = -x + \frac{x^4}{12} - \frac{x^7}{252} + \frac{x^{10}}{6048} + \frac{x^{13}}{157248} + \frac{37x^{16}}{158505984} \tag{16}$$

$$+ \frac{25x^{19}}{3011613696} + \frac{73x^{22}}{252975550464} - \frac{449x^{25}}{45535599083520} \tag{17}$$

$$+ \frac{16531x^{28}}{4972487419920384 0} - O(x^{31}). \tag{18}$$

The authors El-Sayed et al. [10] a series solution was obtained for the integro-differential equations (11) by using the ADM [15] which is the same with one we presented in equation (16) up to $O(x^{13})$. However, higher ordered terms are slightly different from the ones we obtained as shown in the following Table 1.

Table 1. Comparison of WGM, ADM, and DTM method

X	WGM	ADM	DTM
0	0	0	0
0.0312	-0.0312	-0.0311999	-0.0311999210
0.0625	0.0625	0.0624987	0.0624987284
0.0938	0.0937	0.0937935	0.0937935492
0.1562	0.1562	0.1561500	0.1561504020
0.2188	0.2186	0.2186090	0.2186091064
0.2812	0.2807	0.2806800	0.2806795005
0.3125	0.3117	0.3117060	0.3117064249
0.3750	0.3734	0.3733560	0.3733561800
0.4062	0.4040	0.4039390	0.4039385134
0.4688	0.4648	0.4647950	0.4647946327
0.5000	0.4948	0.4948230	0.4948225080
0.5938	0.5835	0.5835420	0.5835419298
0.6250	0.6124	0.6124310	0.6124306816
0.6875	0.6692	0.6691670	0.6691672304
0.7188	0.6969	0.6969410	0.6969414638
0.7512	0.7509	0.7508550	0.7251900069
0.8125	0.7771	0.7770900	0.7770901037
0.8750	0.8277	0.8276670	0.8276674429
0.9062	0.8520	0.8519340	0.8519341746
0.9688	0.8984	0.8984520	0.8984522567
1.0000	0.9205	0.9204760	0.9204757107

2.1. Fredholm Integro-Differential Equations

In this section we apply the differential transformed method for Fredholm integro-differential equations. Consider the Fredholm integro-differential equation

$$u'(x) + f(x)u(x) + \int_a^b k(x,t)u(t)dt = g(x), \tag{19}$$

$$\text{with boundary conditions} \quad u(a) = u_0, \tag{20}$$

where $u(x)$ is an unknown function, the function $f(x), g(x)$ and $k(x,t)$ are given real functions of a, b and the equation (19) and (20) called initial boundary value problem for Fredholm integro-differential equations.

One-dimensional differential transform

Here we will define the basic one-dimensional differential transformation are introduced as follows:

Definition 1. *If $f(x)$ is analytic in the time domain T, then*

$$\phi(x,k) = \frac{\partial^k f(x)}{\partial x^k}\bigg|_{t=t_i}, \quad \forall x \in T \tag{21}$$

where k belongs to the set of non-negative integer, denoted as the K-domain. Therefore, equation (21) *can be written as*

$$F(k) = \frac{1}{k!}\phi(x_i,k) = \frac{1}{k!}\left[\frac{\partial^k f(x)}{\partial x^k}\right]\bigg|_{t=t_i}, \quad \forall k \in K, \tag{22}$$

where $F(k)$ is called the spectrum of $f(x)$ at $t = t_i$ in the K-domain. If $f(x)$ can be represented by Taylor's series, expansion as follows:

$$f(x) = \sum_{k=0}^{\infty}(x-x_i)^k F(k) \equiv D^{-1}F(k). \tag{23}$$

The equation (23) is known as inverse of $f(x)$ with the symbol D which denotes the differential transformation process. The particular case of equation (23) when $x_i = 0$ is referred to as the Maclaurin series of $f(x)$ and is expressed as

$$f(x) = \sum_{k=0}^{\infty}(x)^k F(k) \equiv D^{-1}F(k). \tag{24}$$

Using this transformation a differential equation in the domain of interest can be transformed to an algebraic equation in K- domain and $f(x)$ can be obtained by finite term Taylor series expansion and remainder as

$$f(x) = \sum_{k=0}^{\infty}(x-x_i)^k F(k) + R_{n+1}(x). \tag{25}$$

Lemma 1. *If $w(t) = u_1(t)u_2(t)...u_{n-1}(t)u_n(t)$, then*

$$W(k) = \sum_{l_{n-1}=0}^{k}\sum_{l_{n-2}=0}^{l_{n-1}}\cdots\sum_{l_2=0}^{3}\sum_{l_1=0}^{2} U_1(l_1)U_2(l_2-l_1)...U_n(k-l_{n-1}). \tag{26}$$

The fundamental operations for one-dimensional DTM are represented in the following Table 2.

Table 2. The Fundamental operations of one-dimensional DTM

Time Function	Transformed Function
$w(t) = \alpha u(t) \pm \beta v(t)$	$W(k) = \alpha U(K) \pm \beta V(k)$
$w(t) = \dfrac{\partial^m u(t)}{\partial t^m}$	$W(k) = \frac{(k+m)!}{k!} U(k+m)$
$w(t) = u(t)v(t)$	$W(k) = \sum_{l=0}^{k} U(l)V(k-l)$
$w(t) = t^n$	$W(k) = \delta(k-m) = \begin{cases} 1, & \text{k =m;} \\ 0, & k \neq m. \end{cases}$
$w(t) = exp(t)$	$W(k) = \dfrac{k}{k!}$
$w(t) = sin(\omega t + \alpha)$	$W(k) = \frac{\omega^k}{k!} sin(\frac{k\pi}{2} + \alpha)$
$w(t) = cos(\omega t + \alpha)$	$W(k) = \frac{\omega^k}{k!} cos(\frac{k\pi}{2} + \alpha)$

Proof. By using Definition (21), we have

$$\begin{aligned}
W(0) =& \frac{1}{0!} [u_1(t), u_2(t), ..., u_{n-1}(t), u_n(t)] \Big|_{t=t_0} \\
=& U_1(0)U_2(0)...U_{n-1}(0)U_n(0), \\
W(1) =& \frac{1}{1!}\frac{\partial}{\partial t} [u_1(t), u_2(t), ..., u_{n-1}(t), u_n(t)] \Big|_{t=t_0} \\
=& [u_1'(t)u_2(t), ..., u_{n-1}(t), u_n(t) + u_1(t), u_2'(t), ..., u_{n-1}(t), u_n(t) + ... \\
=& + u_1(t), u_2(t), ..., u_{n-1}'(t), u_n(t) + u_1(t), u_2(t), ..., u_{n-1}(t), u_n'(t)] \Big|_{t=t_0} \\
=& U_1(1)U_2(0)...U_{n-1}(0)U_n(0) + U_1(0)U_2(1)...U_{n-1}(0)U_n(0) + ... \\
& + U_1(0)U_2(0)...U_{n-1}(0)U_n(0) + U_1(0)U_2(0)...U_{n-1}(0)U_n(1), \\
W(2) =& U_1(1)U_2(1)U_3(0)...U_{n-1}(0)U_n(0) + U_1(0)U_2(1)U_3(1)... \\
& \times U_{n-2}(0)U_{n-1}(0)U_n(0) + ... + U_1(0)U_2(0)U_3(0)... \\
& ...U_{n-2}(1)U_{n-1}(1)U_n(0) \\
& + U_1(0)U_2(0)U_3(0)...U_{n-2}(0)U_{n-1}(1)U_n(1).
\end{aligned}$$

In the general form we have

$$W(k) = \sum_{l_{n-1}=0}^{k} \sum_{l_{n-2}=0}^{l_{n-1}} ... \sum_{l_2=0}^{3} \sum_{l_1=0}^{2} U_1(l_1)U_2(l_2 - l_1)...U_n(k - l_{n-1}). \tag{27}$$

Two-dimensional differential transform
The two dimensional differential transformation is introduced as

Definition 2.

$$W(k,h) = \frac{1}{k!h}\frac{\partial^{k+h}}{\partial x^k \partial y^h} w(x,y)\Big|_{x=0,y=0} \tag{28}$$

where

$$W(x,y) = \sum_{k=0}^{\infty}\sum_{h=0}^{\infty}\frac{1}{k!h}\left[\frac{\partial^{k+h}}{\partial x^k \partial y^h} w(x,y)\right]\Big|_{x=0,y=0}, \quad x^k y^h \equiv D^{-1}[W(k,h)] \tag{29}$$

Definition 3. *If* $w(x,y) = u(x,y)v(x,y), u(x,y) = D^{-1}[U(k,h)],\ v(x,y) = D^{-1}[V(k,h)]$ *and* $\otimes$ *denote the convolution, then*

$$D[w(x,y)] = D[u(x,y)v(x,y)] = U(k.h)\otimes V(k,h).$$

In the following Table 3 the operations for two-dimensional DTM are presented.

Table 3. The fundamental operations of two-dimensional DTM

Time Function	Transformed Function
$w(x,y) = \alpha u(x,y) \pm \beta v(x,y)$	$W(k,h) = \alpha U(k,h) \pm V(k,h)$
$w(x,y) = \frac{\partial^{a+b} u(x,y)}{\partial x^a y^b}$	$W(k,h) = (k+1)(k+2)....(k+a)(h+1)(h+2)...$ (a+b)U(k+a,h+b)
w(x,y) = u(x,y)v(x,y)	$W(k,h) = \sum_{a=0}^{k}\sum_{b=0}^{h} U(a,h-a)V(k-b,b)$
$w(x,y) = x^m y^n$	$W(k,h) = \delta(k-m,h-n) = \delta(k-m)\delta(h-n)$

2.2. Higher-Order Integro-Differential Equations

This section will focus on higher order integro-differential equations by using differential transformed method.

Consider the integro-differential equation

$$u^n(t) + \sum_{i=1}^{n} f_i(t) y^{n-i}(t) + \int_a^b w(t,x)y(x)dx = g(t) \tag{30}$$

We convert a equation (30) of nth order integro-differential equation with the initial conditions

$$u(a) = \alpha_1, \quad u'(a) = \alpha_2, u''(a) = \alpha_3, ..., u^{n-1}(a) = \alpha_n \ ,$$

into a system of first-order integro-differential equations, let $y_1(t) = u(t)$, $y_2(t) = u'(t), ..., y_n(t) = u^{n-1}(t)$.

By using this notation, we obtain the first-order system

$$\begin{aligned}
y_1'(t) &= y_2(t), \\
y_2'(t) &= y_3(t), \\
y_3'(t) &= y_4(t), \\
&. \\
&. \\
y_{n-1}'(t) &= y_n(t) \\
y_n'(t) &= g(t) - \sum_{i=1}^{n} f_i(t) y_{n+1-i}(t) - \int_a^b w(t,x) y_1(t) dt
\end{aligned} \tag{31}$$

with initial conditions

$$y_1(a) = \alpha_1, y_2(a) = \alpha_2, ..., y_n(t) = \alpha_n(t).$$

Now, using the differential transformation of system (31) and using the relations, we get

$$\begin{aligned}
Y_1(k+1) &= \frac{1}{k+1} Y_2(k), \\
Y_2(k+1) &= \frac{1}{k+1} Y_3(k), \\
Y_3(k+1) &= \frac{1}{k+1} Y_4(k), \\
&. \\
&. \\
Y_{n-1}(n-1) &= \frac{1}{k+1} Y_n(k), \\
Y_n(k+1) &= \frac{1}{k+1} \left[G(k) - \sum_{i=1}^{n} F_1(k) \otimes Y_{n+1-i}(k) + (b-a) \sum_{i=0}^{h} W(k, h-l) Y_1(k) \right]
\end{aligned}$$

Using the initial conditions we have,

$$Y_1(0) = \alpha_1 \ Y_2(0) = \alpha_2, Y_3(0) = \alpha_3, ..., Y_n(0) = \alpha_n.$$

2.3. Numerical Approach of Differential Transform Method

Here, we developed the numerical version of differential transform method. In the first sub-domain $y(t)$ can be described by $y_0(t)$ can be expressed by using Taylor's series expansion near $a = x_0$ as

$$y_0(t) = \sum_{j=0}^{n} Y_{10}(j)(x-a)^j,$$

where the subscript "0" denotes that the Taylor's polynomial near a.

$$y(t_1) \cong y_0(t_1) = \sum_{j=0}^{n} Y_{10}(j)(h)^j, \quad h = x_1 - x_0.$$

The final value $y_0(x_1) = U_{11}(0)$

$$y(t_2) \cong y_1(t_2) = \sum_{j=0}^{n} Y_{11}(j)(h)^j, \quad h = x_2 - x_1,$$

where $a = x_1$. In the same manner, $y(t_3)$, can be represented as

$$y(t_3) \cong y_2(t_3) = \sum_{j=0}^{n} Y_{12}(j)(h)^j, \quad h = x_2 - x_1.$$

Hence, the solution on the grid points t_{i+1} can be obtained from

$$y(t_{i+1}) \cong y_i(t_{i+1}) = \sum_{j=0}^{n} Y_{1i}(j)(h)^j, h = x_{i+1} - x_i, i = 0, 1, 2, ..., N-1. \tag{32}$$

3. Numerical Results

As an application of differential transform method, we present the couple problems are given below.

Example 3. *Consider the integro-differential equations*

$$y^{'}(t) = 1 - \frac{t}{3} + \int_0^1 xty(x)dx, \tag{33}$$

$$u(0) = 0. \tag{34}$$

Let $h = 0.1$ and $m = 10$. The integro-differential equation of the system between t_i and t_{i+1} can be present as

$$y^{'}(t^*) = 1 - \frac{t_i}{3} - \frac{t^*}{3} + \int_0^1 (y^* + y_i)xy(x)dx, \tag{35}$$

where $t^* = t - t_i$. Taking the differential transformation, we obtain

$$(k+1)Y_i(k+1) = \left(1 - \frac{t_i}{3}\right)\delta(k) - \frac{1}{3}\delta(k-1) + t_i\delta(k-1) \otimes Y(k) + \delta(k-1, h-1) \otimes Y(h)$$

Set $k = h$ and by using tables, we obtain

$$Y_i(k+1) = \frac{1}{(k+1)}\left[(1 - \frac{t_i}{3})\delta(k) - \frac{1}{3}\delta(k-1) + t_i \sum_{j=0}^{k} Y_i(j)\delta(k-j-1)\right] + \sum_{j=0}^{k} Y_i(j)\delta(k-j-1)\delta(k-1). \quad (36)$$

With $Y_0(0) = 0$, the approximation of $y(t)$ at the grid points can be obtained. The following table shows the comparison of numerical solution y_i and exact solution $y(t_i)$. The following Table 4 exhibits the error analysis.

Table 4. Represent the solution with its error

t	y_i	$y(t_i)$	$\|\|y(t_i) - y_i\|\|$
0.0	0.0	0.0	0.0
0.1	0.09833333	0.10000000	1.66666667e-03
0.2	0.19390611	0.20000000	6.09388620e-03
0.3	0.28679821	0.30000000	1.32017875e-02
0.4	0.37708594	0.40000000	2.29140636e-02
0.5	0.46484216	0.50000000	3.51578404e-02
0.6	0.55013645	0.60000000	6.69648304e-02
0.8	0.71360162	0.80000000	8.63983845e-02
0.9	0.79189609	0.90000000	1.08103910e-01
1.0	0.86797601	1.00000000	1.32023989e-01

In the following Fig. 2 we have given the graphical representation of solution of boundary value problem (33).

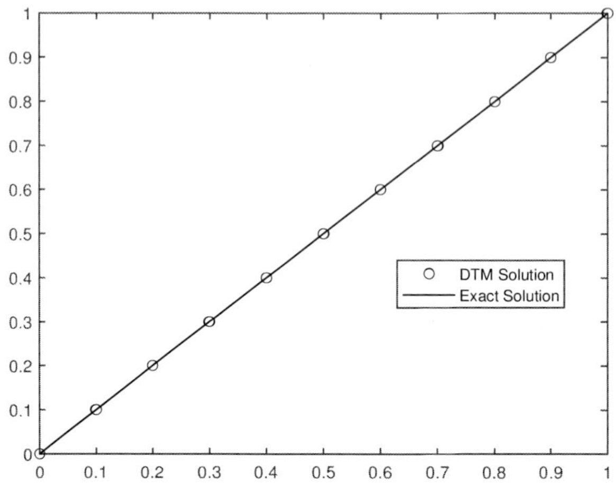

Figure 2. Graphical solution of boundary value problem. (33)

Example 4. *Consider the integro-differential equation*

$$y'(t) = te^t + e^t - t + \int_0^t ty(t)dt \quad y(0) = 0. \tag{37}$$

having exact solution $y(t) = te^t$.

Let $h = 0.1$ and $n = 10$. The integro-differential equation of the system between t_i and t_{i+1} can be represented as

$$y'(t^*) = (x^* + t_i)e^{t^*+t_i} - (t^* + t_i) + \int_0^1 t^* + t_i y(x)dx, \tag{38}$$

where $t^* = t - t_i$. applying the differential transformation of equation (39), we get

$$Y_i(k+1) = \frac{1}{(k+1)}\left[\frac{1}{k!}(t_i+1)e^{t_i} - t_i\delta(k) - \delta(k-1) + t_iY_i(k) + \sum_{j=0}^{k} Y_i(j)\delta(k-j-1)\right] + \sum_{j=0}^{k}\frac{1}{j!}\delta(k-j-1). \tag{39}$$

By using $Y_0(0) = 0$, the approximation of $y(t)$ on the grid points can be obtained. The following Table 5 exhibits the errors between the exact solution and solution obtained by DTM.

Table 5. Represent the solution along with its error.

t	y_i	$y(t_i)$	$\|y(t_i) - y_i\|$
0.0	0.0	0.0	0.0
0.0	0.00000000	0.000000000	0.00000000
0.1	0.10050526	0.11051709	1.00118319e-02
0.2	0.21641542	0.24428055	2.78651355e-02
0.3	0.35408455	0.40495764	5.08730892e-02
0.4	0.52119424	0.59672988	7.55356316e-02
0.5	0.72717178	0.82436064	9.71888592e-02
0.6	0.98371957	1.09327128	1.09551714e-01
0.7	1.30549366	1.40962690	1.04133232e-01
0.8	1.71098147	1.78043274	0.94512700e-02
0.9	2.22364623	2.21364280	1.00034260e-02
1.0	2.87342954	2.71828183	1.55147712e-01

The following Fig. 3 shows the solution obtained by differential transform method.

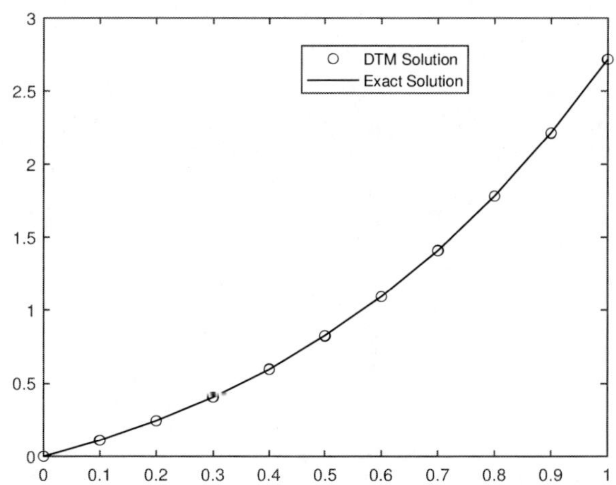

Figure 3. Graphical solution of boundary value problem (37).

4. The Adomain Decomposition Method

Consider a Volterra intergro-differential equation of the second kind given by

$$g''(x) = h(x) + \int_0^x k(x,y)g(y)dy, \;\; g(0) = g_0, \;\; g'(0) = g_1 \rightarrow (1)$$

Integrating w.r.t. $'x'$ from 0 to x, we get

$$g'(x) - g'(0) = \int_0^x h(y)\,dy + \int_0^x \left[\int_0^x k(x,y)g(y)dy\right] dx$$

Again integrating w.r.t. $'x'$ from 0 to x, we get

$$g(x) - g_0 - g_1 x = \int_0^x \int_0^x h(y)\,dydy + \int_0^x \int_0^x \left[\int_0^x k(x,y)g(y)dy\right] dx\,dx.$$

$$\Rightarrow g(x) = g_0 + g_1 x + \int_0^x \int_0^x h(x)\,dxdx + \int_0^x \int_0^x \left[\int_0^x k(x,y)g(y)dy\right] dx\,dx. \rightarrow (2)$$

Now using the decomposition series,

$$g(x) = \sum_{n=0}^{\infty} g_n(x) = g_0(x) + g_1(x) + g_2(x) + \ldots.$$

Substituting $g(x)$ as a series in (2), we get

$$\sum_{n=0}^{\infty} g_n(x) = g_0 + g_1 x + \int_0^x \int_0^x h(x)dxdx + \int_0^x \int_0^x \left[\int_0^x k(x,y)\left(\sum_{n=0}^{\infty} g_n(y)\right)dy\right] dx\,dx$$

$$\Rightarrow g_0(x) + g_1(x) + g_2(x) + \ldots = g_0 + g_1 x + \int_0^x \int_0^x h(x)dx\,d +$$

$$\int_0^x \int_0^x \left[\int_0^x [k(x,y)(g_0(y) + g_1(y) + g_2(y) + \ldots.)]dy\right] dx\,dx.$$

$$\Rightarrow \quad g_0(x) = g_0 + g_1(x) + \int_0^x \int_0^x h(x)dxdx.$$

$$g_1(x) = \int_0^x \int_0^x \left[\int_0^x k(x,y)\,g_0(y)dy\right] dx\,dx.$$

$$g_2(x) = \int_0^x \int_0^x \left[\int_0^x k(x,y)\,g_1(y)dy\right] dx\,dx.$$

...

The components $g_0(x),\ g_1(x),\ g_2(x), \ldots.$ of the solution $g(x) = \sum_{n=0}^{\infty} g_n(x)$ is determined from the following recurrence relations.

$$g_0(x) = g_0 + g_1(x) + \int_0^x \int_0^x h(x)dxdx.$$

$$g_{n+1}(x) = \int_0^x \int_0^x \left[\int_0^x k(x,y)\,g_n(y)dy\right] dx\,dx,\ n = 0, 1, 2, 3, \ldots$$

Example 5. *Consider the following Voltera integro- differential equation*

$$g''(x) = 1 + \int_0^x (x+t)g(t)\,dt, \quad g(0) = 1, \quad g'(0) = 2.$$

By comparing it with the standard form of Volterra integro-differential equation.

$$\begin{aligned} g''(x) &= h(x) + \int_0^x K(x,y)\, g(y))dy, g(0) = g_0,\; g'(0) = g_1, \text{we get} \quad (40)\\ h(x) &= 1,\; k(x,t) = x+t,\; g_0 = 1,\; g_1 = 2. \quad (41)\end{aligned}$$

Let $g(x) = \sum_{n=0}^{\infty} g_n(x)$ be a solution of (1).

Let us find $g_0(x), g_1(x), ...b_n$ by Adomain-Decomposition method.

$$\begin{aligned} g_0(x) &= g_0 + g_1(x) + \int_0^x \int_0^x h(x)\,dx\,dx\\ &= 1+2x+\int_0^x \int_0^x l\,dx\,dx = 1+2x+\int_0^x (x)_0^x dx\\ &= 1+2x+\int_0^x x = dx = 1+2x+\frac{x^2}{2}.\end{aligned}$$

We obtain the recurrence formula as fallow

$$\begin{aligned} g_{n+1}(x) &= \int_0^x \int_0^x \left[\int_0^x K(x,y)g_n(y)dy\right] dxdx, \; n = 0,1,2,...\\ g_1(x) &= \int_0^x \int_0^x \left[\int_0^x (x+y)\left(1+2y+\frac{y^2}{2}\right)dy\right] dxdx,\\ &= \frac{x^4}{8} + \frac{x^5}{24} + \frac{7}{720}x^6,\end{aligned}$$

and so on.

So the solution in series form is

$$g(x) = 1+2x+\frac{x^2}{2}+\frac{x^4}{8}+\frac{1}{24}x^5+\frac{7}{720}x^6+...$$

5. The Laplace Decomposition Method

Consider the nonlinear integro- differential equation of the type

$$U'(x) = f(x) + \int_0^x K(t,U(t),U'(t))\,dt, \to (1)$$

$U(x_0) = \alpha,\; 0 \le x \le 1$, $f(x)$ is the source term, $K(t,U(t),U'(t))$ is a linear or non-linear function.

Applying the Laplace transform to (1), we get

$$L\{U'\} = L\{f\} + L\left\{\int_0^x K(t,U(t),U'(t))dt\right\}.$$

$$\Rightarrow s\,L\{U\} - U(0) = F(s) + L\left\{\int_0^x K(t,U(t),U'(t))dt\right\}.$$

$$\Rightarrow s\,\bar{U}(s) - \alpha = F(s) + L\left\{\int_0^x K(t,U(t),U'(t))dt\right\}.$$

In the Laplce decomposition method, we assume the solution as an infinite series as follows:

$U = \sum_{n=0}^{\infty} U_n$, where the terms U_n are to be determined recursively.

Also the nonlinear term $K(t,U(t),U'(t))$ is decomposed as an infinite series of Adomain polynomials as follows:

$K(t,U(t),U'(t)) = \sum_{n=0}^{\infty} A_n$, where $A_n = A_n(U_0,U_1,U_2...,U_n)$ are to be determined by recursive relation:

$$A_n = \frac{1}{n!}\left[\frac{d^n}{d\lambda^n}\left[K\left(\sum_{i=0}^{n} U_i\,\lambda^i\right)\right]\right]_{\lambda=0}.$$

Then $S\,L\{U\} = \alpha + F(s) + L\left\{\int_0^x K(t,U(t),U'(t))dt\right\}$ becomes

$$L\{U\} = \frac{\alpha}{s} + \frac{1}{s}F(s) + \frac{1}{s}L\left[\int_0^x\left[\sum_{n=0}^{\infty} A_n\right]dt\right]$$

$$\Rightarrow L\left\{\sum_{n=0}^{\infty} U_n\right\} = \frac{\alpha}{s} + \frac{1}{s}F(s) + \frac{1}{s}\int_0^x \sum_{n=0}^{\infty} L[A_n]dt$$

Define the following iterative algorithm:

$$L\{U_0\} = \frac{\alpha}{s} + L\{f(x)\};$$

$$L\{U_1\} = \frac{1}{s} + L\left[\int_0^x A_0\,dt\right].$$

In general,

$$L\{U_{n+1}\} = \frac{1}{s}L\left[\int_0^x A_n\, dt\right], n \geq 1.$$

with these, the components $U_0, U_1, U_2...$, are identified and the series solution is entirely determined.

In many cases, the exact solution in closed form may also be obtained.

From numerical point of view, the approximation

$$U(x) = \lim_{n\to\infty}\left(\sum_{k=0}^{n-1} U_k(x)\right).$$

5.1. Laplace Transform Method

Consider a Volterra -integro differential equation of the form

$$g^{(n)}(x) = h(x) + \lambda \int_0^x K(x-t)\, g(t)\, dt,$$
$$g(0) = g_0,\ g'(0) = g_1, ..., g^{(n-1)}_{(0)} = g_{n-1}.$$

Applying Laplace transform, and using the given initial conditions, find $L\{g(x)\}$ in terms of s, by using convolution theorem.

Then by inverse Laplace transform, find $g(x)$.

Example 6. *Consider the following Volterra integro- differential equation*
$U'(x) = x - \int_0^x U(t)dt, U(0) = 0.$

Applying Laplace transform, and using the initial condition $U(0) = 0$, we get

$$S\,\bar{U}(s) - 0 = \frac{1}{s^2} - \frac{1}{s}.U(s).$$

$$\Rightarrow \left(s + \frac{1}{s}\right)U(s) = \frac{1}{s^2}$$

$$\Rightarrow \frac{s^2+1}{s}U(s) = \frac{1}{s^2} \quad \Rightarrow U(s) = \frac{1}{s(s^2+1)}$$

$$\Rightarrow U(s) = \frac{1}{s} - \frac{s}{s^2+1}$$

Applying inverse Laplace transform, we get

$U(x) = 1 - cos\, x$, is the solution.

6. The Variational Iteration Method

Consider the jth order integro -differential equation

$$g^{(i)}_{(x)} = h(x) + \int_0^x K(x,y)g(y)dy,$$

Where $g^{(i)}_{(x)} = \frac{d^i g}{dx^j}$, $g(0) = g_0, g'(0) = g_1, \ldots\ldots,$

$g^{(j-1)}_{(0)} = g_{(j-1)}$, are the initial conditions.

The correction functional for the integro differential equation is

$$g^{(x)}_{n+1} = g_n(x) + \int_0^x \lambda(\varsigma)\left(g^{(i)}_n(\varsigma) - h(\varsigma) - \int_0^\varsigma K(\varsigma,r)F(g_n(r))dr\right)d\varsigma,$$

where λ is a Lagrange multiplier.

For λ :

$$g' + h(g(\varsigma), g'(\varsigma)) = 0,\ \lambda = -1.$$

$$g'' + h(g(\varsigma), g'(\varsigma), g''(\varsigma)) = 0,\ \lambda = \varsigma - x.$$

$$\begin{aligned} g''' &+ h(g(\varsigma), g'(\varsigma), g''(\varsigma), g'''(\varsigma)) = 0,\ \lambda = -\frac{1}{2},(\varsigma - x)^2. \\ g^{(n)} &+ h(g(\varsigma), g'(\varsigma), g''(\varsigma), \ldots., g^{(n)}(\varsigma)) = 0, \\ \lambda &= (-1)^n \frac{1}{(n-1)!}(\varsigma - x)^{n-1}. \end{aligned}$$

Example 7. *Consider the Volterra integro-differential equation*

$$g'(x) = -1 + \int_0^x g(t)dt = 1.$$

The correction functional for this equation is given by

$$g^{(x)}_{n+1} = g_n(x) - \int_0^x \left(g'_n(\varsigma) + 1 - \int_0^{(\varsigma)} g_n(r)dr\right)d\varsigma$$

where $\lambda = -1$ for first order integro -differential equation.

Use initial condition to select $g_0(x) = g_0(0) = 1$.
Now the successive approximations :

$$
\begin{aligned}
g_0(x) &= 1 \\
g_1(x) &= g_0(x) - \int_0^x (g_0'(\varsigma) + 1 - \int_0^\varsigma g_0(r)dr)d\varsigma \\
&= 1 - \int_0^x (0 + 1 - \int_0^\varsigma dr)d\varsigma \\
&= 1 - \int_0^x (1 - \varsigma dr)d\varsigma) = 1 - x + \frac{x^2}{2} \\
g_2(x) &= g_1(x) - \int_0^x (g_1'(\varsigma) + 1 - \int_0^\varsigma g_1(r)dr)d\varsigma \\
&= 1 - x + \frac{x^2}{2} - \int_0^x (-1 + \varsigma + 1 - \int_0^\varsigma (1 - r + \frac{r^2}{2})dr)d\varsigma \\
&= 1 - x + \frac{x^2}{2} - \int_0^x (\varsigma - (\varsigma - \frac{\varsigma^2}{2} + \frac{\varsigma^3}{6})d\varsigma \\
&= 1 - x + \frac{x^2}{2} - \frac{x^3}{6} + \frac{x^4}{24} - - - - - - \\
g_3(x) &= g_2(x) - \int_0^x (g_1'(\varsigma) + 1 - \int_0^\varsigma g_2(r)dr)d\varsigma \\
&= 1 - x + \frac{x^2}{2} - \frac{x^3}{3!} + \frac{x^4}{4!} - \ldots\ldots
\end{aligned}
$$

and so on.

Now by VIM, $U(x) = \lim_{n \to oo} U_n(x) = e^{-x}$ is the exact solution.

Conclusion

The DTM, ADM, LDM and Variational iteration methods are successfully applied to find the solutions of linear as well as nonlinear integro-differential equations. The efficiency and accuracy of the proposed methods are demonstrated by test problems. These methods may be successfully used to solve other linear as well as nonlinear integro-differential equations.

References

[1] A. Avudainayagam, C. Vani, Wavelet-Galerkin method for integro-differential equations, *Appl. Numer. Math.* 32 (2000) 247-254.

[2] M. T. Rashed, Lagrange interpolation to compute the numerical solutions of differential,integral and integro-differential equations, *Appl. Math. Comput.* 151(3) (2004) 869-878.

[3] S. M. Hosseini, S. Shahmorad, Tau numerical solution of Fredholm integro-differential equations with arbitrary polynomial bases, *Appl. Math. Model.* 27 (2003) 145-154.

[4] R. P. Kanwal, K. C. Liu, A Taylor expansion approach for solving integral equations, *Int. J. Edu. Sci. Technol.* 3 (1989) 411-414.

[5] K. Maleknejad, Y. Mahmoudi, Taylor polynomial solution of high-order nonlinear Volterra-Fredholm integro-differential equations, *Appl. Math. Comput.* 145 (2003) 641-653.

[6] K. Maleknejad, F. Mirzaee, S. Abbasbandy, Solving linear integro-differential equations system by using rationalized Haar functions method, *Appl. Math. Comput.* 155 (2), (2004) 317-328.

[7] I. H. Abdel-Halim Hassan, Differential transformation technique for solving higher-order initial value problems, *Appl. Math. Comput.* 154 (2004)299-311.

[8] C. K. Chen, S. H. Ho, Solving partial differential equations by two-dimensional differential transform method, *Appl. Math. Comput.* 106 (1999) 171-179.

[9] P. Darania, Ali Ebadian, A method for the numerical solution of the integro-differential equations, *Appl. Math. Comput.* 188 (1), (2007), 657-668.

[10] S. M. El-Sayed, M. R. Abdel-Aziz, A Comparison of Adomian decompositon method and Wavelet-Galerkin method for solving integro-differential equations, *Appl. Math. Comput.* 136 (2003) 151-159.

[11] M. J. Jang, C. L. Chen, Y. C. Liu, Two-dimensional differential transform for partial differential equations, *Appl. Math. Comput.* 121 (2001) 261-271.

[12] G. Adomian, *Nonlinear Stochastic Operator Equations*, Academic Press, San Diego, (1986).

[13] G. Adomian, *Solving Frontier Problems of Physics, The Decomposition Method*, Kluwer, Boston, (1994).

[14] G. Adomian and R. Rach, Noise terms in decomposition series solution, *Comput. Math. Appl.*, 24 (1992) 61-64.

[15] A.-M. Wazwaz, A reliable algorithm for solving boundary value problems for higher-order integro-differential equations, *Appl. Math. Comput.* 118 (2001) 327-342.

[16] A. M. Wazwaz, *A First Course in Integral Equations*, World Scientific, Singapore, (1997), 212.

[17] A. M. Wazwaz, The variational iteration method; a reliable tool for solving linear and nonlinear wave equations, *Comput. Math. Appl.*, 54 (20007) 926-932.

[18] M. Sezer, Taylor polynomial solution of Volterra integral equations, *Int. J. Math. Educ. Sci. Technol.* 5 (1994) 625-633.

[19] S. Yalcinbas, Taylor polynomial solution of nonlinear Volterra-Fredholm integral equations, *Appl. Math. Comput.* 127 (2002) 195-206.

[20] J. K. Zhou, *Differential Transformation and Its Application for Electrical Circuits*, Huazhong University Press, Wuhan, China, 1986.

In: Understanding Integro-Differential Equations ISBN: 979-8-89113-040-1
Editors: J. Vasundhara Devi et al. © 2023 Nova Science Publishers, Inc.

Chapter 5

Operational Matrices for Solving Fractional Order Integral and Integro-Differential Equations

***H. Tajadodi*[1]**
***R. M. Ganji*[2]**
***S. M. Narsale*[3]**
***and H. Jafari*[4*]**
[1]Department of Mathematics,
University of Sistan and Baluchestan, Zahedan, Iran
[2]Department of Applied Mathematics,
University of Mazandaran, Babolsar, Iran
[3]Symbiosis Institute of Technology,
Symbiosis International (Deemed University),
Pune, India
[4]Department of Mathematical Sciences,
University of South Africa,
South Africa

Abstract

In this chapter a numerical method is presented to obtain numerical solutions of fractional-order differential equations, as well as fractional

*Corresponding Author's Email: jafari.usern@gmail.com

integro-differential equations. In this method, operational matrices are computed and used to convert the main equations into a system of algebraic equations. After solving the algebraic equations, the values are substituted in the approximate function to obtain a solution of the governing equation. The operational matrices are obtained by using Legendre polynomials as basic functions. Further the existence, uniqueness, stability and convergence of solutions are discussed.

Keywords: fractional-order differential equations, operational matrices, fractional integro-differential equations, Legendre polynomials

AMS Subject Classification: 53D, 37C, 65P

1. Introduction

During the past three decades fractional calculus had many applications in the area of physics and engineering. The applications of fractional calculus used in many fields such as electrical networks, control theory of dynamical systems, probability and statistics, electrochemistry of corrosion, chemical physics, optics and signal processing can be successfully modelled by linear or nonlinear fractional differential equations (FDEs).

Finding approximate or exact solutions of FDEs is an important task. Except for a limited number of these equations, we have difficulty in finding their analytical solutions. Therefore, there have been attempts to develop new methods for obtaining numerical solutions which reasonably approximate the exact solutions.

The operational matrix method using orthogonal functions has been considered important to deal with various problems. The main feature of this technique is that it reduces these equations into a system of algebraic equations and thus greatly simplifies the problem. In this method, a truncated orthogonal series is used for numerical integration of differential equations, with the goal of obtaining efficient computational solutions. A lot of work has been done using shifted Legendre polynomials in the literature.

In this chapter, we have mainly focused on the application of Legendre polynomials to create operational matrices for solving a class of fractional-order differential equations and fractional integro-differential equations.

The chapter is organized as follows: In Section 2., we have introduced some necessary definitions and mathematical preliminaries of the fractional calculus theory. In Section 3. definition and properties of Legendre polynomials are discussed. In Section 4. the Legendre polynomials are used to approximate the function. Section 5. operational matrices of the fractional derivative and integration are constructed. In Section 6. the proposed method is applied to several examples. Section 7. consists of existence, uniqueness, stability and convergence of the method. Also a conclusion is given in Section 7.3..

2. Fractional Calculus

In this section, several common definitions and properties of fractional integrals and derivatives will be introduced.

2.1. Riemann-Liouville Fractional Integral and Derivative

The Riemann-Liouville fractional order integral and derivative are presented in the following definitions [1–3].

Definition 1. *Suppose a is a positive real number and the function $f(t)$ is defined on the interval $[a,b]$. The α-th order fractional integral of the function $f(t)$ is defined as*

$$^{RL}I_t^{\alpha} f(t) = \frac{1}{\Gamma(\alpha)} \int_a^t (t-\tau)^{\alpha-1} f(\tau) d\tau,$$

where $t \in [a,b]$ and $\Gamma(x)$ is the Gamma function defined by

$$\Gamma(x) = \int_0^{\infty} e^{-t} t^{x-1} dt, \quad Re(x) > 0.$$

Some of the important properties of the Riemann-Liouville fractional integral are given as:

1. $^{RL}I_t^{\alpha}\, {}^{RL}I_t^{\beta} f(t) = {}^{RL}I_t^{\beta}\, {}^{RL}I_t^{\alpha} f(t),$
2. $^{RL}I_t^{\alpha}\, {}^{RL}I_t^{\beta} f(t) = {}^{RL}I_t^{\alpha+\beta} f(t),$
3. $^{RL}I_t^{\alpha} (\lambda f(t) + \mu g(t)) = \lambda\, {}^{RL}I_t^{\alpha} f(t) + \mu\, {}^{RL}I_t^{\alpha} g(t),$
4. $^{RL}I_t^{\alpha} C = \frac{C}{\Gamma(\alpha+1)} t^{\alpha}$, C is a constant,

5. ${}^{RL}I_t^{\alpha}(t-a)^p = \frac{\Gamma(p+1)}{\Gamma(\alpha+p+1)}(t-a)^{\alpha+p}, \ Re(p) > -1.$

Definition 2. *Suppose a is a positive real number, $n-1 \le \alpha < n$ with n a positive integer and the function $f(t)$ is defined on the interval $[a,b]$. The α-th order Riemann-Liouville (RL) fractional derivative of function $f(t)$ is defined as*

$$ {}^{RL}D_t^{\alpha} f(t) = \frac{1}{\Gamma(n-\alpha)} \frac{d^n}{dt^n} \int_a^t \frac{f(\tau)}{(t-\tau)^{\alpha-n+1}} d\tau = \frac{d^n}{dt^n} \left({}^{RL}I_t^{n-\alpha} f(t) \right). $$

In particular if $n = 1$, then the Riemann-Liouville fractional derivative of order $\alpha \in (0, 1)$ is defined as:

$$ {}^{RL}D_t^{\alpha} f(t) = \frac{1}{\Gamma(1-\alpha)} \frac{d}{dt} \int_a^t \frac{f(\tau)}{(t-\tau)^{\alpha}} d\tau = \frac{d}{dt} \left({}^{RL}I_t^{1-\alpha} f(t) \right). $$

A direct calculation gives

$$ {}^{RL}D_t^{\alpha}(t-a)^p = \frac{\Gamma(p+1)}{\Gamma(p-\alpha+1)}(t-a)^{p-\alpha}, \quad Re(p) > -1. $$

It can be proved that

1. $\frac{d^m}{dt^m} {}^{RL}D_t^{\alpha} f(t) = {}^{RL}D_t^{m+\alpha} f(t), \quad \alpha > 0, \ m > 0, \ m \in \mathbb{Z}^+,$
2. ${}^{RL}D_t^{\alpha} \, {}^{RL}I_t^{\alpha} f(t) = f(t).$

2.2. Caputo Fractional Derivative

Although the Riemann-Liouville definition of fractional derivatives seems to play an important role in the development of fractional calculus, several authors including Caputo (1967, 1969) [1–3] realized that the Riemann-Liouville definition needs revision because the applied problems in viscoelaticity, solid mechanics and in rheology require physically interpretable initial conditions such as $f(a), f'(a), f''(a)$.

Definition 3. *Suppose a is a positive real number, $n-1 \le \alpha < n$ with n a positive integer and the function $f(t)$ is defined on the interval $[a,b]$. The α-th order Caputo (C) fractional derivative of function $f(t)$ is defined as*

$$ {}^{C}D_t^{\alpha} f(t) = \frac{1}{\Gamma(n-\alpha)} \int_a^t \frac{f^{(n)}(\tau)}{(t-\tau)^{\alpha-n+1}} d\tau, $$

where $t \in [a,b]$.

It is clear that

$$^{C}D_t^{\alpha} f(t) = {}^{RL}I_t^{n-\alpha}\left(f^{(n)}(t)\right).$$

A direct calculation gives

$$^{C}D_t^{\alpha}(t-a)^p = \frac{\Gamma(p+1)}{\Gamma(p-\alpha+1)}(t-a)^{p-\alpha}, \quad Re(p) > n-1 \geq 0.$$

It can be proved that

1. $^{C}D_t^{\alpha} C = 0, \quad C$ is a constant,
2. $\lim_{\alpha\to n} {}^{C}D_t^{\alpha} f(t) = f^{(n)}(t)$,
3. $^{C}D_t^{\alpha}\left(\frac{d^m}{dt^m} f(t)\right) = {}^{C}D_t^{m+\alpha} f(t), \quad n-1 < \alpha < n, \; m \in \mathbb{N}$,
4. $^{C}D_t^{\alpha} f(t) = {}^{RL}D_t^{\alpha} f(t) - \sum_{k=0}^{n-1} \frac{f^{(k)}(0^+)}{\Gamma(k-\alpha+1)}(t-a)^{k-\alpha}$,
5. $^{RL}I_t^{\alpha}\, {}^{C}D_t^{\alpha} f(t) = f(t) - \sum_{k=0}^{n-1} \frac{f^{(k)}(0^+)}{k!}(t-a)^k$.

3. Legendre Polynomials and Related Results

Polynomials play the most important role in spectral methods, so it is necessary to have a thorough study of their relevant properties. The well-known Legendre polynomials $L_i(x), i = 0, 1, \cdots$, are the eigenfunctions of the singular Sturm-Liouville problem. Consider the Sturm-Liouville problem in the interval $(-1,1)$ as

$$-\left(pu'\right)' + qu = \lambda wu,$$

the coefficients p, q and w are three given, real-valued functions such that: p is continuously differentiable, strictly positive in $(-1,1)$ and continuous at $x = \pm 1$; q is continuous, nonnegative and bounded in $(-1,1)$; the weight function w is continuous, nonnegative and integrable over $(-1,1)$. Now let us suppose $p(x) = 1-x^2$, $q(x) = 0$ and $w(x) = 1$, then

$$\left(\left(1-x^2\right)L_i'(x)\right)' + i(i+1)L_i(x) = 0.$$

$L_i(x)$ is even if i is even and odd if i is odd. If $L_i(x)$ is normalized so that $L_i(1) = 1$, then for any i:

$$L_i(x) = \frac{1}{2^i}\sum_{\ell=0}^{\left[\frac{i}{2}\right]}(-1)^{\ell}\binom{i}{\ell}\binom{2i-2\ell}{i}x^{i-2\ell},$$

where $\left[\frac{i}{2}\right]$ denotes the integral part of $\frac{i}{2}$. The Legendre polynomials satisfy the recursion relation [4]:

$$L_{i+1}(x) = \frac{2i+1}{i+1}xL_i(x) - \frac{i}{i+1}L_{i-1}(x), \quad i = 1,2,3,\cdots,$$

where $L_0(x) = 1$ and $L_1(x) = x$. Relevant properties are

1. $|L_i(x)| \leq 1, \qquad -1 \leq x \leq 1,$
2. $L_i(\pm 1) = (\pm 1)^i,$
3. $|L_i'(x)| \leq \frac{1}{2}i(i+1), \quad -1 \leq x \leq 1,$
4. $L_i'(\pm 1) = (\pm 1)^{i+1}\frac{1}{2}i(i+1),$
5. $\int_{-1}^{1} L_i^2(x)dx = \left(i+\frac{1}{2}\right)^{-1}.$

In order to use these polynomials on the interval $t \in [0,1]$ we define the so-called shifted Legendre polynomials by introducing the change of variable $x = 2t-1$. Let the shifted Legendre polynomials $L_i(2t-1)$ be denoted by $L_i^*(t)$. Then $L_i^*(t)$ can be obtained as follows:

$$L_i^*(t) = L_i(2t-1), \quad i = 0,1,2,\cdots.$$

The analytic form of the shifted Legendre polynomial of degree i is defined by

$$L_i^*(t) = \sum_{k=0}^{i} l_{i,k}\, t^k, \tag{1}$$

where

$$l_{i,k} = \frac{(-1)^{i+k}(i+k)!}{(i-k)!(k!)^2}. \tag{2}$$

The orthogonality property is satisfied for these polynomials with respect to the weight function $w(t) = 1$, as follows

$$\int_0^1 w(t)\, L_m^*(t)\, L_n^*(t)\, dt = \begin{cases} \dfrac{1}{2m+1}, & m = n, \\ 0, & m \neq n. \end{cases}$$

4. Approximation of Functions in Terms of Legendre Polynomials

Any arbitrary function $y \in L^2(0,1)$ can be expanded in terms of the Legendre polynomials, $L_i^*(t)$, described in the previous section by the following infinite series [5]:

$$y(t) = \sum_{i=0}^{\infty} y_i\, L_i^*(t), \tag{3}$$

where

$$y_i = \frac{\langle y(t), L_i^*(t)\rangle}{\langle L_i^*(t), L_i^*(t)\rangle} = \frac{1}{\langle L_i^*(t), L_i^*(t)\rangle} \int_0^1 y(\tau)\, L_i^*(\tau)\, d\tau. \tag{4}$$

By taking only the first $M+1$ terms in (3), $y(t)$ can be approximated as

$$y(t) \simeq y_M(t) = \sum_{i=0}^{M} y_i\, L_i^*(t) = Y^T\, \varphi(t),$$

where

$$\varphi(t) = [L_0^*(t), L_1^*(t), \cdots, L_M^*(t)]^T, \tag{5}$$

and

$$Y = [y_0, y_1, \cdots, y_M]^T,$$

which Y can be calculated by

$$Y = Q^{-1} \int_0^1 y(\tau)\varphi(\tau)\, d\tau, \tag{6}$$

with

$$Q = \int_0^1 \varphi(\tau)\varphi^T(\tau) d\tau.$$

The basis vector $\varphi(t)$ can be written in terms of the Taylor basis functions as

$$\varphi(t) = A\,T_M(t),$$

where

$$T_M(t) = [1, t, \cdots, t^M]^T,$$

is the Taylor basis vector and $A = [a_{i,j}]$, $i, j = 0, 1, \cdots, M$, with

$$a_{i,j} = \begin{cases} l_{i,j}, & i \geq j, \\ 0, & i < j, \end{cases}$$

where $l_{i,j}$ is given by (2).

Similarly, we can expand a function $y(x,t)$ in $L^2((0,1) \times (0,1))$ in terms of the Legendre polynomials as

$$y(x,t) = \sum_{i=0}^{\infty} \sum_{j=0}^{\infty} y_{i,j}\, L_i^*(x)\, L_j^*(t). \tag{7}$$

By considering only the first $(M+1) \times (M+1)$ terms in (7), $y(x,t)$ can be approximated as

$$y(x,t) \simeq \sum_{i=0}^{M} \sum_{j=0}^{M} y_{i,j}\, L_i^*(x)\, L_j^*(t) = \varphi^T(x)\, Y\, \varphi(t),$$

where Y is an $(M+1) \times (M+1)$ matrix and is given by

$$Y = Q^{-1} \langle \varphi(x), \langle y(x,t), \varphi(t) \rangle \rangle\, Q^{-1}. \tag{8}$$

5. Operational Matrices Based on Legendre Polynomials

In this section, we obtain the operational matrices of the fractional integral and the product of basis function based on the shifted Legendre Polynomials.

5.1. The Operational Matrix of Riemann-Liouville Fractional Integral

Let $\varphi(t)$ be the shifted Legendre polynomials vector given by (5), then the Riemann-Liouville fractional integral of $\varphi(t)$ can be approximated as:

$$^{RL}I_t^{\alpha}\varphi(t) \simeq \mathbf{I}_{RL}^{\alpha}\varphi(t),$$

where $\mathbf{I}_{RL}^{\alpha}$ is the operational matrix of Riemann-Liouville fractional integral. To evaluate this matrix, first we apply the Riemann-Liouville fractional integral, $^{RL}I_t^{\alpha}$, onto $L_i^*(t)$, $i = 0, 1, \cdots, M$, as

$$^{RL}I_t^{\alpha}L_i^*(t) = {}^{RL}I_t^{\alpha}\left(\sum_{k=0}^{i} l_{i,k}t^k\right) = \sum_{k=0}^{i} l_{i,k}\left({}^{RL}I_t^{\alpha}t^k\right) = \sum_{k=0}^{i}\frac{\Gamma(k+1)l_{i,k}}{\Gamma(k+\alpha+1)}t^{k+\alpha}.$$

By approximating the function $t^{k+\alpha}$ in terms of the the shifted Legendre polynomials, we have

$$t^{k+\alpha} \simeq \sum_{s=0}^{M} d_{k,s}L_s^*(t), \tag{9}$$

where $d_{k,s}$ can be calculated by (4). In view of (9) and for $i = 0, 1, \cdots, M$, we get

$$\begin{aligned} ^{RL}I_t^{\alpha}L_i^*(t) &\simeq \sum_{k=0}^{i}\frac{\Gamma(k+1)l_{i,k}}{\Gamma(k+\alpha+1)}\left(\sum_{s=0}^{M} d_{k,s}L_s^*(t)\right) = \sum_{s=0}^{M}\left(\sum_{k=0}^{i}\frac{\Gamma(k+1)l_{i,k}d_{k,s}}{\Gamma(k+\alpha+1)}\right)L_s^*(t) \\ &= \sum_{s=0}^{M}\left(\sum_{k=0}^{i}\sigma_{i,s,k}\right)L_s^*(t). \end{aligned}$$

Therefore, for $i = 0, 1, \cdots, M$, we can write

$$^{RL}I_t^{\alpha}\varphi(t) \simeq \mathbf{I}_{RL}^{\alpha}\varphi(t),$$

where

$$\mathbf{I}_{RL}^{\alpha} = \begin{bmatrix} \sigma_{0,0,0} & \sigma_{0,1,0} & \cdots & \sigma_{0,M,0} \\ \sum_{k=0}^{1}\sigma_{1,0,k} & \sum_{k=0}^{1}\sigma_{1,1,k} & \cdots & \sum_{k=0}^{1}\sigma_{1,M,k} \\ \vdots & \vdots & \cdots & \vdots \\ \sum_{k=0}^{M}\sigma_{M,0,k} & \sum_{k=0}^{M}\sigma_{M,1,k} & \cdots & \sum_{k=0}^{M}\sigma_{M,M,k} \end{bmatrix},$$

with

$$\sigma_{i,s,k} = \frac{\Gamma(k+1) l_{i,k} d_{k,s}}{\Gamma(k+\alpha+1)}.$$

5.2. The Operational Matrix of the Product

Suppose that $C = [c_0, c_1, \cdots, c_M]^T$ is an arbitrary $(M+1) \times 1$ matrix, then $\widehat{C}$ is an $(M+1) \times (M+1)$ operational matrix of product based on the shifted Legendre polynomials whenever

$$\varphi(t)\,\varphi^T(t)\,C \simeq \widehat{C}\varphi(t).$$

To evaluate the product of $\varphi(t)$ and $\varphi^T(t)$, that will give the operational matrix of the product based on the shifted Legendre basis, first, we write

$$L_i^*(t)\,L_j^*(t) = \sum_{r=0}^{i+j} a_r\,L_r^*(t). \tag{10}$$

The coefficients a_r, $r = 0, 1, \ldots, i+j$, are computed in the following manner. Multiplying both sides of equation (10) by $L_m^*(t), m = 0, 1, \cdots, i+j$, and integrating the result yields

$$\int_0^1 L_i^*(\tau)\,L_j^*(\tau)\,L_m^*(\tau)\,dt = \sum_{r=0}^{i+j} a_r \int_0^1 L_r^*(\tau)\,L_m^*(\tau)d\tau = \frac{1}{2m+1} a_m. \tag{11}$$

Using (11) and the analytic form of the shifted Legendre polynomial given by (1), we obtain

$$\begin{aligned} a_m &= (2m+1)\int_0^1 L_i^*(\tau)\,L_j^*(\tau)\,L_m^*(\tau)\,d\tau \\ &= (2m+1)\sum_{k=0}^{i}\sum_{\iota=0}^{j}\sum_{s=0}^{m} l_{i,k}\,l_{j,\iota}\,l_{m,s}\int_0^1 \tau^{k+\iota+s}\,d\tau \\ &= (2m+1)\Delta_{i,j,m}, \end{aligned}$$

where

$$\Delta_{i,j,m} = \sum_{k=0}^{i}\sum_{l=0}^{j}\sum_{s=0}^{m} \frac{l_{i,k}\,l_{j,\iota}\,l_{m,s}}{k+\iota+s+1}.$$

By substituting a_m into equation (10), we have

$$L_i^*(t)\,L_j^*(t) = \sum_{m=0}^{i+j} (2m+1)\Delta_{i,j,m}. \tag{12}$$

then Eq. (12) can be written as

$$\varphi(t)\,\varphi^T(t)\,C \simeq \widehat{C}\varphi(t),$$

where $\widehat{C}$ is given by

$$\widehat{C} = [\widehat{c}_{i,j}], \quad i,j = 0,1,\cdots,M,$$

where

$$\widehat{c}_{i,j} = (2j+1)\sum_{m=0}^{M} \Delta_{i,j,m}\, c_m.$$

6. Applications

6.1. Solving Fractional Differential Equation

Consider the inhomogeneous Bagley-Torvik equation as a multi-order fractional differential equations (FDEs) [6–8]

$$\begin{aligned} &y''(t) + {}^{C}D_t^{\frac{3}{2}} y(t) + y(t) = 1+t, \quad 0 < t < 1, \\ &y(0) = 1, \quad y'(t) = 1. \end{aligned}$$

For solving Bagley Torvik equation using the operational matrix, first we apply the integral operator, I^2, twice on both sides of this equation. It gives the following relation

$$y(t) - (1+t) + {}^{RL}I_t^{\frac{1}{2}} y(t) - \left(\frac{t^{\frac{1}{2}}}{\Gamma\left(\frac{3}{2}\right)} + \frac{t^{\frac{3}{2}}}{\Gamma\left(\frac{5}{2}\right)} \right) + I^2 y(t) = \frac{t^2}{2} + \frac{t^3}{6}.$$

This implies that

$$y(t) + {}^{RL}I_t^{\frac{1}{2}} y(t) + I^2 y(t) = \frac{t^2}{2} + \frac{t^3}{6} + 1 + t + \frac{t^{\frac{1}{2}}}{\Gamma\left(\frac{3}{2}\right)} + \frac{t^{\frac{3}{2}}}{\Gamma\left(\frac{5}{2}\right)}. \tag{13}$$

Now, we approximate $y(t)$ and the right side of the above equation by the Legendre basis as follows:

$$y(t) \simeq \sum_{i=0}^{M} y_i L_i^*(t) = Y^T \varphi(t), \tag{14}$$

where

$$Y^T = [y_0, y_1, \cdots, y_M],$$
$$\varphi(t) = [L_0^*(t), L_1^*(t), \cdots, L_M^*(t)]^T,$$

and

$$\frac{t^2}{2} + \frac{t^3}{6} + 1 + t + \frac{t^{\frac{1}{2}}}{\Gamma\left(\frac{3}{2}\right)} + \frac{t^{\frac{3}{2}}}{\Gamma\left(\frac{5}{2}\right)} \simeq F^T \varphi(t), \tag{15}$$

where F can be calculated by (6) as follows:

$$F^T \simeq Q^{-1} \int_0^1 \left(\frac{\tau^2}{2} + \frac{\tau^3}{6} + 1 + \tau + \frac{\tau^{\frac{1}{2}}}{\Gamma\left(\frac{3}{2}\right)} + \frac{\tau^{\frac{3}{2}}}{\Gamma\left(\frac{5}{2}\right)} \right) \varphi(\tau)\, d\tau.$$

In view of (13), we need to approximate $I^{\frac{1}{2}} y(t)$ and $I^2 y(t)$. To do that, we use the operational matrix of the integral. Applying this matrix on (14) yields:

$$\begin{aligned} {}^{RL}I_t^{\frac{1}{2}} y(t) &\simeq {}^{RL}I_t^{\frac{1}{2}} \left(Y^T \varphi(t)\right) = Y^T \underbrace{\left({}^{RL}I_t^{\frac{1}{2}} \varphi(t)\right)}_{\mathbf{I}_{RL}^{\frac{1}{2}} \varphi(t)} = Y^T \mathbf{I}_{RL}^{\frac{1}{2}} \varphi(t), \\ I^2 y(t) &\simeq I^2 \left(Y^T \varphi(t)\right) = Y^T \underbrace{\left(I^2 \varphi(t)\right)}_{\mathbf{I}_{RL}^{2} \varphi(t)} = Y^T \mathbf{I}_{RL}^2 \varphi(t). \end{aligned} \tag{16}$$

Substituting (14), (15) and (16) into (13), yields

$$Y^T \varphi(t) + Y^T \mathbf{I}_{RL}^{\frac{1}{2}} \varphi(t) + Y^T \mathbf{I}_{RL}^2 \varphi(t) = F^T \varphi(t).$$

Consequently, we can write

$$Y^T + Y^T \mathbf{I}_{RL}^{\frac{1}{2}} + Y^T \mathbf{I}_{RL}^2 = F^T.$$

Finally,

$$Y^T = F^T \left(I + \mathbf{I}_{RL}^{\frac{1}{2}} + \mathbf{I}_{RL}^{2} \right)^{-1}, \tag{17}$$

where I is an $(M+1)\times(M+1)$ identity matrix. By solving (17), we can calculate Y. Then, the approximation $y(t)$ can be computed using (14).

Now, let us suppose in (14), $M = 2$, then we calculate the following results

$$\begin{aligned}
&Y = [y_0, y_1, y_2]^T, && \varphi(t) = [1, -1+2t, 1-6t+6t^2]^T,\\
&Q = \begin{bmatrix} 1 & 0 & 0 \\ 0 & \frac{1}{3} & 0 \\ 0 & 0 & \frac{1}{5} \end{bmatrix}, && Q^{-1} = \begin{bmatrix} 1 & 0 & 0 \\ 0 & 3 & 0 \\ 0 & 0 & 5 \end{bmatrix},\\
&\mathbf{I}_{RL}^{\frac{1}{2}} = \begin{bmatrix} \frac{4}{3\sqrt{\pi}} & \frac{4}{5\sqrt{\pi}} & -\frac{4}{21\sqrt{\pi}} \\ -\frac{4}{15\sqrt{\pi}} & \frac{4}{7\sqrt{\pi}} & \frac{4}{9\sqrt{\pi}} \\ -\frac{4}{105\sqrt{\pi}} & -\frac{4}{15\sqrt{\pi}} & \frac{100}{231\sqrt{\pi}} \end{bmatrix}, && \mathbf{I}_{RL}^{2} = \begin{bmatrix} \frac{1}{6} & \frac{1}{4} & \frac{1}{12} \\ -\frac{1}{12} & -\frac{1}{10} & 0 \\ \frac{1}{60} & 0 & -\frac{1}{42} \end{bmatrix},\\
&F = \left[\frac{41}{24} + \frac{28}{15\sqrt{\pi}}, \frac{33}{40} + \frac{52}{35\sqrt{\pi}}, \frac{1}{8} - \frac{4}{63\sqrt{\pi}} \right]^T,
\end{aligned} \tag{18}$$

then, according to (18), Y is calculated as

$$Y^T = \left[\frac{3}{2}, \frac{1}{2}, 0 \right],$$

when put into (14) gives the exact solution $(y(t) = 1 + t)$.

6.2. Solving Fractional Integro-Differential Equation

Consider the following fractional integro-differential equation (FIDE) :

$$\begin{cases} {}^{C}D_t^{\frac{1}{2}} y(t) = \int_0^1 t\,\tau\, y(\tau)\, d\tau + \int_0^t (t+\tau)\, y(\tau)\, d\tau + \frac{4\sqrt{t}}{\sqrt{\pi}} - \frac{17t}{12} + \frac{8t^{\frac{3}{2}}}{3\sqrt{\pi}} - \frac{3t^2}{2} - \frac{5t^3}{3} - \frac{7t^4}{12}, \\ y(0) = 1, \end{cases} \tag{19}$$

For solving this equation using the operational matrix, first we approximate ${}^{C}D_t^{\frac{1}{2}} y(t)$ as

$${}^{C}D_t^{\frac{1}{2}} y(t) \simeq \sum_{i=0}^{M} y_i L_i^*(t) = Y^T \varphi(t). \tag{20}$$

By taking the Riemann-Liouville integral, ${}^{RL}I_t^{\frac{1}{2}}$, of (20) and using the initial condition, we have

$$y(t) \simeq Y^T \mathbf{I}_{RL}^{\frac{1}{2}} \varphi(t) + \mathbf{1}.$$

Approximating $\mathbf{1} \simeq Y_0^T \varphi(t)$, where Y_0 can be calculated by (6), yields

$$y(t) \simeq \gamma\varphi(t), \tag{21}$$

where $\gamma = Y^T \mathbf{I}_{RL}^{\frac{1}{2}} + Y_0^T$. Now, approximating $t\,\tau$ and $t+\tau$ as

$$\begin{cases} t\,\tau \simeq \varphi^T(t)\,\mathcal{K}_1\,\varphi(\tau), \\ t+\tau \simeq \varphi^T(t)\,\mathcal{K}_2\,\varphi(\tau), \end{cases} \tag{22}$$

where $\mathcal{K}_1$ and $\mathcal{K}_2$ can be computed by (8), and using (21), the integral parts are approximated by

$$\begin{aligned} \int_0^1 t\,\tau\, y(\tau)\, d\tau &\simeq \int_0^1 \varphi^T(t)\,\mathcal{K}_1\,\varphi(\tau)\,\varphi^T(\tau)\,\gamma^T\, d\tau \\ &= \varphi^T(t)\,\mathcal{K}_1 \Bigg(\underbrace{\int_0^1 \varphi(\tau)\,\varphi^T(\tau)\, d\tau}_{Q} \Bigg) \gamma^T \\ &= \varphi^T(t)\,\mathcal{K}_1 Q \gamma^T, \end{aligned} \tag{23}$$

$$\begin{aligned} \int_0^t (t+\tau)\, y(\tau)\, d\tau &\simeq \int_0^t \varphi^T(t)\,\mathcal{K}_2\varphi(\tau)\,\varphi^T(\tau)\,\gamma^T\, d\tau \\ &= \varphi^T(t)\,\mathcal{K}_2 \int_0^t \underbrace{\varphi(\tau)\,\varphi^T(\tau)\,\gamma^T}_{\widehat{\gamma^T}\varphi(\tau)}\, d\tau \\ &= \varphi^T(t)\,\mathcal{K}_2 \int_0^t \widehat{\gamma^T}\,\varphi(\tau)\, d\tau \\ &= \varphi^T(t)\,\mathcal{K}_2\widehat{\gamma^T} \underbrace{\int_0^t \varphi(\tau)\, d\tau}_{\mathbf{I}_{RL}^1\varphi(t)} \\ &= \varphi^T(t)\,\mathcal{K}_2\widehat{\gamma^T}\,\mathbf{I}_{RL}^1\varphi(t). \end{aligned} \tag{24}$$

Now, it is enough to approximate the term $\frac{4\sqrt{t}}{\sqrt{\pi}} - \frac{17t}{12} + \frac{8t^{\frac{3}{2}}}{3\sqrt{\pi}} - \frac{3t^2}{2} - \frac{5t^3}{3} - \frac{7t^4}{12}$. Therefore,

$$\frac{4\sqrt{t}}{\sqrt{\pi}} - \frac{17t}{12} + \frac{8t^{\frac{3}{2}}}{3\sqrt{\pi}} - \frac{3t^2}{2} - \frac{5t^3}{3} - \frac{7t^4}{12} = F^T\varphi(t). \tag{25}$$

Substituting (20) and (23)-(25) in equation (19), we have

$$Y^T \varphi(t) = \varphi^T(t)\, \mathcal{K}_1 Q \gamma^T + \varphi^T(t)\, \mathcal{K}_2 \widehat{\gamma^T} \mathbf{I}^1_{RL} \varphi(t) + F^T \varphi(t). \tag{26}$$

For solving (26), there are many methods, such as, **Collocation Method**, **Standard Least Squares Method** or other methods. By computing Y and substituting into (21), the approximation $y(t)$ can be computed.

Now, let us suppose in (20), $M = 2$, then we calculate the following results

$$Y = [y_0, y_1, y_2]^T, \qquad \varphi(t) = [1, -1+2t, 1-6t+6t^2]^T,$$

$$Q = \begin{bmatrix} 1 & 0 & 0 \\ 0 & \frac{1}{3} & 0 \\ 0 & 0 & \frac{1}{5} \end{bmatrix}, \qquad \gamma = \begin{bmatrix} 1 + \frac{4y_0}{3\sqrt{\pi}} - \frac{4y_1}{15\sqrt{\pi}} - \frac{4y_2}{105\sqrt{\pi}} \\ \frac{4y_0}{5\sqrt{\pi}} + \frac{4y_1}{7\sqrt{\pi}} - \frac{4y_2}{15\sqrt{\pi}} \\ -\frac{4y_0}{21\sqrt{\pi}} + \frac{4y_1}{9\sqrt{\pi}} + \frac{100y_2}{231\sqrt{\pi}} \end{bmatrix},$$

$$\mathcal{K}_1 = \begin{bmatrix} \frac{1}{4} & \frac{1}{4} & 0 \\ \frac{1}{4} & \frac{1}{4} & 0 \\ 0 & 0 & 0 \end{bmatrix}, \qquad \mathcal{K}_2 = \begin{bmatrix} 1 & \frac{1}{2} & 0 \\ \frac{1}{2} & 0 & 0 \\ 0 & 0 & 0 \end{bmatrix},$$

$$\mathbf{I}^1_{RL} = \begin{bmatrix} \frac{1}{2} & \frac{1}{2} & 0 \\ -\frac{1}{6} & 0 & \frac{1}{6} \\ 0 & -\frac{1}{10} & 0 \end{bmatrix}, \quad \mathbf{I}^{\frac{1}{2}}_{RL} = \begin{bmatrix} \frac{4}{3\sqrt{\pi}} & \frac{4}{5\sqrt{\pi}} & -\frac{4}{21\sqrt{\pi}} \\ -\frac{4}{15\sqrt{\pi}} & \frac{4}{7\sqrt{\pi}} & \frac{4}{9\sqrt{\pi}} \\ -\frac{4}{105\sqrt{\pi}} & -\frac{4}{15\sqrt{\pi}} & \frac{100}{231\sqrt{\pi}} \end{bmatrix},$$

$$F = \left[-\frac{209}{120} + \frac{56}{15\sqrt{\pi}} \quad 3\left(-\frac{293}{360} + \frac{104}{105\sqrt{\pi}}\right) \quad 5\left(-\frac{1}{6} - \frac{8}{315\sqrt{\pi}}\right) \right]^T,$$

$$\widehat{\gamma^T} = \begin{bmatrix} \frac{105\sqrt{\pi}+140y_0-28y_1-4y_2}{105\sqrt{\pi}} & \frac{4(21y_0+15y_1-7y_2)}{105\sqrt{\pi}} & \frac{4(-33y_0+77y_1+75y_2)}{693\sqrt{\pi}} \\ \frac{4(21y_0+15y_1-7y_2)}{315\sqrt{\pi}} & \frac{3465\sqrt{\pi}+4356y_0-308y_1+468y_2}{3465\sqrt{\pi}} & \frac{8(21y_0+15y_1-7y_2)}{315\sqrt{\pi}} \\ \frac{4(-33y_0+77y_1+75y_2)}{3465\sqrt{\pi}} & \frac{8(21y_0+15y_1-7y_2)}{525\sqrt{\pi}} & \frac{24255\sqrt{\pi}+31020y_0-3388y_1+207}{24255\sqrt{\pi}} \end{bmatrix},$$

then, according to (26) and using the above results, the residual function is written as

$$\begin{aligned} &R(t, y_0, y_1, y_2) = \\ &\frac{1}{69300\sqrt{\pi}} \big(-44000 + 9240\sqrt{\pi} - 464640t - 42735\sqrt{\pi}t + 52800t^2 - 884y_2 \\ &- 5280t^2 y_2 + 242550\sqrt{\pi}t^2 + 132\left(-9 + 525\sqrt{\pi} - 272t - 840t^2 - 420t^3\right) y_0 \\ &+ 220\left(-13 + 112t + 216t^2 - 180t^3 + 315\sqrt{\pi}(-1+2t)\right) y_1 + 18480t^3 y_2 \\ &+ 69300\sqrt{\pi} y_2 + 3008t y_2 - 415800\sqrt{\pi} t y_2 + 415800\sqrt{\pi} t^2 y_2\big). \end{aligned}$$

Finally, substituting the collocation points $t_\ell = \frac{\ell}{M+2}$, $\ell = 1, 2, \cdots, M+1$, into the residual function and solving the obtained system, the unknown parameters are computed as

$$y_0 = 2.11973, \quad y_1 = 1.62850, \quad y_2 = -0.06854.$$

Therefore, putting these parameters into (21), the approximate solution is calculated as

$$y_2(t) \simeq 1.02277 + 2.00127t + 0.98287t^2.$$

With a similar process, and setting $M = 3, 4$ and 5, the approximate solutions are calculated as

$$y_3(t) \simeq 1.00358 + 1.92942t + 1.19507t^2 - 0.141377t^3,$$
$$y_4(t) \simeq 1.00303 + 1.93623t + 1.29922t^2 - 0.481707t^3 + 0.248103t^4,$$
$$y_5(t) \simeq 1.00201 + 1.93709t + 1.45666t^2 - 1.25618t^3 + 1.45187t^4 - 0.594764t^5.$$

7. Existence, Uniqueness, Stability and Convergence Results

Let us consider a general form of the fractional integro-differential equation as follows:

$$^{C}D_t^{\alpha} y(t) = \lambda_1 \int_0^1 K_1(t,\tau)\Psi_1(\tau, y(\tau))\, d\tau + \lambda_2 \int_0^t K_2(t,\tau)\Psi_2(\tau, y(\tau))\, d\tau + f(t, y(t)), \tag{27}$$

with initial condition

$$y(0) = y_0, \tag{28}$$

where $0 < \alpha \leq 1$, the parameters λ_1 and λ_2 are real constants, K_1, K_2, Ψ_1, Ψ_2 and f are given known functions.

In this section, we prove the existence and uniqueness of a solution to problem (27)–(28). Furthermore, Ulam-Hyers stability of this problem is discussed. Finally, the convergence of the approximate solution is considered.

7.1. Existence and Uniqueness

In the following theorem, we suggest some conditions to have a unique solution for problem (27)–(28).

Theorem 1. *Suppose that* ${}^{C}D_t^{\alpha}y \in C[0,1]$ *and* $K_1, K_2 \in C([0,1] \times [0,1])$. *Let* $\Psi_1, \Psi_2, f \in C([0,1] \times \mathbb{R})$ *satisfy the Lipschitz conditions*

$$|\Psi_1(t,y_1) - \Psi_1(t,y_2)| \le L_1 |y_1 - y_2|,$$
$$|\Psi_2(t,y_1) - \Psi_2(t,y_2)| \le L_2 |y_1 - y_2|,$$
$$|f(t,y_1) - f(t,y_2)| \le L_3 |y_1 - y_2|.$$

Define

$$I_1 y(t) = \int_0^1 K_1(t,\tau)\,\Psi_1(\tau, y(\tau))\,d\tau,$$

$$I_2 y(t) = \int_0^t K_2(t,\tau)\,\Psi_2(\tau, y(\tau))\,d\tau,$$

$$M_1 = \max_{0\le t,\tau \le 1} |K_1(t,\tau)|,$$

$$M_2 = \max_{0\le t,\tau \le 1} |K_2(t,\tau)|.$$

If $\lambda_1 I_1 y(0) + f(0,y(0)) = 0$ *and* $\frac{5}{4}(|\lambda_1| L_1 M_1 + |\lambda_2| L_2 M_2 + L_3) < 1$, *then, problem* (27)–(28) *has a unique solution* $y \in C[0,1]$.

Proof. First, we prove that $y(t)$ is a solution of problem (27)–(28) if and only if it is a solution of the following integral equation

$$y(t) = y_0 + {}^{RL}I_t^{\alpha}\left(\lambda_1 I_1 y(t) + \lambda_2 I_2 y(t) + f(t, y(t))\right). \tag{29}$$

To this aim, let $y(t)$ be a solution of problem (27)–(28). By applying the Riemann-Liouville integral operator of order α to both sides of equation (27), we get

$$y(t) - y(0) = {}^{RL}I_t^{\alpha}\left(\lambda_1 I_1 y(t) + \lambda_2 I_2 y(t) + f(t, y(t))\right).$$

Employing $y(0) = y_0$ in the above equation leads to (29). On the other hand, suppose that y is a solution of (29), then, by applying the Caputo derivative operator of order α to both sides of (29), we obtain equation (27). Furthermore, by setting $t = 0$ in (29) and using the condition $\lambda_1 I_1 y(0) + f(0, y(0)) = 0$, we get the initial condition $y(0) = y_0$. Therefore, y is a solution of problem

(27)–(28).

Now, we prove that (29) has a unique solution. To do this, we define the operator $A : C[0,1] \to C[0,1]$ by

$$Ay(t) = y_0 + {}^{RL}I_t^{\alpha}\left(\lambda_1 I_1 y(t) + \lambda_2 I_2 y(t) + f(t, y(t))\right).$$

Then, finding a solution of problem (27)–(28) in $C[0,1]$ is equivalent to finding a unique fixed point of the operator A. For $y, z \in C[0,1]$, we have

$$\begin{aligned} |Ay(t) - Az(t)| &= \left|{}^{RL}I_t^{\alpha}\left(\lambda_1 I_1 y(t) + \lambda_2 I_2 y(t) + f(t, y(t))\right) - {}^{RL}I_t^{\alpha}\left(\lambda_1 I_1 z(t) + \lambda_2 I_2 z(t) + f(t, z(t))\right)\right| \\ &\le \left|\lambda_1 I_1 (y-z)(t) + \lambda_2 I_2 (y-z)(t) + f(t, y(t)) - f(t, z(t))\right| \left({}^{RL}I_t^{\alpha} 1\right). \end{aligned} \tag{30}$$

Also, using the Lipschitz condition for the operators I_1 and I_2 and taking into account the maximum of K_1 and K_2, we have

$$\begin{aligned} |I_1 (y-z)(t)| &\le L_1 M_1 |y-z|, \\ |I_2 (y-z)(t)| &\le L_2 M_2 |y-z|. \end{aligned} \tag{31}$$

By utilizing (31) and the Lipschitz condition for the function f, (30) yields

$$|Ay(t) - Az(t)| \le \left(|\lambda_1| L_1 M_1 + |\lambda_2| L_2 M_2 + L_3\right) \frac{t^{\alpha}}{\Gamma(\alpha+1)} |y-z|.$$

Since $0 < \alpha \le 1$, we get $\dfrac{t^{\alpha}}{\Gamma(\alpha+1)} \le \dfrac{5}{4}$ for $t \in [0,1]$ [9]. Therefore, we obtain

$$|Ay(t) - Az(t)| \le \frac{5}{4}\left(|\lambda_1| L_1 M_1 + |\lambda_2| L_2 M_2 + L_3\right) |y(t) - z(t)| < |y(t) - z(t)|,$$

which yields

$$\|Ay - Az\| < \|y - z\|,$$

where $\|\cdot\|$ is the max-norm in the space $C[0,1]$. The above inequality implies that the operator A is a contraction. Consequently, it has a unique fixed point by the Banach fixed-point theorem. □

7.2. Ulam-Hyers Stability

Theorem 2. *[10] Suppose that $\alpha > 0$, $h_1(t)$ is nonnegative, nondecreasing and locally integrable on $a < t < b < \infty$, $h_2(t) \leq K$ and $y(t)$ is nonnegative and locally integrable on $[a,b)$ with*

$$y(t) \leq h_1(t) + h_2(t)\left({}^{RL}I_t^{\alpha} y(t)\right),$$

then,

$$y(t) \leq h_1(t) E_{\alpha}(h_2(t)(t-a)^{\alpha}), \tag{32}$$

where E_{α} is the Mittag-Leffler function defined by $E_{\alpha}(t) = \sum_{k=0}^{\infty} \frac{t^k}{\Gamma(\alpha k+1)}$.

Remark 1. *[11] If in Theorem 2, $h_1(t)$ is not necessarily nondecreasing then the inequality* (32) *takes the most general form*

$$y(t) \leq h_1(t) + \int_a^t \sum_{k=1}^{\infty} \left(\frac{(h_2(t))^k}{\Gamma(\alpha k)} (t-\tau)^{\alpha k - 1} \right) d\tau, \quad a \leq t < b.$$

Definition 4. *Equation* (27) *is said to be Ulam-Hyers stable if for each $\varepsilon > 0$ and $\forall z(t)$ satisfying the inequality*

$$\left|{}^{C}D_t^{\alpha} z(t) - (\lambda_1 I_1 z(t) + \lambda_2 I_2 z(t) + f(t, z(t)))\right| < \varepsilon, \tag{33}$$

there exists a solution $y(t)$ of equation (27) *satisfying*

$$|z(t) - y(t)| < d\varepsilon, \qquad d \in \mathbb{R}.$$

Theorem 3. *According to Theorem 1, equation* (27) *is Ulam-Hyers stable.*

Proof. If $z(t)$ satisfies (33), there exists a function $g(t)$ satisfying $|g(t)| < \varepsilon$ such that

$${}^{C}D_t^{\alpha} z(t) - (\lambda_1 I_1 z(t) + \lambda_2 I_2 z(t) + f(t, z(t))) = g(t),$$

which is equivalent to

$$z(t) - z(0) - {}^{RL}I_t^{\alpha}(\lambda_1 I_1 z(t) + \lambda_2 I_2 z(t) + f(t, z(t))) = {}^{RL}I_t^{\alpha} g(t).$$

Therefore, we have

$$|z(t) - z(0) - {}^{RL}I_t^{\alpha}(\lambda_1 I_1 z(t) + \lambda_2 I_2 z(t) + f(t, z(t)))| = |{}^{RL}I_t^{\alpha} g(t)| \quad \leq |g(t)| \left({}^{RL}I_t^{\alpha} 1\right) \leq \frac{5}{4}\varepsilon.$$

Now, let $y(t)$ be the solution of equation (27) satisfying $y(0) = z(0) = y_0$. Then, we have

$$y(t) = z(0) + {}^{RL}I_t^{\alpha}\left(\lambda_1 I_1 y(t) + \lambda_2 I_2 y(t) + f(t, y(t))\right).$$

Note that the existence and uniqueness of $y(t)$ is assured by Theorem 1. We have

$$\begin{aligned}
&|z(t) - y(t)| \\
&= \left|z(t) - z(0) - {}^{RL}I_t^{\alpha}\left(\lambda_1 I_1 y(t) + \lambda_2 I_2 y(t) + f(t, y(t))\right)\right| \\
&= \Big|z(t) - z(0) - {}^{RL}I_t^{\alpha}\left(\lambda_1 I_1 z(t) + \lambda_2 I_2 z(t) + f(t, z(t))\right) \\
&+ {}^{RL}I_t^{\alpha}\left(\lambda_1 I_1 z(t) + \lambda_2 I_2 z(t) + f(t, z(t))\right) - {}^{RL}I_t^{\alpha}\left(\lambda_1 I_1 y(t) + \lambda_2 I_2 y(t) + f(t, y(t))\right)\Big| \\
&\leq \frac{5}{4}\varepsilon + \left(|\lambda_1| L_1 M_1 + |\lambda_2| L_2 M_2 + L_3\right)\left({}^{RL}I_t^{\alpha}|z(t) - y(t)|\right).
\end{aligned}$$

Now, using the Gronwall's inequality with $h_1(t) = \frac{5}{4}\varepsilon$ and $h_2(t) = |\lambda_1| L_1 M_1 + |\lambda_2| L_2 M_2 + L_3$, we get

$$\begin{aligned}
|z(t) - y(t)| &\leq \frac{5}{4}\varepsilon E_{\alpha}\left(\left(|\lambda_1| L_1 M_1 + |\lambda_2| L_2 M_2 + L_3\right) t^{\alpha}\right) \\
&\leq \frac{5}{4}\varepsilon E_{\alpha}\left(|\lambda_1| L_1 M_1 + |\lambda_2| L_2 M_2 + L_3\right),
\end{aligned}$$

which gives $|z(t) - y(t)| \leq d\varepsilon$, with

$$d = \frac{5}{4} E_{\alpha}\left(|\lambda_1| L_1 M_1 + |\lambda_2| L_2 M_2 + L_3\right).$$

Hence, equation (27) is Ulam-Hyers stable. □

7.3. Convergence Analysis

Here, we consider the convergence of the approximate solution obtained by the proposed scheme in the Section 6. to the analytical solution of problem (27)–(28).

To do this, we suppose that (a, b) is a bounded interval in the real line ($\mathbb{R}$) and $H^m(a, b)$ (m is an integer number greater than or equal to zero) is the following vector space

$$H^m(a, b) = \left\{ y : (a, b) \longrightarrow \mathbb{R} : \text{for } 0 \leq j \leq m,\ y^{(j)}(t) \in L^2(a, b) \right\},$$

where $y^{(j)}$ denotes the j-th order derivative of the function y. The Sobolev space $H^m(a,b)$ is a Hilbert space equipped with the following norm

$$\|y\|_{H^m(a,b)} = \left(\sum_{j=0}^{m} \|y^{(j)}\|^2_{L^2(a,b)} \right)^{\frac{1}{2}}.$$

To continue the error discussion, the following lemma from [4] is recalled.

lemma 1 (See [4]). *Suppose that $y \in H^m(-1,1)$ with $m \geq 0$. Let $P_M(y) = \sum_{i=0}^{M} y_i L_i(t)$ be the truncated Legendre series of y. Then,*

$$\|y - P_M(y)\|_{L^2(-1,1)} \leq CM^{-m} |y|_{H^{m;M}(-1,1)},$$

where

$$|y|_{H^{m;M}(-1,1)} = \left(\sum_{j=\min\{m,M+1\}}^{m} \|y^{(j)}\|^2_{L^2(-1,1)} \right)^{\frac{1}{2}},$$

and C is a positive constant independent of function y and integer M.

lemma 2. *Let $y : (0,1) \longrightarrow \mathbb{R}$ be a function in $H^m(0,1)$. Suppose that function $\bar{y} : (-1,1) \longrightarrow \mathbb{R}$ is defined by $\bar{y}(t) = y(\frac{1}{2}(t+1))$ for all $t \in (-1,1)$. Then, for $0 \leq j \leq m$*

$$\|\bar{y}^{(j)}\|_{L^2(-1,1)} = 2^{\frac{1}{2}-j} \|y^{(j)}\|_{L^2(0,1)}.$$

Proof. Using the change of variable $\tau = \frac{1}{2}(t+1)$, we have

$$\begin{aligned}
\|\bar{y}^{(j)}\|^2_{L^2(-1,1)} &= \int_{-1}^{1} |\bar{y}^{(j)}(t)|^2 dt \\
&= \int_{-1}^{1} \left| y^{(j)} \left(\frac{1}{2}(t+1) \right) \right|^2 dt \\
&= \int_{0}^{1} 2^{-2j} \left| y^{(j)}(\tau) \right|^2 2d\tau \\
&= 2^{1-2j} \|y^{(j)}\|^2_{L^2(0,1)},
\end{aligned}$$

Then, taking the square root from its both sides of this equation, gives the desired result. □

Theorem 4. *Suppose $m \geq 0$ and $y \in H^m(0,1)$. Let $y_M(t) = \sum_{i=0}^{M} y_i L_i^*(t)$ be the approximate solution obtained by the operational matrix method in the Section 6.. Then,*

$$\|y - y_M\|_{L^2(0,1)} \leq CM^{-m} |y|_{H^{m;M;0}(0,1)},$$

where

$$|y|_{H^{m;M;0}(0,1)} = \left(\sum_{k=\min\{m,M+1\}}^{m} \left(\frac{1}{2}\right)^{2k} \|y^{(k)}\|^2_{L^2(0,1)} \right)^{\frac{1}{2}}.$$

Proof. With the help Lemma 2, we obtain

$$\begin{aligned}
\|y - y_M\|^2_{L^2(0,1)} &= \frac{1}{2} \|\bar{y} - P_M(\bar{y})\|^2_{L^2(-1,1)} \\
&\leq \frac{1}{2} C' M^{-2m} \sum_{k=\min\{m,M+1\}}^{m} \|\bar{y}^{(k)}\|^2_{L^2(-1,1)} \\
&= C' M^{-2m} \sum_{k=\min\{m,M+1\}}^{m} \left(\frac{1}{2}\right)^{2k} \|y^{(k)}\|^2_{L^2(0,1)}.
\end{aligned}$$

By definition

$$|y|_{H^{m;M;0}(0,1)} = \left(\sum_{k=\min\{m,M+1\}}^{m} \left(\frac{1}{2}\right)^{2k} \|y^{(k)}\|^2_{L^2(0,1)} \right)^{\frac{1}{2}},$$

which completes the proof. □

Theorem 5. *Let $y_M(t)$ be the approximate solution of problem* (27)–(28) *obtained by the operational matrix method in Section 6., $y(t)$ be its analytical solution and $R_M(t)$ be the residual error for the approximate solution. Also, the conditions of Theorem 1 are satisfied. Then, if $y(t)$ satisfies the conditions of Theorem 4, $R_M(t)$ tends to zero when $M \to \infty$.*

Proof. By applying the Riemann-Liouville integral, ${}^{RL}I_t^{\alpha}$, to both sides of equation (27), we can rewrite equation (27) as follows:

$$y(t) = y_0 + {}^{RL}I_t^{\alpha} \left(\lambda_1 I_1 y(t) + \lambda_2 I_2 y(t) + f(t, y(t))\right).$$

where

$$I_1 y(t) = \int_0^1 K_1(t,\tau)\Psi_1(\tau, y(\tau))\, d\tau,$$
$$I_2 y(t) = \int_0^t K_2(t,\tau)\Psi_2(\tau, y(\tau))\, d\tau.$$

So, $y_M(t)$ satisfies the following equation

$$y_M(t) = y_0 + {}^{RL}I_t^{\alpha}\left(\lambda_1 I_1 y_M(t) + \lambda_2 I_2 y_M(t) + f(t, y_M(t))\right) + R_M(t),$$

where $R_M(t)$ is the residual function given by

$$R_M(t) = y_M(t) - y(t) +$$
$$+ {}^{RL}I_t^{\alpha}\left(\lambda_1 I_1(y(t) - y_M(t)) + \lambda_2 I_2(y(t) - y_M(t)) + f(t, y(t)) - f(t, y_M(t))\right).$$

Then,

$$\|R_M(t)\| =$$
$$\left\| y_M(t) - y(t) + {}^{RL}I_t^{\alpha}\left(\lambda_1 I_1(y(t) - y_M(t)) + \lambda_2 I_2(y(t) - y_M(t)) + f(t, y(t)) - f(t, y_M(t))\right)\right\|.$$

According to Theorem 1, we have

$$\|R_M(t)\| \leq \left(1 + \frac{5}{4}\left(|\lambda_1| K_1 L_1 + |\lambda_2| K_2 L_2 + L_3\right)\right) \|y_M(t) - y(t)\|.$$

Now applying Theorem 4, yields

$$\|R_M(t)\| \leq C M^{-m}\left(1 + \frac{5}{4}\left(|\lambda_1| K_1 L_1 + |\lambda_2| K_2 L_2 + L_3\right)\right) |y|_{H^{m;M;0}(0,1)}.$$

Therefore, it is clear that $R_M(t)$ tends to zero when $M \to \infty$. □

Conclusion

A numerical method using operational matrices based on the shifted Legendre polynomials has been used to obtain numerical solutions of fractional-order differential equations, as well as fractional integro-differential equations. The main problems are converted into a system of algebraic equations by using operational matrices. After solving the algebraic equations, the values are substituted in the approximate function to obtain a solution of the governing equation. The operational matrices could be obtained with other basic functions. Finally, stability and convergence of the given method are discussed.

References

[1] K. S. Miller, B. Ross (1993). *An Introduction to the Fractional Calculus and Fractional Differential Equations.* Wiley, New York.

[2] I. Podlubny (1999): *Fractional Differential Equations.* Academic Press, San Diego.

[3] S. G. Samko, A. A. Kilbas and O. I. Marichev (1993): *Fractional Integrals and Derivatives: Theory and Applications.* Gordon and Breach, Yverdon.

[4] C. Canuto, M. Y. Hussaini, A. Quarteroni, T. A. Zang (2006): *Spectral Methods. Scientific Computation*, Springer-Verlag, Berlin.

[5] E. Kreyszig (1989): *Introductory Functional Analysis with Applications.* John Wiley and Sons.

[6] R. L. Bagley, P. J. Torvik (1983): Fractional calculus: a different approach to the analysis of viscoelastically damped structures, *AIAA J.*, 21(5), 741–748.

[7] H. Jafari (2016): Numerical Solution of Time-Fractional Klein–Gordon Equation by Using the Decomposition Methods. *ASME. J. Comput. Nonlinear Dynam.*, 11(4).

[8] A. Saadatmandi, M. Dehghan (2010): A new operational matrix for solving fractional-order differential equations. *Computers and Mathematics with Applications*, 59, 1326–1336.

[9] R. M. Ganji, H. Jafari and S. Nemati, A new approach for solving integro-differential equations of variable order, *Journal of Computational and Applied Mathematics*, 379 (2020), 112940.

[10] H. Ye, J. Gao and Y. Ding, A generalized Gronwall inequality and its application to a fractional differential equation, *Journal of Mathematical Analysis and Applications*, 328(2) (2007), 1075–1081.

[11] Y. Adjabi, F. Jarad and T. Abdeljawadc, On generalized fractional operators and a Gronwall type inequality with applications, *Filomat*, 31(17) (2017), 5457–5473.

In: Understanding Integro-Differential Equations ISBN: 979-8-89113-040-1
Editors: J. Vasundhara Devi et al. © 2023 Nova Science Publishers, Inc.

Chapter 6

ON A SECOND ORDER NONLINEAR INTEGRO-DIFFERENTIAL EQUATION OF FREDHOLM TYPE IN A COMPLEX PLANE

Lemita Samir[1,*]
Selim Raja[2,†]
and Belahbib Zeineb[2,‡]
[1]Ecole normale supérieure de Ouargla, Algeria;
Laboratoire de Mathématiques Appliquées et de Modélisation,
Université 8 Mai 1945 Guelma, Guelma, Algeria
[2]Ecole normale supérieure de Ouargla, Algeria

Abstract

In this paper, we deal with the study of a nonlinear Fredholm integro-differential equation of second order in the complex plane. At first, under appropriate conditions, the existence and uniqueness of the solution of the proposed equation have been demonstrated by using the Banach fixed point theorem. After that, we have resorted to construct a numerical process based on the combination between Nyström and Picard iterative methods in order to obtain an approximate solution of our equation. Our study has been supported by illustrative examples to confirm the efficiency of the used numerical process.

*Corresponding Author's Emails: lemita.samir@ens-ouargla.dz; lem.samir@gmail.com.
†Corresponding Author's Email: jijisweet99@gmail.com.
‡Corresponding Author's Email: belahbibens98@gmail.com.

Keywords: nonlinear Fredholm equation, integro-differential equation, fixed point theorem, Nyström method, Picard iterative method, complex plane

MSC (2010): 45B05, 47G20, 30D05, 37C25

1. Introduction

Integral and integro-differential equations in complex analysis are considered to be an important branch of mathematics. They are used for modeling fundamental problems and have applications in diverse scientific fields, such as, electrical engineering, heat and mass transfer, elasticity, oscillation, fluid dynamics, control, etc [1, 2]. For this reason, various articles have discussed linear, nonlinear, integral and integro-differential equations in the complex plane [3, 4, 5, 6], and have chosen appropriate techniques to obtain solutions.
In our work, we propose a different form of Fredholm integral equation. This equation is a second order nonlinear integro-differential in the complex plane, where all derivatives (first and second) of the unknown function appearing inside of the nonlinear integral operator. Consider the following proposed equation:

$$\psi(s) = g(s) + \int_a^b k(s,r,\psi(r),\psi'(r),\psi''(r))\,dr, \quad \forall s \in [a,b], \tag{1}$$

where, $g(s) \in C^2([a,b],\ \mathbb{C})$ and $k \in C^2([a,b]^2 \times \mathbb{C}^3,\ \mathbb{C})$ are some given complex-valued continuously differentiable functions, and $\psi(s)$ is the unknown function in $C^2([a,b],\ \mathbb{C})$.

Our work is concerned, first, with the treatment of existence and uniqueness of solutions. Then, we present a numerical process to approach this solution. Furthermore, we prove the convergence of the suggested numerical process and give some illustrative examples.

2. Preliminaries

We introduce some notions and concepts, which we will use in the next section. First, we differentiate twice both sides of the proposed equation (1) to get more information about the unknown function $\psi(s)$ and obtain the following system in the complex plane:

$$\begin{aligned}
\psi(s) &= g(s)+\int_a^b k(s,r,\psi(r),\psi'(r),\psi''(r))\,dr, &(2)\\
\psi'(s) &= g'(s)+\int_a^b \frac{\partial k}{\partial s}(s,r,\psi(r),\psi'(r),\psi''(r))\,dr, &(3)\\
\psi''(s) &= g''(s)+\int_a^b \frac{\partial^2 k}{\partial s^2}(s,r,\psi(r),\psi'(r),\psi''(r))\,dr. &(4)
\end{aligned}$$

For a complex variable $z = x+iy$, any complex-valued function $f(z)$ can be represented as:

$$f(z) = f(x+iy) = f_1(x,y)+if_2(x,y), \textit{(real and imaginary parts)}.$$

This allows us to give the following representations:

$$\begin{aligned}
\psi(s) &= x(s)+iy(s), \quad g(s) = g_1(s)+ig_2(s),\\
\psi'(s) &= x'(s)+iy'(s), \quad g'(s) = g_1'(s)+ig_2'(s),\\
\psi''(x) &= x''(s)+iy''(s), \quad g''(s) = g_1''(s)+ig_2''(s),
\end{aligned}$$

where, $x(s), y(s), g_1(s)$ and $g_2(s) \in C^2([a,b], \mathbb{R})$. Also, we obtain:

$$\begin{aligned}
k(s,r,\psi,\psi',\psi'') &= \phi_1(s,r,x,y,x',y',x'',y'')+i\phi_2(s,r,x,y,x',y',x'',y''),\\
\frac{\partial k}{\partial s}(s,r,\psi,\psi',\psi'') &= \frac{\partial \phi_1}{\partial s}(s,r,x,y,x',y',x'',y'')+i\frac{\partial \phi_2}{\partial s}(s,r,x,y,x',y',x'',y''),\\
\frac{\partial^2 k}{\partial s^2}(s,r,\psi,\psi',\psi'') &= \frac{\partial^2 \phi_1}{\partial s^2}(s,r,x,y,x',y',x'',y'')+i\frac{\partial^2 \phi_2}{\partial s^2}(s,r,x,y,x',y',x'',y''),
\end{aligned}$$

where, $\phi_1, \phi_2 \in C^2([a,b]^2 \times \mathbb{R}^6, \mathbb{R})$. As well as, system of equations (2)-(4), can be transformed as follows:

$$\begin{aligned}
x(s) &= g_1(s) + \int_a^b \phi_1(s,r,x(r),y(r),x'(r),y'(r),x''(r),y''(r))\,dr, && (5)\\
y(s) &= g_2(s) + \int_a^b \phi_2(s,r,x(r),y(r),x'(r),y'(r),x''(r),y''(r))\,dr, && (6)\\
x'(s) &= g_1'(s) + \int_a^b \frac{\partial \phi_1}{\partial s}(s,r,x(r),y(r),x'(r),y'(r),x''(r),y''(r))\,dr, && (7)\\
y'(s) &= g_2'(s) + \int_a^b \frac{\partial \phi_2}{\partial s}(s,r,x(r),y(r),x'(r),y'(r),x''(r),y''(r))\,dr, && (8)\\
x''(s) &= g_1''(s) + \int_a^b \frac{\partial^2 \phi_1}{\partial s^2}(s,r,x(r),y(r),x'(r),y'(r),x''(r),y''(r))\,dr, && (9)\\
y''(s) &= g_2''(s) + \int_a^b \frac{\partial^2 \phi_2}{\partial s^2}(s,r,x(r),y(r),x'(r),y'(r),x''(r),y''(r))\,dr. && (10)
\end{aligned}$$

However, to rewrite the previous system in a simple formula, we consider the Banach product space $\mathscr{P} = \prod_{i=1}^6 C([a,b],\mathbb{R})$ equipped with the following norm:

$$\forall\, V = (v_1, v_2, v_3, v_4, v_5, v_6) \in \mathscr{P}, \; \|V\|_{\mathscr{P}} = \sum_{i=1}^{6} \|v_i\|_\infty.$$

Then, our system of equations (5)-(10) can be reformulated as:

$$\forall s \in [a,b], \forall\, V = (x(s), y(s), x'(s), y'(s), x''(s), y''(s)) \in \mathscr{P},$$
$$V = T_G(V),$$

where, the functional T_G is defined from $\mathscr{P}$ into itself by:

$$\begin{aligned}
T_G :\; &\mathscr{P} \longrightarrow \mathscr{P}\\
&V \longmapsto T_G(V) = \left(T_{G_1}(V), T_{G_2}(V), T'_{G_1}(V), T'_{G_2}(V), T''_{G_1}(V), T''_{G_2}(V)\right),
\end{aligned}$$

where,

$$\begin{aligned}
T_{G_1}(V) &= g_1(s) + \int_a^b \phi_1(s,r,x(r),y(r),x'(r),y'(r),x''(r),y''(r))\,dr,\\
T_{G_2}(V) &= g_2(s) + \int_a^b \phi_2(s,r,x(r),y(r),x'(r),y'(r),x''(r),y''(r))\,dr,
\end{aligned}$$

and $T'_{G_1}(V), T'_{G_2}(V), T''_{G_1}(V), T''_{G_2}(V)$ are the derivatives of $T_{G_1}(V)$ and $T_{G_2}(V)$, respectively.

3. Existence and Uniqueness

The current section shows us how to take the conditions that are ensuring the existence and uniqueness of solutions of system (5)-(10). At first, we give the following necessary definitions and hypothesis:

Definition 1. *(6-Lipschitzian function)*
We say that the function $K(s,r,x_1,x_2,x_3,x_4,x_5,x_6)$ is $6-Lipschitzian$ with constants $l_1,l_2,l_3,l_4,l_5,l_6 \in \mathbb{R}^+$, if it satisfies the following condition: $\forall s,r \in [a,b], \forall x_i,y_i \in \mathbb{R}, i=1:6$, then:

$$|K(s,r,x_1,x_2,x_3,x_4,x_5,x_6)-K(s,r,y_1,y_2,y_3,y_4,y_5,y_6)| \leq \sum_{i=1}^{6} l_i|x_i-y_i|.$$

For short, we write: K is $6-Lipsch$.

Hypothesis $\mathscr{H}$:
We suppose that kernels: $\phi_1, \phi_2, \frac{\partial \phi_1}{\partial s}, \frac{\partial \phi_1}{\partial s}, \frac{\partial^2 \phi_1}{\partial s^2}$ and $\frac{\partial^2 \phi_2}{\partial s^2}$ are $6-Lipsch$ with constants $l_{2i+1}, l_{2i+2}, m_{2i+1}, m_{2i+2}, t_{2i+1}, t_{2i+2}$ for $i=0:5$, respectively. Moreover, let θ be a positive parameter that verifies:

$$\max\left(\sum_{i=1}^{2} l_i, \sum_{i=3}^{4} l_i, \sum_{i=5}^{6} l_i, \sum_{i=7}^{8} l_i, \sum_{i=9}^{10} l_i, \sum_{i=11}^{12} l_i\right)(b-a) \leq \theta < \frac{1}{3},$$

$$\max\left(\sum_{i=1}^{2} m_i, \sum_{i=3}^{4} m_i, \sum_{i=5}^{6} m_i, \sum_{i=7}^{8} m_i, \sum_{i=9}^{10} m_i, \sum_{i=11}^{12} m_i\right)(b-a) \leq \theta < \frac{1}{3},$$

$$\max\left(\sum_{i=1}^{2} t_i, \sum_{i=3}^{4} t_i, \sum_{i=5}^{6} t_i, \sum_{i=7}^{8} t_i, \sum_{i=9}^{10} t_i, \sum_{i=11}^{12} t_i\right)(b-a) \leq \theta < \frac{1}{3}.$$

Theorem 3..1. *Let hypothesis $(\mathscr{H})$ be satisfied. Then, the functional T_G is a contraction from $\mathscr{P}$ into itself. As well as, the system $V = T_G(V)$ has a one solution $V \in \mathscr{P}$.*

Proof. For $V_1=(x_1,y_1,x_1',y_1',x_1'',y_1''),V_2=(x_2,y_2,x_2',y_2',x_2'',y_2'')\in\mathscr{P}$, we have:

$$\begin{aligned}
\sum_{i=1}^{2} | & T_{G_i}(V_1)-T_{G_i}(V_2)| \\
\leq & \int_a^b |\phi_1(s,r,x_1,y_1,x_1',y_1',x_1'',y_1'')-\phi_1(s,r,x_2,y_2,x_2',y_2',x_2'',y_2'')|\,dr \\
+ & \int_a^b |\phi_2(s,r,x_1,y_1,x_1',y_1',x_1'',y_1'')-\phi_2(s,r,x_2,y_2,x_2',y_2',x_2'',y_2'')|\,dr, \\
\leq & \int_a^b (l_1|x_1-x_2|+l_3|y_1-y_2|+l_5|x_1'-x_2'|+l_7|y_1'-y_2'|+l_9|x_1''-x_2''|+l_{11}|y_1''-y_2''|)\,dr \\
+ & \int_a^b (l_2|x_1-x_2|+l_4|y_1-y_2|+l_6|x_1'-x_2'|+l_8|y_1'-y_2'|+l_{10}|x_1''-x_2''|+l_{12}|y_1''-y_2''|)dr, \\
\leq & (b-a)\left(\sum_{i=1}^{2}(l_i)\|x_1-x_2\|_\infty+\sum_{i=3}^{4}(l_i)\|y_1-y_2\|_\infty+\sum_{i=5}^{6}(l_i)\|x_1'-x_2'\|_\infty\right) \\
+ & (b-a)\left(\sum_{i=7}^{8}(l_i)\|y_1'-y_2'\|_\infty+\sum_{i=9}^{10}(l_i)\|x_1''-x_2''\|_\infty+\sum_{i=11}^{12}(l_i)\|y_1''-y_2''\|_\infty\right), \\
\leq & \max\left(\sum_{i=1}^{2}l_i,\sum_{i=3}^{4}l_i,\sum_{i=5}^{6}l_i,\sum_{i=7}^{8}l_i,\sum_{i=9}^{10}l_i,\sum_{i=11}^{12}l_i\right)(b-a)\|V_1-V_2\|_{\mathscr{P}}, \\
\leq & \theta\|V_1-V_2\|_{\mathscr{P}}.
\end{aligned}$$

Similarly, we get:

$$\begin{aligned}
\sum_{i=1}^{2} | & T_{G_i}'(V_1)-T_{G_i}'(V_2)| \\
\leq & \max\left(\sum_{i=1}^{2}m_i,\sum_{i=3}^{4}m_i,\sum_{i=5}^{6}m_i,\sum_{i=7}^{8}m_i,\sum_{i=9}^{10}m_i,\sum_{i=11}^{12}m_i\right)(b-a)\|V_1-V_2\|_{\mathscr{P}}, \\
\leq & \theta\|V_1-V_2\|_{\mathscr{P}}.
\end{aligned}$$

Also,

$$\begin{aligned}
\sum_{i=1}^{2} | & T_{G_i}''(V_1)-T_{G_i}''(V_2)| \\
\leq & \max\left(\sum_{i=1}^{2}t_i,\sum_{i=3}^{4}t_i,\sum_{i=5}^{6}t_i,\sum_{i=7}^{8}m_i,\sum_{i=9}^{10}t_i,\sum_{i=11}^{12}t_i\right)(b-a)\|V_1-V_2\|_{\mathscr{P}}, \\
\leq & \theta\|V_1-V_2\|_{\mathscr{P}}.
\end{aligned}$$

Thus,

$$\|T_G(V_1) - T_G(V_2)\|_{\mathscr{P}} \leq 3\theta \|V_1 - V_2\|_{\mathscr{P}}. \tag{11}$$

Since $0 \leq \theta < \frac{1}{3}$ and by applying Banach fixed point theorem, the functional T_G has a one fixed point $V \in \mathscr{P}$, which represents the solution of the system (5)-(10). □

In practice, the solution of the system $V = T_G(V)$ cannot be found exactly. For this reason, we must construct it according the Picard successive iterative method, that we will see in the next theorem.

Theorem 3..2. *Consider the sequences $V_n = (x_n, y_n, x'_n, y'_n, x''_n, y''_n) \in \mathscr{P}$ defined by: for $V_0 \in \mathscr{P}$:*

$$V_n = T_G(V_{n-1}), \qquad n \geq 1.$$

Then, V_n converges to V:

$$V_n \longrightarrow V, \quad as \quad n \longrightarrow \infty \quad in\ the\ space \quad \mathscr{P}.$$

Further, we have the following estimate:

$$\|V - V_n\|_{\mathscr{P}} \leq \frac{(3\theta)^n}{1 - 3\theta} \|V_1 - V_0\|_{\mathscr{P}}.$$

Proof. First, the induction on $n \in \mathbb{N}$ for inequality (11) gives us:

$$\|T_G^n(V_1) - T_G^n(V_2)\|_{\mathscr{P}} \leq (3\theta)^n \|V_1 - V_2\|_{\mathscr{P}}. \tag{12}$$

Then, according to the definition of V_n and inequality (12), we have that: for $V_0 \in \mathscr{P}$:

$$V_n = T_G^n(V_0) \longrightarrow V, \quad as \quad n \longrightarrow \infty \quad in\ the\ space \quad \mathscr{P}.$$

However, we have:

$$\begin{aligned} \|V - V_n\|_{\mathscr{P}} = \|V - T_G^n(V_0)\|_{\mathscr{P}} &\leq \sum_{k=n}^{\infty} (3\theta)^k \|T_G(V_0) - V_0\|_{\mathscr{P}}, \\ &\leq \frac{(3\theta)^n}{1 - 3\theta} \|V_1 - V_0\|_{\mathscr{P}}. \end{aligned} \tag{13}$$

□

4. Numerical Study

First, as we have seen above in Theorem 3..2, we have constructed a sequence $V_n \in \mathscr{P}$, to approach solutions of the system (5)-(10) using Picard iterative method. However, we can see that it is very hard to calculate exactly each integral term in the previous iterated sequences. Therefore, we must apply the well-known Nyström method [7], as a second approximate part, in order to approach all integral terms of the iterated sequences V_n. We describe that process in this section.
Let $N \in \mathbb{N}^*$, consider the following equidistance partition δ_N:

$$\delta_N = \left\{ s_i = a + ih, h = \frac{b-a}{N}, \ i = 0, \ldots, N \right\}.$$

According to the partition δ_N, we construct our numerical process based on the conjunction between Nyström and Picard iterative methods. This numerical procedure permits us to give approximate solutions $\widetilde{V}_n = (\widetilde{x}_n, \widetilde{y}_n, \widetilde{x}'_n, \widetilde{y}'_n, \widetilde{x}''_n, \widetilde{y}''_n) \in \mathscr{P}$ of the system (5)-(10), which are given by the following formulas: For all $i = 0, 1, \ldots, N$, for $n \geq 1$:

$$\widetilde{x}_n(s_i) = g_1(s_i) + \sum_{j=0}^{N} w_j \phi_1(s_i, r_j, \widetilde{x}_{n-1}(r_j), \widetilde{y}_{n-1}(r_j), \widetilde{x}'_{n-1}(r_j), \widetilde{y}'_{n-1}(r_j), \widetilde{x}''_{n-1}(r_j), \widetilde{y}''_{n-1}(r_j)),$$

$$\widetilde{y}_n(s_i) = g_2(s_i) + \sum_{j=0}^{N} w_j \phi_2(s_i, r_j, \widetilde{x}_{n-1}(r_j), \widetilde{y}_{n-1}(r_j), \widetilde{x}'_{n-1}(r_j), \widetilde{y}'_{n-1}(r_j), \widetilde{x}''_{n-1}(r_j), \widetilde{y}''_{n-1}(r_j)),$$

$$\widetilde{x}'_n(s_i) = g'_1(s_i) + \sum_{j=0}^{N} w_j \frac{\partial \phi_1}{\partial s}(s_i, r_j, \widetilde{x}_{n-1}(r_j), \widetilde{y}_{n-1}(r_j), \widetilde{x}'_{n-1}(r_j), \widetilde{y}'_{n-1}(r_j), \widetilde{x}''_{n-1}(r_j), \widetilde{y}''_{n-1}(r_j)),$$

$$\widetilde{y}'_n(s_i) = g'_2(s_i) + \sum_{j=0}^{N} w_j \frac{\partial \phi_2}{\partial s}(s_i, r_j, \widetilde{x}_{n-1}(r_j), \widetilde{y}_{n-1}(r_j), \widetilde{x}'_{n-1}(r_j), \widetilde{y}'_{n-1}(r_j), \widetilde{x}''_{n-1}(r_j), \widetilde{y}''_{n-1}(r_j)),$$

$$\widetilde{x}''_n(s_i) = g''_1(s_i) + \sum_{j=0}^{N} w_j \frac{\partial^2 \phi_1}{\partial s^2}(s_i, r_j, \widetilde{x}_{n-1}(r_j), \widetilde{y}_{n-1}(r_j), \widetilde{x}'_{n-1}(r_j), \widetilde{y}'_{n-1}(r_j), \widetilde{x}''_{n-1}(r_j), \widetilde{y}''_{n-1}(r_j)),$$

$$\widetilde{y}''_n(s_i) = g''_2(s_i) + \sum_{j=0}^{N} w_j \frac{\partial^2 \phi_2}{\partial s^2}(s_i, r_j, \widetilde{x}_{n-1}(r_j), \widetilde{y}_{n-1}(r_j), \widetilde{x}'_{n-1}(r_j), \widetilde{y}'_{n-1}(r_j), \widetilde{x}''_{n-1}(r_j), \widetilde{y}''_{n-1}(r_j)),$$

where, $\{w_j\}_{j=0}^{N}$ are real weights of the used Nyström method, such that: $\sup_{N \geq 2} \sum_{j=0}^{N} |w_j| < \infty$.

The next theorem, confirms the convergence of the proposed numerical process.

Theorem 4..1. *Let hypothesis* $(\mathscr{H})$ *be satisfied. Then, the approximate solutions* $\widetilde{V}_n$ *converge to* V*:*

$$\widetilde{V}_n \longrightarrow V, \quad \text{as} \quad n \longrightarrow \infty \quad \text{and} \quad N \longrightarrow \infty \quad \text{in the space} \quad \mathscr{P}.$$

Further, we have the following estimate:

$$\|V-\widetilde{V}_n\|_{\mathscr{P}} \leq \frac{(3\theta)^n}{1-3\theta}\|V_1-V_0\|_{\mathscr{P}} + \sum_{p=1}^{n}(3\theta)^{n-p}\varepsilon_p,$$

where, $\lim\limits_{N\longrightarrow\infty}\varepsilon_p = 0,\ p = 1,...,n.$

Proof. . By applying the triangle inequality twice and from (13), we obtain:

$$\begin{aligned}\|V-\widetilde{V}_n\|_{\mathscr{P}} &\leq \|V-V_n\|_{\mathscr{P}} + \|V_n-\widetilde{V}_n\|_{\mathscr{P}},\\ &\leq \tfrac{(3\theta)^n}{1-3\theta}\|V_1-V_0\|_{\mathscr{P}} + \|T_G(V_{n-1})-T_G(\widetilde{V}_{n-1})\|_{\mathscr{P}} + \|T_G(\widetilde{V}_{n-1})-\widetilde{V}_n\|_{\mathscr{P}}.\end{aligned}$$

In the same way of inequality (11), we can write:

$$\|T_G(V_{n-1})-T_G(\widetilde{V}_{n-1})\|_{\mathscr{P}} \leq (3\theta)\|V_{n-1}-\widetilde{V}_{n-1}\|_{\mathscr{P}}. \tag{14}$$

On the other hand, from [7], we have the following estimate:

$$\|T_G(\widetilde{V}_{n-1})-\widetilde{V}_n\|_{\mathscr{P}} \leq \varepsilon_n, \tag{15}$$

where, ε_n represents the convergence's order of the used Nystrom method. In addition, inequalities (14) and (15) give us:

$$\|V_n-\widetilde{V}_n\|_{\mathscr{P}} \leq (3\theta)\|V_{n-1}-\widetilde{V}_{n-1}\|_{\mathscr{P}} + \varepsilon_n.$$

Repeating the last inequality n-times, we get the requested estimate:

$$\|V-\widetilde{V}_n\|_{\mathscr{P}} \leq \frac{(3\theta)^n}{1-3\theta}\|V_1-V_0\|_{\mathscr{P}} + \sum_{p=1}^{n}(3\theta)^{n-p}\varepsilon_p.$$

□

5. Extended Study Overview

Consider the general form of the equation proposed above:

$$\psi(s) = g(s) + \int_a^b k(s, r, \psi(r), \psi'(r), \ldots, \psi^{(m)}(r))\, dr, \quad \forall s \in [a,b], \quad for \quad m > 2 \in \mathbb{N},$$

where, $\psi(s),\ g(s) \in C^m([a,b],\ \mathbb{C})$ and $k \in C^m([a,b]^2 \times \mathbb{C}^{m+1},\ \mathbb{C})$. In the same way that we have shown before we can get the following system:

$$\begin{aligned}
x(s) &= g_1(s) + \int_a^b \phi_1(s, r, x(r), y(r), \ldots, x^{(m)}(r), y^{(m)}(r))\, dr,\\
y(s) &= g_2(s) + \int_a^b \phi_2(s, r, x(r), y(r), \ldots, x^{(m)}(r), y^{(m)}(r))\, dr,\\
&\ldots\\
&\ldots\\
x^{(m)}(s) &= g_1^{(m)}(s) + \int_a^b \frac{\partial^m \phi_1}{\partial s^m}(s, r, x(r), y(r), \ldots, x^{(m)}(r), y^{(m)}(r))\, dr,\\
y^{(m)}(s) &= g_2^{(m)}(s) + \int_a^b \frac{\partial^m \phi_2}{\partial s^m}(s, r, x(r), y(r), \ldots, x^{(m)}(r), y^{(m)}(r))\, dr.
\end{aligned}$$

However, the extended vision of Definition 1 and hypothesis $(\mathscr{H})$, which confirm the existence and uniqueness of solutions of this system will be taken as follows:

Suppose that kernels: $\phi_1, \phi_2, \frac{\partial \phi_1}{\partial s}, \frac{\partial \phi_2}{\partial s}, \ldots, \frac{\partial^m \phi_1}{\partial s^m}, \frac{\partial^m \phi_2}{\partial s^m}$ are $(2m+2) - Lipsch$ with constants:

$$l_{2i+1}^j,\ l_{2i+2}^j \quad for \quad i = 0 : 2m+1,\ j = 1 : m+1.$$

Moreover, choose the positive parameter θ as follows: for $j = 1 : m+1$ as follows:

$$\max\left(\sum_{i=1}^{2} l_i^j, \sum_{i=3}^{4} l_i^j, \ldots, \sum_{i=4m+1}^{4m+2} l_i^j, \sum_{i=4m+3}^{4m+4} l_i^j\right)(b-a) \le \theta < \frac{1}{m+1}.$$

The numerical process, based on Nyström and Picard iterative methods, gives us an approximate solution $\widetilde{V}_n = (\widetilde{x}_n, \widetilde{y}_n, \widetilde{x}'_n, \widetilde{y}'_n, \ldots, \widetilde{x}_n^{(m)}, \widetilde{y}_n^{(m)})$ of the generalized previous system by: For all $i = 0, 1, \ldots, N$, for $n \geq 1$:

$$\begin{aligned}
\widetilde{x}_n(s_i) &= g_1(s) + \sum_{j=0}^{N} w_j \phi_1(s_i, r_j, \widetilde{x}_{n-1}(r_j), \widetilde{y}_{n-1}(r_j), \ldots, \widetilde{x}_{n-1}^{(m)}(r_j), \widetilde{y}_{n-1}^{(m)}(r_j)),\\
\widetilde{y}_n(s_i) &= g_2(s) + \sum_{j=0}^{N} w_j \phi_2(s_i, r_j, \widetilde{x}_{n-1}(r_j), \widetilde{y}_{n-1}(r_j), \ldots, \widetilde{x}_{n-1}^{(m)}(r_j), \widetilde{y}_{n-1}^{(m)}(r_j)),\\
&\ldots\\
&\ldots\\
\widetilde{x}_n^{(m)}(s_i) &= g_1^{(m)}(s) + \sum_{j=0}^{N} w_j \frac{\partial^m \phi_1}{\partial s^m}(s_i, r_j, \widetilde{x}_{n-1}(r_j), \widetilde{y}_{n-1}(r_j), \ldots, \widetilde{x}_{n-1}^{(m)}(r_j), \widetilde{y}_{n-1}^{(m)}(r_j)),\\
\widetilde{y}_n^{(m)}(s_i) &= g_2^{(m)}(s) + \sum_{j=0}^{N} w_j \frac{\partial^m \phi_2}{\partial s^m}(s_i, r_j, \widetilde{x}_{n-1}(r_j), \widetilde{y}_{n-1}(r_j), \ldots, \widetilde{x}_{n-1}^{(m)}(r_j), \widetilde{y}_{n-1}^{(m)}(r_j)).
\end{aligned}$$

Therefore, the approximate solution $\widetilde{V}_n$ converges to the exact solution V, in the product Banach space $\mathscr{P} = \prod_{i=1}^{2m+2} C([a,b], \mathbb{R})$, with the following estimated error:

$$\|V - \widetilde{V}_n\|_{\mathscr{P}} \leq \frac{((m+1)\theta)^n}{1-(m+1)\theta} \|V_1 - V_0\|_{\mathscr{P}} + \sum_{p=1}^{n} ((m+1)\theta)^{n-p} \varepsilon_p,$$

where, $\lim\limits_{N \to \infty} \varepsilon_p = 0,\ p = 1, \ldots, n.$

6. Simulation Results

In this section, we apply our numerical process to approach the solutions of two illustrative examples. These examples illustrate the effectiveness of the proposed process.

Example 1. *Consider the following integro-differential equation:*

$$\psi(s) = sin^3(2\pi s) + i\cos^3(2\pi s) - \frac{t^2}{3} + \int_0^1 \frac{\alpha(s,r)}{1 + (\psi(r) + \psi'(r) + \psi''(r))^2} dr, \quad s \in [0,1],$$

where,

$$\begin{aligned}
\alpha(s,r) &= s^2 r^2 (1 + [(1 - 12\pi^2)(sin^3(2\pi r) + i\cos^3(2\pi r)) + 3\pi \sin(4\pi r)[sin(2\pi r)\\
&+ 4\pi \cos(2\pi r) - i(\cos(2\pi r) + 4\pi \sin(2\pi r))]]^2).
\end{aligned}$$

The exact solution of this example is: $\psi(s) = sin^3(2\pi s) + i\cos^3(2\pi s),\ s \in [0,1].$

Table 1. Computed error E for examples (1) and (2), respectively

N	30	50	100	250	500
E : for example (1)	9.30E-4	3.35E-4	8.37E-5	1.34E-5	3.35E-6
E : for example (2)	2.28E-5	2.96E-6	1.85E-7	4.80E-9	3.54E-10

Example 2. *Consider the second equation:*

$$\psi(s) = cos(2\pi s) + i\sin(4\pi s) - \frac{t^3}{8} + \int_0^1 \frac{\beta(s,r)}{2 + e^{\psi(r)+\psi'(r)+\psi''(r)+\psi'''(r)}} dr, \quad s \in [0,1],$$

where,

$$\beta(s,r) = s^3 r^7 (2 + e^{[(1-4\pi^2)(cos(2\pi r)+(8\pi^3-2\pi)\sin(2\pi r))+i[(1-16\pi^2)sin(4\pi r)+(4\pi-64\pi^3)\cos(4\pi r)]]})$$

with exact solution: $\psi(s) = cos(2\pi s) + i\sin(4\pi s)$, $s \in [0,1]$.

Using the numerical process described above, we give an approximate solution for examples (1) and (2), respectively. We mention that we have taken the zero vector $\widetilde{V}_0 = 0$ as a starting point in the iterative scheme and we have chosen the following stopped condition: $\|\widetilde{V}_{n-1} - \widetilde{V}_n\|_{\mathscr{P}} < 10^{-9}$. We also calculate the absolute error E, between the approximate and exact solutions by the formula: $E = \|V - \widetilde{V}_n\|_{\mathscr{P}}$.

According to different values of steps N, Table (1) shows us the absolute error E for examples (1) and (2), which is close to 0. Therefore, the accuracy of the error increases with N. On the other hand, Figures (1)-(7) display a graphical comparison between the exact and approximate solution and their derivatives, which appear to be almost identical. So, from these simulation results, the performance and efficacy of the presented process, have been proved.

Conclusion

In this work, we have suggested a numerical process which is based on the following two procedures: Picard iterative scheme coupled with Nyström method. We have applied this numerical process to obtain approximate solutions of a type of Fredholm equation with a second order integro-differential in the complex plane. However, the illustrative examples give us very interesting results,

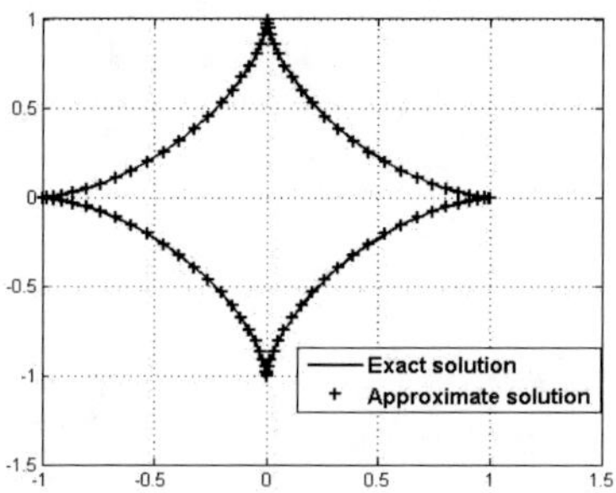

Figure 1. ψ versus $\widetilde{\psi}_n$ with $N = 50$, for Example (1).

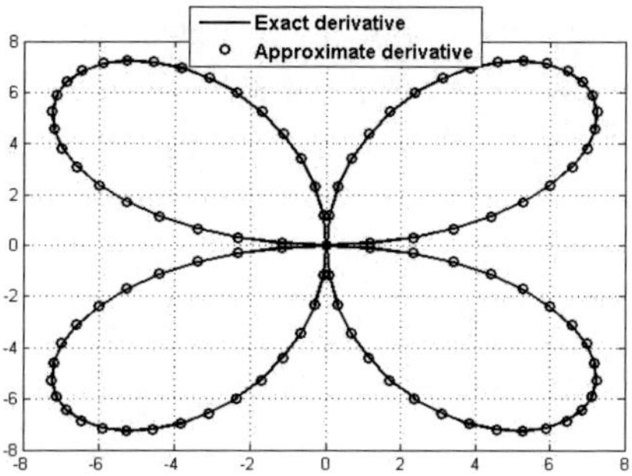

Figure 2. ψ' versus $\widetilde{\psi'_n}$ with $N = 50$, for Example (1).

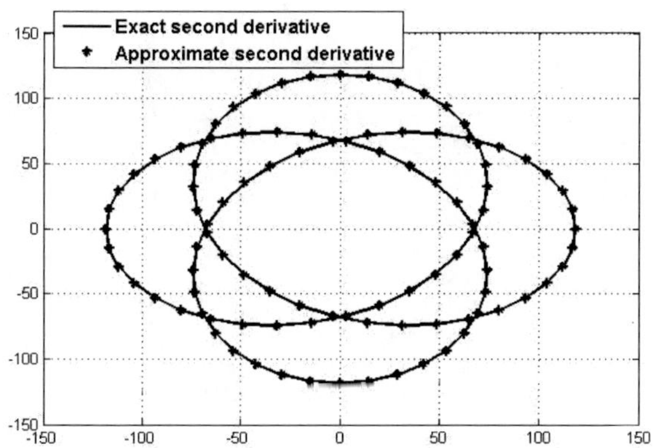

Figure 3. ψ'' versus $\widetilde{\psi''_n}$ with $N = 50$, for Example (1).

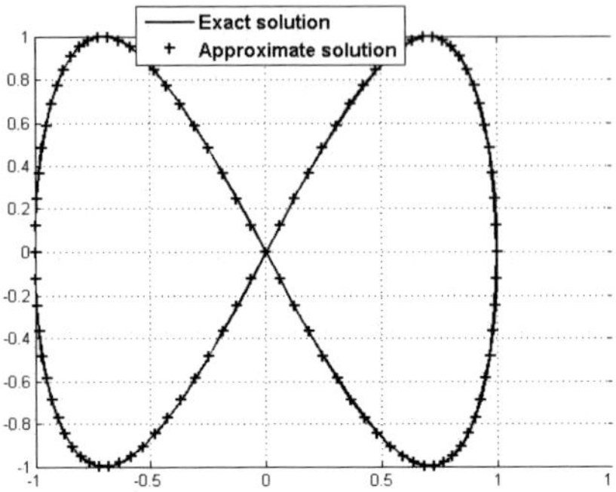

Figure 4. ψ versus $\widetilde{\psi_n}$ with $N = 50$, for Example (2).

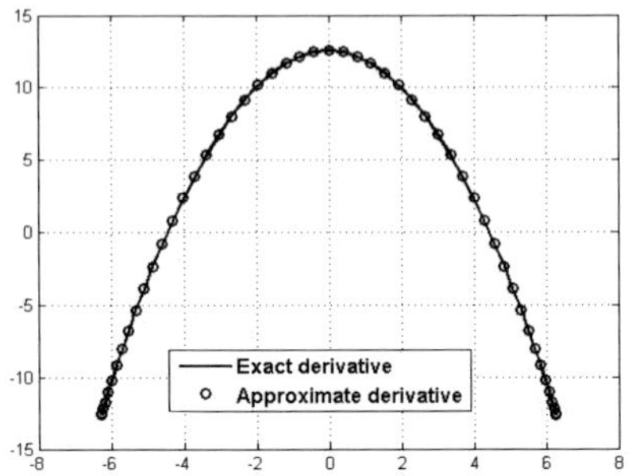

Figure 5. ψ' versus $\widetilde{\psi_n'}$ with $N = 50$, for Example (2).

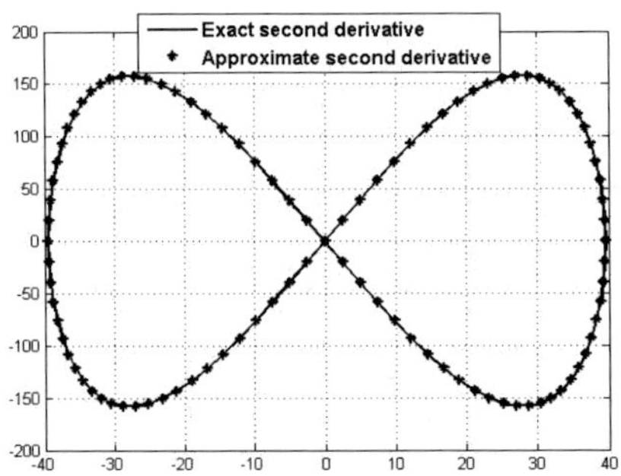

Figure 6. ψ'' versus $\widetilde{\psi_n''}$ with $N = 50$, for Example (2).

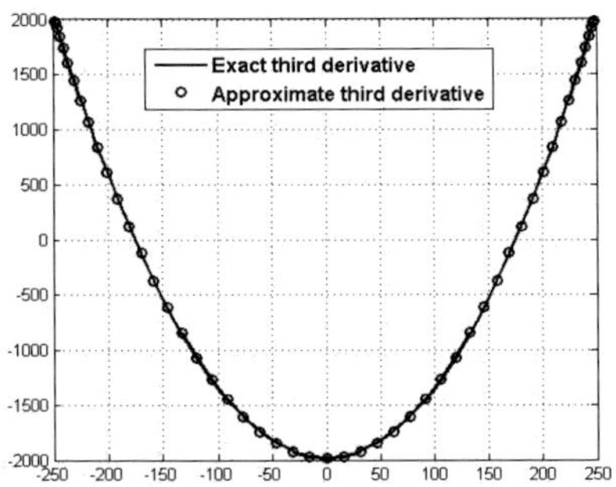

Figure 7. ψ''' versus $\widetilde{\psi_n'''}$ with $N = 50$, for Example (2).

which affirm the effectiveness of this numerical scheme. In future, we seek to develop this numerical process that will be suitable for equations with general kernels, for example: Integral equations with weakly singular kernels described in [8].

References

[1] Kwok, Y.K. (2010). *Applied Complex Variables for Scientists and Engineers.* Cambridge University Press, Cambridge.

[2] Kythe, P.K. and Puri, P. (1992). *Computational methods for linear integral equations.* University of New Orleans, New Orleans.

[3] Erfanian, M. and Akrami, A. (2017). RH wavelet bases to approximate solution of nonlinear Fredholm-Hammerstein integral equations in complex plane. *Mathematical Modelling of Systems*, 4(1), 1-8.

[4] Erfanian, M. (2018). The approximate solution of nonlinear mixed Volterra-Fredholm-Hammerstein integral equations with RH wavelet bases in a complex plane. *Mathematical Methods in the Applied Sciences*, 41(18), 8942-8952.

[5] Erfanian, M. and Zeidabadi, H. (2019). Solving of nonlinear Fredholm integro-differential equation in a complex plane with rationalized Haar wavelet bases. *Asian-European Journal of Mathematics*, 12(04), 1950055.

[6] Erfanian, M., Zeidabadi, H. and Parsamanesh, M. (2020). Using of PQWs for solving NFID in the complex plane. *Advances in Difference Equations*, 2020(1), 1-13.

[7] Atkinson, K. and Han, H. (2001). *Theoretical numerical analysis: a functional analysis framework.* Springer, New York.

[8] Touati, S., Lemita, S., Ghiat, M. and Aissaoui, M.Z. (2019). Solving a nonlinear Volterra-Fredholm integro-differential equation with weakly singular kernels. *Fasciculi Mathematici*, 62, 1-14.

In: Understanding Integro-Differential Equations ISBN: 979-8-89113-040-1
Editors: J. Vasundhara Devi et al. © 2023 Nova Science Publishers, Inc.

Chapter 7

LINEAR AND NONLINEAR PARTIAL INTEGRO-DIFFERENTIAL EQUATIONS ARISING FROM FINANCE

José Cruz[1]
Maria Grossinho[1]
Daniel Ševčovič[2,*]
and Cyril Udeani[2]
[1]ISEG, University of Lisbon, Portugal
[2]Comenius University in Bratislava, Slovakia

Abstract

The purpose of this review chapter is to present our recent results on nonlinear and nonlocal mathematical models arising from modern financial mathematics. It is based on our four papers written jointly by J. Cruz, M. Grossinho, D. Ševčovič, and C. Udeani [1], [2], [3], [4], as well as parts of the PhD thesis by J. Cruz [5]. We investigated linear and nonlinear partial integro-differential equations (PIDEs) arising from option pricing and portfolio selection problems and studied the systematic relationships between the PIDEs with option pricing theory and Black–Scholes models. First, we relax the liquid and complete market assumptions and extend the models

*Corresponding Author's Email: sevcovic@fmph.uniba.sk.

that study the market illiquidity to the case where the underlying asset price follows a Lévy stochastic process with jumps. Then, we establish the corresponding PIDE for option pricing under suitable assumptions. The qualitative properties of solutions to nonlocal linear and nonlinear PIDE are presented using the theory of abstract semilinear parabolic equation in the scale of Bessel potential spaces. The existence and uniqueness of solutions to the PIDE for a general class of the so-called admissible Lévy measures satisfying suitable growth conditions at infinity and origin are also established in the multidimensional space. Additionally, the qualitative properties of solutions to the generalized PIDE are investigated by considering a general shift function arising from nonlinear option pricing models, which takes into account a large trader stock-trading strategy with the underlying asset price following the Lévy process. For the portfolio management problem, we present the existence and uniqueness results of the fully nonlinear Hamilton-Jacobi-Bellman equation arising from the stochastic dynamic optimization problem in Sobolev spaces using the theory of monotone operator technique, which can also be viewed as PIDE in some sense. Furthermore, a stable, convergent, and consistent numerical scheme that can give an approximate solution to such PIDE is presented, and various numerical experiments are conducted to illustrate the influence of a large trader and the intensity of jumps on the option price.

Keywords: Lévy measure, option pricing, partial integro-differential equation, Hamilton-Jacobi-Bellman equation, maximal monotone operator, dynamic stochastic portfolio optimization

1. Introduction

This review chapter contains our recent advances in research focused on nonlinear and nonlocal mathematical models arising from modern financial mathematics. The main parts of this chapter are based on our four chapters jointly written by J. Cruz, M. Grossinho, D. Ševčovič, and C. Udeani [1], [2], [3], [4], as well as parts of the PhD thesis by J. Cruz [5].

The classical Black–Scholes model has been widely used in financial industry because of its simplicity and the existence of analytical formula for pricing derivative securities. This model relies on restrictive assumptions, such as completeness, frictionlessness of the market, and the assumption that the underlying asset price follows a geometric Brownian motion. However, the assumption that an investor can trade a large amount of assets without affecting the underlying asset price is generally not satisfied, especially in illiquid markets. It is also known that the fully nonlinear Hamilton–Jacobi–Bellman (HJB) equation plays an essential role in finance. For instance, it gives the necessary and sufficient condition for a control with respect to the value function. Therefore, this chapter investigates linear and nonlinear partial integro-differential equations (PIDEs) arising from the option pricing and portfolio selection problem. We investigate the systematic relationships of the PIDEs with option pricing theory and Black–Scholes models. First, we relax the liquid and complete market assumptions and extend the models that study market illiquidity to the case where the underlying asset price follows a Lévy stochastic process with jumps. Then, we establish the corresponding PIDE for option pricing under suitable assumptions. The qualitative properties of solutions to nonlocal linear and nonlinear PIDEs are presented using the theory of abstract semilinear parabolic equations in the scale of Bessel potential spaces. The existence and uniqueness of solutions to the PIDE for a general class of the so-called admissible Lévy measures satisfying suitable growth conditions at infinity and origin are also established in the multidimensional space. Additionally, the qualitative properties of solutions to the generalized PIDE are investigated by considering a general shift function arising from nonlinear option pricing models, which takes into account a large trader stock-trading strategy with the underlying asset price following the Lévy process. For the portfolio management problem, we present the existence and uniqueness results to the fully nonlinear HJB equation arising from stochastic dynamic optimization problem in Sobolev spaces using the theory of monotone operator technique, which can also be viewed as PIDE in some sense. Furthermore, a stable, convergent and consistent numerical scheme is presented that can efficiently approximate the solution of this PIDE, and various numerical experiments are conducted to illustrate the influence of a large trader and the intensity of jumps on the option price.

The Black–Scholes model and the HJB equation have been widely used in financial markets. However, evidence from the stock market observation shows that the Black–Scholes model is not the most realistic one because it depends on some restrictive assumptions, such as the liquidity, completeness, and frictionless of the market. Additionally, the linear Black–Scholes equation provides a solution that corresponds to a perfectly replicated portfolio, which is not a desirable property. For this reason, several attempts have been made to generalize and relax some of these assumptions. Some authors relaxed these assumptions by (i) considering the presence of transaction costs (see Kwok [6] and Avellaneda and Paras [7]), (ii) feedback and illiquid market effects due to large traders choosing given stock-trading strategies (Schönbucher and Willmott [8], Frey and Patie [9], Frey and Stremme [10]), and (iii) the risk from the unprotected portfolio (Jandačka and Ševčovič [11]). In these generalizations, the constant volatility was replaced by a nonlinear function based on the second derivative of the option price. Frey and Stremme derived a nonlinear Black–Scholes model that plays an essential role in the class of the generalized Black–Scholes equation with such a nonlinear diffusion function [9, 12, 11]). In this model, the asset dynamics considers the presence of feedback effects due to a large trader choosing his/her stock-trading strategy [8]. Another important direction in generalizing the original Black–Scholes equation arises from the fact that the sample paths of a Brownian motion are continuous; however, the realized stock price of a typical company exhibits random jumps over the intraday scale, making the price trajectories discontinuous. In the classical Black–Scholes model, the logarithm of the price process has a normal distribution. However, the empirical distribution of stock returns exhibits fat tails. Meanwhile, when calibrating the theoretical prices to the market prices, the implied volatility is not constant as a function of strike price nor as a function of time to maturity, contradicting the prediction of the Black–Scholes model. However, the models with jumps and diffusion can solve the problems inherent to the Black–Scholes model. Jump models also play an essential role in the option market. In the Black–Scholes model, the market is complete, implying that every payoff can exactly be replicated; meanwhile, there is no perfect hedge in jump models, making the way of options not redundant.

Market illiquidity has been widely studied in the literature [13, 14, 15, 16, 17]. The first major contribution was made by Robert Jarrow, in 1994, who studied the market manipulation strategies that may arise in illiquid markets. The author also studied option pricing theory in discrete time when there is a large trader. The pricing-arbitrage condition was used to ensure that no market manipulation strategy is used by the large trader and large trader's optimality conditions; thus, replacing the usual free-arbitrage argument. Then, Frey (1998) extended Jarrow's analysis to the continuous time case and established the existence and uniqueness of solution of a nonlinear partial differential equation (PDE) satisfied by the large trader's hedging strategy. Additionally, Platten and Schweizer (1998) proposed an explanation for the smile and skewness of the implied volatilities and showed that hedging strategies followed by large traders can lead to option price bias. Sircar and Papanicolaou (1998) also proposed a model where the derivative security price is characterized by a nonlinear PDE that becomes the Black–Scholes equation when there is no feedback. When the program traders are a small fraction of the economy, numerical and analytical methods can be used to analyze the nonlinear PDE through perturbation. This equation is derived using an argument similar to the one used in deriving the classical Black–Scholes equation. Consequently, they obtained that this model also predicts increased implied volatilities as in Platten and Schweizer. Furthermore, Schonbucher and Willmott (2000) analyzed the feedback effects from the presence of hedging strategies. They also derived a nonlinear PDE for an option replication strategy and studied these effects for a put option. The effects are more pronounced in markets with low liquidity, which can induce discontinuities in the price process. However, none of these studies that investigated jump models [18, 19, 20, 21, 22, 23, 24] considered the market illiquidity. Meanwhile, investors and risk managers have realized that financial models based on the assumption that an investor can trade large amounts of an asset without affecting its price are no longer true in markets that are not liquid. Therefore, in this chapter, we relax the liquid and complete market hypothesis and extend the models that study market illiquidity to the case where the underlying asset price follows a Lévy stochastic process with jumps to obtain a model for pricing European and American call and put options on an underlying asset characterized by a Lévy measure. In this way, it is assumed that trading strategies affect the stock price and the possibility to account for sudden jumps that might occur when the market is under stress .

Recently, the relationships between more general nonlocal operators and jump processes have been widely investigated. For instance, there is an actual connection between the solution to PIDEs and properties of the corresponding Markov jump process (cf. Abels and Kassmann [25]; Florescu and Mariani [26]). In recent decades, the role of PIDEs has been investigated in various fields, such as pure mathematics, biological sciences, and economics [27, 28, 29]. PIDE problems arising from financial mathematics, especially option pricing models, have been of great interest to many researchers. In most cases, standard methods for solving these problems lead to the study of parabolic equations. Mikulevičius and Pragaraustas [30] investigated solutions of the Cauchy problem to the parabolic PIDE with variable coefficients in Sobolev spaces. They used their results to obtain solutions of the corresponding martingale problem. Crandal *et al.* [31] employed the notion of a viscosity solution to investigate the qualitative results. Soner *et al.* [32] and Barles *et al.* [33] extended and generalized their results for the first and second order operators, respectively. Florescu and Mariani [26] employed the Schaefer fixed point argument to establish existence of a weak solution of the generalized PIDE. Amster *et al.* [34] used the notion of upper and lower solutions to obtain the solution of such PIDEs. They proved the existence of solutions in a general domain for multiple assets and the regime switching jump-diffusion model. Cont *et al.* [35] investigated the actual connection between option pricing in exponential Lévy models and the corresponding PIDEs for European options and those with single or double barriers. They discussed and established the conditions for which the prices of the option are the classical solution of the corresponding PIDE. In this chapter, we obtain a certain PIDE for option pricing in an illiquid market by assuming a certain dynamics for the stock price. The existence of a solution and the localization results of the associated PIDE are also established. We investigated and established the qualitative properties of solutions to the nonlocal linear and nonlinear PIDE in the scale of Bessel potential spaces using the theory of abstract semilinear parabolic equation. Furthermore, we present the existence and uniqueness results for nonlinear parabolic equations using the monotone operator technique, Fourier transform, and Banach fixed point argument. We considered the fully nonlinear HJB equation arising from the portfolio selection problem, where the goal of an investor is to optimize the conditional expected value of the terminal utility of the portfolio. Such a nonlinear parabolic equation is presented in an abstract setting, which can also be viewed as a nonlinear PIDE. Many previous studies have developed numerical methods for PIDEs, such as finite difference and finite element methods. However, the equation corresponding to the case of illiquid markets is more difficult. Therefore, this chapter also presents a stable, convergent, and consistent numerical scheme that can give an approximate solution of such PIDEs. Various experiments are presented to illustrate the influence of a large trader and the intensity of jumps on the option price.

2. Background and Motivation

Based on the classical theory developed by Black, Scholes, and Merton, the price $V(t,S)$ of an option in a stylized market at time $t \in [0,T]$ and the underlying asset price S can be calculated as a solution to the following linear Black–Scholes parabolic equation:

$$\frac{\partial V}{\partial t}(t,S)+\frac{1}{2}\sigma^2 S^2\frac{\partial^2 V}{\partial S^2}(t,S)+rS\frac{\partial V}{\partial S}(t,S)-rV(t,S)=0,\ t\in[0,T), S>0. \tag{1}$$

Here, $\sigma > 0$ is the historical volatility of the underlying asset driven by the geometric Brownian motion, and $r > 0$ is the risk-free interest rate of zero-coupon bond. The solution of the above equation is subject to the terminal payoff condition $V(T,S) = \Phi(S)$ at maturity $t = T$. Meanwhile, evidence from stock market observations indicates that the model is not the most realistic one because it assumes that the market is liquid, complete, frictionless and without transaction costs. It is also known that the linear Black–Scholes equation provides a solution corresponding to a perfectly replicated portfolio, which need not be a desirable property. To solve these problems, several attempts have been made to generalized the linear Black–Scholes equation (1) by replacing the constant volatility σ with a nonlinear function $\tilde{\sigma}(S\partial_S^2 V)$ depending on the second derivative $\partial_S^2 V$ of the option price. In this regard, Frey and Stremme derived a nonlinear Black–Scholes model, which plays an essential role in the class of generalized Black–Scholes equation with such a nonlinear diffusion function [9, 12]). They considered the case in which asset dynamics takes into account the presence of feedback effects due to a large trader choosing his/her stock-trading strategy (see also [8]). The diffusion coefficient is non-constant, and it is given by

$$\tilde{\sigma}(S\partial_S^2 V)^2 = \sigma^2 \left(1 - \varrho S \partial_S^2 V\right)^{-2}, \tag{2}$$

where σ and $\varrho > 0$ are constants.

Furthermore, several researchers have attempted to generalize the original Black–Scholes equation, which arises from the fact that the sample paths of a Brownian motion are continuous. However, the realized stock price of a typical company exhibits random jumps over the intraday scale, making the price trajectories discontinuous. The underlying asset price process is usually assumed to follow a geometric Brownian motion in the classical Black–Scholes model. However, the empirical distribution of stock returns exhibits fat tails. The models with jumps and diffusion can solve the problems inherent to the linear Black–Scholes model and play an essential role in options pricing. It is well known that the market is complete in the Black–Scholes model, illustrating that each payoff can be perfectly replicated; however, there is no perfect hedge in jump-diffusion models, making the options not redundant. It turns out that the option price can be computed from the solution $V(t,S)$ to the following PIDE Black–Scholes equation [1]:

$$\begin{aligned}&\frac{\partial V}{\partial t}(t,S) + \frac{1}{2}\sigma^2 S^2 \frac{\partial^2 V}{\partial S^2}(t,S) + rS\frac{\partial V}{\partial S}(t,S) - rV(t,S)\\ &+ \int_{\mathbb{R}} V(t, S + H(z,S)) - V(t,S) - H(z,S)\frac{\partial V}{\partial S}(t,S)\nu(\mathrm{d}z) = 0,\end{aligned} \tag{3}$$

where $H(z,S) = S(e^z - 1)$, and ν is the so-called Lévy measure characterizing the underlying asset process with random jumps in time and space. It is worth noting that (3) reduces to the classical linear Black–Scholes equation (1) if $\nu = 0$.

In this chapter, we consider both directions of generalization of the Black–Scholes equation. First, we relax the assumption of a liquid market following the Frey–Stremme model by assuming that the underlying asset price follows a Lévy stochastic process with jumps and establish the corresponding PIDEs. Then, we present the existence and uniqueness results to the linear and nonlinear nonlocal PIDE in the framework of Bessel potential spaces for the multidimensional case. A more generalized nonlinear nonlocal PIDE is also presented by considering a shift function $\xi = \xi(\tau, x, z)$ depending on variables $x, z \in \mathbb{R}^n$. In addition, we derive, analyze and perform numerical computations of the model. We also show that the corresponding nonlinear PIDE has the following form:

$$\frac{\partial V}{\partial t} + \frac{1}{2}\frac{\sigma^2}{(1-\varrho S\partial_S\phi)^2}S^2\frac{\partial^2 V}{\partial S^2} + rS\frac{\partial V}{\partial S} - rV$$
$$+ \int_{\mathbb{R}} V(t, S + H(t,z,S)) - V(t,S) - H(t,z,S)\frac{\partial V}{\partial S}\nu(\mathrm{d}z) = 0. \quad (4)$$

It is worth noting that the function $H(t, z, S)$ may depend on the large trader strategy function $\phi = \phi(t, S)$ and the delta $\partial_S V$ of the price V if $\varrho > 0$.

We consider a stylized economy with two traded assets: a riskless asset (a bond with a price B_t taken as numeraire) and a risky asset (stock with a price S_t). Here, we assume that the bond market is perfectly elastic, since it is more liquid than stocks, and consider two types of traders: reference and program traders. Program traders are also known as portfolio insurers because they use dynamic hedging strategies to hedge the portfolio against jumps in stock prices. They are classified as single traders or a group of traders acting together. It is assumed that their trades influence the stock price equilibrium. On the contrary, reference traders can be considered as representative traders of many small agents. We assume that they act as price takers. Generally, $\tilde{D}(t, Y_t, S_t)$ represents the demand function of the reference trader that depends on the income process Y_t or some other fundamental state variable that influences the demand of the reference trader. The aggregate demand of program traders is denoted by $\varphi(t, S_t) = \xi\phi(t, S_t)$, where ξ is the number of identical written securities that program traders are trying to hedge, and $\phi(t, S_t)$ is the demand per unit of the security being hedged. For simplicity, we assume that ξ is the same for every program trader. See [17] for a more general case where different securities are considered. Suppose that the supply of a stock with the price $\tilde{S}_0$ is constant, and let $D(t, Y, S) = \frac{\tilde{D}(t,Y,S)}{\tilde{S}_0}$ be the quantity demanded by a reference trader per unit supply. Then, the total demand relative to the supply at time t is given by $G(t, Y, S) = D(t, Y, S) + \rho\phi(t, S)$, where $\rho = \frac{\xi}{\tilde{S}_0}$, and $\rho\phi(t, S)$ is the proportion of the total supply of the stock traded by the program traders. Therefore, to obtain the market equilibrium, the variables Y and

S should satisfy $G(t, Y, S) = 1$. Assume that the function G is monotone with respect to the variables Y and S, and it is sufficiently smooth. Then, we can solve the implicit equation $G(t, Y_t, S_t) = 1$ to obtain $S_t = \psi(t, Y_t)$, where ψ is a sufficiently smooth function. Using the approach in [17], we assume that the stochastic process Y_t has the following dynamics:

$$\mathrm{d}Y_t = \mu(t, Y_t)\mathrm{d}t + \eta(t, Y_t)\mathrm{d}W_t.$$

Then, using Itô's lemma for the process $S_t = \psi(t, Y_t)$, we have

$$\mathrm{d}S_t = \left(\partial_t\psi + \mu\partial_y\psi + \frac{\eta^2}{2}\partial_y^2\psi\right)\mathrm{d}t + \eta\partial_y\psi\mathrm{d}W_t \equiv b(t, S_t)S_t\mathrm{d}t + v(t, S_t)S_t\mathrm{d}W_t. \tag{5}$$

It means that S_t follows a geometric Brownian motion with a nonconstant volatility function $v(t, S) = \eta(t, Y)\partial_Y\psi(t, Y)/\psi(t, Y)$, where $Y = \psi^{-1}(t, S)$. Thus, we follow the argument used in the derivation of the original Black–Scholes equation to obtain a generalization of the Black–Scholes PDE with a nonconstant volatility function $\sigma = v(t, S)$. We employ the Frey–Stremme's approach (cf. [9, 12]) to prescribe a dynamics for the underlying stock price instead of deriving it using the market equilibrium and dynamics for the income process Y_t as is done in [17]. In this way, Frey and Stremme derived the same stock price dynamics as in [17] corresponding to a situation where the demand function is of logarithmic type, $D(Y, S) = \ln(\frac{Y^\gamma}{S})$, where $\gamma = \frac{\sigma}{\eta_0}$, and the income process Y_t follows a geometric Brownian motion, i.e.,

$$\partial_Y D(Y, S) = \gamma\frac{1}{Y}, \ \partial_S D(Y, S) = -\frac{1}{S}, \ \mathrm{d}Y_t = \mu_0 Y_t\mathrm{d}t + \eta_0 Y_t\mathrm{d}W_t, \tag{6}$$

$$v(t, S) = \eta(t, Y)\frac{\partial_Y\psi(t, Y)}{\psi(t, Y)} = -\frac{\eta_0 Y}{S}\frac{\gamma\frac{1}{Y}}{-\frac{1}{S} + \rho\frac{\partial\phi}{\partial S}} = \frac{\sigma}{1 - \rho S\frac{\partial\phi}{\partial S}}.$$

Assuming the delta hedging strategy with $\phi(t, S) = \partial_S V(t, S)$ and substituting the volatility function $v(t, S)$ in (5), we obtain the generalized Black–Scholes equation with the nonlinear diffusion function of the form (2).

In this chapter, we first generalized the Frey–Stremme model by considering an underlying asset following a Lévy process with jumps. After that, we establish the corresponding PIDE for option pricing. Furthermore, we investigate the existence and uniqueness of solutions to such PIDE in multidimensional spaces.

3. Preliminaries and Definitions

This section presents some basic definitions and properties of Lévy measures and notion of admissible activity Lévy measures. Here,$|\cdot|$ and $\|\cdot\|$ represent the Euclidean norm in $\mathbb{R}^n$ and the norm in an infinite dimensional function space (e.g., $L^p(\mathbb{R}^n), X^\gamma$). In what follows, $a \cdot b$ stands for the usual Euclidean product in $\mathbb{R}^n$ with the norm $|z| = \sqrt{z \cdot z}$.

Definition 1. *[3] A Lévy process on $\mathbb{R}^n$ is a stochastic (right continuous) process $X = \{X_t, t \geq 0\}$ having the left limit with independent stationary increments. It is uniquely characterized by its Lévy exponent ϕ:*

$$\mathbb{E}_x(e^{iy \cdot X_t}) = e^{-t\phi(y)}, \; y \in \mathbb{R}^n.$$

The subscript x in the expectation operator $\mathbb{E}_x$ indicates that the process X_t starts with a given value x at the origin $t = 0$. The L'evy exponent ϕ has the following unique decomposition:

$$\phi(y) = ib \cdot y + \sum_{i,j=1}^{n} a_{ij} y_i y_j + \int_{\mathbb{R}^n} \left(1 - e^{iy \cdot z} + iy \cdot z 1_{|z| \leq 1}\right) \nu(\mathrm{d}z),$$

where $b \in \mathbb{R}^n$ is a constant vector; (a_{ij}) is a constant matrix, which is positive semidefinite; $\nu(\mathrm{d}z)$ is a nonnegative measure in $\mathbb{R}^n \setminus \{0\}$ such that $\int_{\mathbb{R}^n} \min(1, |z|^2) \nu(\mathrm{d}z) < \infty$ (cf. [36]).

3.1. Exponential Lévy Models

Let $X_t, t \geq 0$, be a stochastic process. The Poisson random measure $\nu(A)$ of a Borel set $A \in \mathcal{B}(\mathbb{R})$ is defined by $\nu(A) = \mathbb{E}\left[J_X([0,1] \times A)\right]$, where $J_X([0,t] \times A) = \#\{s \in [0,t] \, . \, \Delta X_s \in A\}$. This measure gives the mean number per unit of time of jumps whose amplitude belongs to the set A. It is worth noting that the Lévy–Itô decomposition provides a representation of X_t interpreted as a combination of a Brownian motion with a drift ω and an infinite sum of independent compensated Poisson processes with variable jump sizes (see [2]), i.e.,

$$\mathrm{d}X_t = \omega \mathrm{d}t + \sigma \mathrm{d}W_t + \int_{|x| \geq 1} x J_X(\mathrm{d}t, \mathrm{d}x) + \int_{|x| < 1} x \widetilde{J}_X(\mathrm{d}t, \mathrm{d}x),$$

where $\widetilde{J}_X([0,t] \times A) = J_X([0,t] \times A) - t\nu(A)$ is the compensation of J_X.

Remark 1. *[2, Remark 1] Note that any Lévy process is a strong Markov process, and the associated semigroup is a convolution. Its infinitesimal generator* $L : u \mapsto L[u]$ *is a nonlocal partial integro-differential operator given by*

$$L[u](x) = \lim_{h\to 0^+} \frac{\mathbb{E}[u(x+X_h)]-u(x)}{h}$$
$$= \frac{\sigma^2}{2}\frac{\partial^2 u}{\partial x^2} + \gamma\frac{\partial u}{\partial x} + \int_{\mathbb{R}} \left[u\left(x+y\right) - u\left(x\right) - y1_{|y|\le 1}\frac{\partial u}{\partial x}\left(x\right)\right]\nu(\mathrm{d}y), \quad (7)$$

which is well-defined for any compactly supported function $u \in C_0^2\left(\mathbb{R}\right)$.

Let $S_t, t \geq 0$, be a stochastic process that represents an underlying asset process in a filtered probability space $(\Omega, \mathcal{F}, \{\mathcal{F}_t\}, \mathbb{P})$. The filtration $\{\mathcal{F}_t\}$ represents the price history up to time t. In an arbitrage-free market, there is an equivalent measure $\mathbb{Q}$ under which the discounted prices of all traded assets are $\mathbb{Q}-$ martingales, which is called the fundamental theorem of asset pricing (see [37]). The measure $\mathbb{Q}$ is also known as the risk-neutral measure. We consider the exponential Lévy model in which the risk-neutral price process S_t under $\mathbb{Q}$ is given by $S_t = e^{rt+X_t}$, where X_t is a Lévy process under $\mathbb{Q}$ with the characteristic triplet (σ, γ, ν). Then, the arbitrage-free market hypothesis states that $\widehat{S}_t = S_t e^{-rt} = e^{X_t}$ is a martingale, which is equivalent to the following conditions imposed on the triplet (σ, γ, ν):

$$\int_{|y|\ge 1} e^y \nu(\mathrm{d}y) < \infty, \;\; \gamma \in \mathbb{R}, \;\; \gamma = -\frac{\sigma^2}{2} - \int_{-\infty}^{+\infty} \left(e^y - 1 - y1_{|y|\le 1}\right)\nu(\mathrm{d}y). \quad (8)$$

The risk-neutral dynamics of S_t under $\mathbb{Q}$ is given by

$$\mathrm{d}S_t = rS_t\mathrm{d}t + \sigma S_t \mathrm{d}W_t + \int_{\mathbb{R}} \left(e^y - 1\right) S_t \widetilde{J}_X\left(dt, dy\right). \quad (9)$$

The exponential price process, $e^{X_t}, t \geq 0$, is also a Markov process with the state space $(0, \infty)$. It has the following infinitesimal generator:

$$L^S[V](S) = \lim_{h\to 0} \frac{\mathbb{E}[V(Se^{X_h})] - V(S)}{h} = rS\frac{\partial V}{\partial S} + \frac{\sigma^2}{2}S^2\frac{\partial^2 V}{\partial S^2} \quad (10)$$
$$+ \int_{\mathbb{R}} \left[V(Se^y) - V(S) - S(e^y - 1)\frac{\partial V}{\partial S}\right]\nu(\mathrm{d}y) \quad (11)$$

(see [37]). A Lévy process with the following representation:

$$\mathrm{d}X_t = \omega \mathrm{d}t + \sigma \mathrm{d}W_t + \int_{|x|\geq 1} K(t,x) J_X(\mathrm{d}t, \mathrm{d}x) + \int_{|x|<1} H(t,x)\tilde{J}_X(\mathrm{d}t, \mathrm{d}x).$$

is called the Lévy type stochastic integral. The following variant of Itô's lemma is an essential results, which will be needed later in this study.

Theorem 1. *[1, Therorem 2.1] Let $f \in C^{1,2}([0,T] \times \mathbb{R})$ and $H, K \in C([0,T] \times \mathbb{R})$. Suppose $X_t, t \geq 0$, is a Lévy stochastic process. Then,*

$$\begin{aligned}
\mathrm{d}f(t,X_t) &= \tfrac{\partial f}{\partial t}\mathrm{d}t + \tfrac{\partial f}{\partial x}\mathrm{d}X_t + \tfrac{1}{2}\tfrac{\partial^2 f}{\partial x^2}\mathrm{d}[X_t, X_t] \\
&+ \textstyle\int_{|x|\geq 1} f(t, X_t + K(t,x)) - f(t,X_t) J_X(\mathrm{d}t, \mathrm{d}x) \\
&+ \textstyle\int_{|x|<1} f(t, X_t + H(t,x)) - f(t,X_t) \tilde{J}_X(\mathrm{d}t, \mathrm{d}x) \\
&+ \textstyle\int_{|x|<1} f(t, X_t + H(t,x)) - f(t,X_t) - H(t,x)\tfrac{\partial f}{\partial x}(t,X_t)\nu(\mathrm{d}x)\mathrm{d}t.
\end{aligned} \tag{12}$$

3.2. Examples of Lévy Processes in Finance

There are two types of exponential Lévy models considered in the literature. The first types are jump-diffusion models, where the log-price is represented as a Lévy process with a nonzero diffusion part ($\sigma > 0$) and a jump process with finite activity (i.e., $\nu(\mathbb{R}) < \infty$). The second types of models are infinite activity pure jump models, where there is no diffusion part and only a jump process with infinite activity (i.e., $\nu(\mathbb{R}) = \infty$). This section presents different types of exponential Lévy models that differ in the choice of the Lévy measure.

3.2.1. Jump-Diffusion Models

A Lévy process with jump-diffusion has the following general form:

$$X_t = \gamma t + \sigma W_t + \sum_{i=1}^{N_t} Y_i,$$

where $\sigma > 0$, and N_t is a Poisson process with intensity λ that counts the jumps of X_t, and $Y_i, i = 1, 2, 3...$ are independent and identically distributed random variables with distribution μ. The Lévy measure ν is $\lambda\mu$, and the drift γ is given by

$$-\frac{\sigma^2}{2} - \int_{\mathbb{R}} \left(e^y - 1 - y1_{|y|\le 1}\right) \nu\,(\mathrm{d}y)\,.$$

Merton's Model

This is the first jump-diffusion model proposed by Merton [22] in the financial application. Random variables $Y_i, i = 1, 2, 3...$, are normally distributed with mean m and variance δ^2. It has the following Lévy density:

$$\nu(\mathrm{d}z) = \lambda \frac{1}{(2\pi\delta^2)^{n/2}} e^{-\frac{|z-m|^2}{2\delta^2}} \mathrm{d}z\,, \tag{13}$$

where the parameters $m \in \mathbb{R}^n, \lambda, \delta > 0$, are given. Merton's measure is a finite activity Lévy measure, i.e., $\nu(\mathbb{R}^n) = \int_{\mathbb{R}^n} \nu(\mathrm{d}z) < \infty$, with finite variation $\int_{|z|\le 1} |z|\nu(\mathrm{d}z) < \infty$. Therefore, the probability density of X_t can be obtain as a series that converges rapidly (see [37]):

$$p_t(x) = \sum_{j=0}^{\infty} e^{-\lambda t} (\lambda t)^j \frac{e^{-\frac{|z-\gamma t - jm|^2}{2(\sigma^2 t + j\delta^2)}}}{j!\sqrt{2\pi(\sigma^2 t + j\delta^2)}}. \tag{14}$$

Thus, the price of an European call option can be expressed as a weighted sum of Black–Scholes prices:

$$C_{Merton}(S_0, K, T, \sigma, r) = e^{-rT} \sum_{j=0}^{\infty} e^{-\lambda t} \frac{(\lambda t)^j}{j!} e^{r_j T} C_{BS}(S_0 e^{\frac{j\delta^2}{T}}, K, T, \sigma_j, r_j), \tag{15}$$

where $r_j = r - \lambda(e^{m+\frac{\delta^2}{2}} - 1) + \frac{jm}{T}$, $\sigma_j = \sqrt{\sigma^2 + \frac{j\delta^2}{T}}$, and $C_{BS}(S, K, T, \sigma, r)$ is the well-known Black–Scholes formula.

3.2.2. Infinite Activity Pure Jump Models

The variance Gamma and normal inverse Gaussian (NIG) processes are obtained by a subordination of a Brownian motion and a tempered α-stable process; variance Gamma and NIG processes correspond to $\alpha = 0$ and $\alpha = 1/2$, respectively. These models are widely used in finance because of the existence of probability density of the subordinator in a closed form for these values of α (see [37]).

Variance Gamma Process

This is a pure discontinuous process of infinite activity and finite variation ($\int_{|x|\leq 1}|x|\nu(\mathrm{d}x)<\infty$) that is widely used in financial modeling. It has the following Lévy measure:

$$\nu(x)=\frac{1}{\kappa|x|}e^{Ax-B|x|} \text{ with } A=\frac{\theta}{\sigma^2} \text{ and } B=\frac{\sqrt{\theta^2+2\frac{\sigma^2}{\kappa}}}{\sigma^2}.$$

Here, σ and θ are parameters related to the volatility and drift of the Brownian motion, respectively; κ is the parameter related to the variance of the subordinator, which is a Gamma process (see [37]). The probability density is given by

$$p_t(x)=Ce^{Ax}|x|^{\frac{t}{k}}K_{\frac{t}{k}-\frac{1}{2}}(|x|),$$

where K is the modified Bessel function of second kind. The characteristic function of $X_t+\gamma t$ is given by

$$\Phi_t(u)=e^{itu\gamma}\phi_t(u)=e^{itu\gamma}\left(1+\frac{\sigma^2u^2\kappa}{2}-i\theta\kappa u\right)^{-t/\kappa},$$

where γ is determined by the martingale condition, and $\phi_t(u)$ is the characteristic function of X_t. Moreover, we have

$$\mathbb{E}[e^{-rT}S_T|\mathbb{F}_t]=e^{-rt}S_t, \tag{16}$$

where

$$S_t=S_0e^{rt+\gamma t+X_t} \tag{17}$$

is the risk-neutral process introduced in [23]. Therefore, $\gamma=\frac{1}{\kappa}log(1-\frac{\sigma^2\kappa}{2}-\theta\kappa)$.

Normal Inverse Gaussian Model

The NIG model is a process of infinite activity and infinite variation without any Brownian component. It has the following Lévy measure [37]

$$\nu(x)=\frac{C}{|x|}e^{Ax}K_1(B|x|)$$

and

$$C = \frac{\sqrt{\theta^2 + \frac{\sigma^2}{\kappa}}}{2\pi\sigma\sqrt{\kappa}}, A = \frac{\theta}{\sigma^2}, B = \frac{\sqrt{\theta^2 + \frac{\sigma^2}{\kappa}}}{\sigma^2},$$

where θ, σ, and κ have the same meaning as in the Variance Gamma process. The probability density is given by

$$p_t(x) = Ce^{Ax} \frac{K_1(B\sqrt{x^2 + \frac{t^2\sigma^2}{\kappa}})}{\sqrt{x^2 + \frac{t^2\sigma^2}{\kappa}}}$$

where K is the modified Bessel function of second kind. The characteristic function is given by

$$\Phi_t(u) = e^{\frac{t}{\kappa} - \frac{t}{\kappa}\sqrt{1+u^2\sigma^2\kappa - 2iu\theta\kappa}}. \tag{18}$$

Generalized Hyperbolic Model

The generalized hyperbolic model is a process of infinite variation without Gaussian part. It has the following characteristic function (see [37]):

$$\phi_t(u) = e^{i\mu u}(\frac{\alpha^2 - \beta^2}{\alpha^2 - (\beta + iu)^2})^{\frac{t}{2\kappa}} \frac{K_{\frac{t}{\kappa}}(\delta\sqrt{\lambda^2 - (\beta + iu)^2})}{K_{\frac{t}{\kappa}}(\delta\sqrt{\alpha^2 - \beta^2})}, \tag{19}$$

where δ is a scale parameter, μ is the shift parameter, and κ has the same meaning as in the variance Gamma process. Parameters λ, α, and β determine the shape of the distribution. The density function

$$p_t(x) = C(\sqrt{\delta^2 + (x-\mu)^2})^{\frac{t}{k} - \frac{1}{2}} K_{\frac{t}{\kappa} - \frac{1}{2}}(\alpha\sqrt{\delta^2 - (x-\mu)^2})e^{\beta(x-\mu)},$$

where K is the modified Bessel function and

$$C = \frac{(\sqrt{\alpha^2 - \beta^2})^{\frac{t}{k}}}{\sqrt{2\pi}\alpha^{\frac{t}{\kappa} - \frac{1}{2}}\delta^{\frac{t}{\kappa}} K_{\frac{t}{\kappa}}(\delta\sqrt{\alpha^2 - \beta^2})}.$$

The variance Gamma process is obtained for $\mu = 0$ and $\delta = 0$. The NIG process corresponds to $\lambda = -\frac{1}{2}$.

3.3. Admissible Activity Lévy Measures

This subsection presents the notion of an admissible activity Lévy measure introduced by Cruz and Ševčovič [1, 2] for the one-dimensional case $n = 1$, which was later extended by Ševčovič and Udeani [3] for the multidimensional case $n \geq 1$.

Definition 2. *[3, Definition 1] A measure ν in $\mathbb{R}^n$ is called an admissible activity Lévy measure if there exists a nonnegative Lebesgue measurable function $h : \mathbb{R}^n \to \mathbb{R}$ such that $\nu(\mathrm{d}z) = h(z)\mathrm{d}z$ with*

$$0 \leq h(z) \leq C_0 |z|^{-\alpha} e^{-D|z| - \mu|z|^2}, \tag{20}$$

for all $z \in \mathbb{R}^n$ and the shape parameters $\alpha, \mu \geq 0, D \in \mathbb{R}$ ($D > 0$ if $\mu = 0$), where $C_0 > 0$ is a positive constant.

Remark 2. *It is worth noting that the additional conditions $\int_{\mathbb{R}} \min(|z|^2, 1)\nu(\mathrm{d}z) < \infty$ and $\int_{|z|>1} e^z \nu(\mathrm{d}z) < \infty$ are satisfied provided that ν is an admissible Lévy measure with shape parameters $\alpha < 3$, and either $\mu > 0, D^\pm \in \mathbb{R}$, or $\mu = 0$ and $D^- + 1 < 0 < D^+$. For the Merton model, we have $\alpha = 0, D^\pm = 0$ and $\mu = 1/(2\delta^2) > 0$. Meanwhile, for the Kou model, we have $\alpha = \mu = 0, D^+ = \lambda^-, D^- = -\lambda^+$. For the variance Gamma process, we have $\alpha = 1, \mu = 0, D^\pm = A \pm B$.*

4. Multidimensional Linear and Nonlinear PIDE

This section focuses on qualitative properties of solutions to the linear and nonlinear nonlocal parabolic PIDE of the form:

$$\begin{aligned} \frac{\partial u}{\partial \tau} &= \frac{\sigma^2}{2}\Delta u + \int_{\mathbb{R}^n} \left[u(\tau, x+z) - u(\tau, x) - z \cdot \nabla_x u(\tau, x)\right] \nu(\mathrm{d}z) + g(\tau, x, u, \nabla_x u) \\ u(0, x) &= u_0(x), \quad x \in \mathbb{R}^n, \tau \in (0, T), \end{aligned} \tag{21}$$

where g is a given sufficiently smooth function; ν is a positive measure on $\mathbb{R}^n$ such that its Radon derivative is a nonnegative Lebesgue measurable function h in $\mathbb{R}^n$, i.e., $\nu(\mathrm{d}z) = h(z)\mathrm{d}z$.

Additionally, we will analyze the solution of the following generalization of the above PIDE, in which the shift function may depend on the variables $\tau > 0, x, z \in \mathbb{R}$:

$$\frac{\partial u}{\partial \tau} = \frac{\sigma^2}{2}\Delta u + \int_{\mathbb{R}^n} [u(\tau, x+\xi) - u(\tau, x) - \xi \cdot \nabla_x u(\tau, x)]\, \nu(\mathrm{d}z) + g(\tau, x, u, \nabla_x u), \quad (22)$$

where $\xi = \xi(\tau, x, z)$ is the shift function. An application of such a general shift function ξ can be found in nonlinear option pricing models considering a large trader stock-trading strategy with the underlying asset price dynamic following the Lévy process (cf., Cruz and Ševčovič [1]). If $\xi(x, z) \equiv z$, then (22) reduces to equation (21). For example, nonlinearity g often arises from applications occurring in pricing XVA derivatives (cf., Arregui *et al.* [28, 38]) or applications of the penalty method for American option pricing under a PIDE model (cf., Cruz and Ševčovič [2]).

4.1. Existence and Uniqueness Results of PIDE

In this section, we present the existence and uniqueness results for the general equation (22) for a class of Lévy measures using the theory of abstract semilinear parabolic equation in the scale of Bessel potential spaces. First, we rewrite the PIDE (22) in high-dimensional space as follows:

$$\tfrac{\partial u}{\partial \tau} + Au = f(u) + g(\tau, x, u, \nabla_x u), \quad u(0, x) = u_0(x), \ x \in \mathbb{R}^n, \tau \in (0, T), \quad (23)$$

where $A = -(\sigma^2/2)\Delta$. The linear nonlocal operator f is defined by

$$f(u)(\cdot) = \int_{\mathbb{R}^n} [u(\cdot + \xi) - u(\cdot) - \xi \cdot \nabla_x u(\cdot)\,]\, \nu(\mathrm{d}z), \quad (24)$$

where $\xi = \xi(\tau, x, z)$ is a given shift function. The function g is assumed to be Hölder and Lipschitz continuous in τ and other variables, respectively. Then, we employ the theory of abstract semilinear parabolic equations presented by Henry [39] to establish the existence, continuation, and uniqueness of a solution. A solution to the PIDE (23) is constructed in the scale of the Bessel potential spaces $\mathscr{L}^p_{2\gamma}(\mathbb{R}^n)$, $\gamma \geq 0$ in high-dimensional space, $n \geq 1$. These spaces can be viewed as a natural extension of the classical Sobolev spaces $W^{k,p}(\mathbb{R}^n)$ for non-integer values of order k. It is worth noting that the nested scale of Bessel potential spaces allows for a finer formulation of existence and uniqueness results than the classical Sobolev spaces.

Definition 3. *[39, Definition 1] An analytic semigroup is a family of bounded linear operators* $\{S(t), t \geq 0\}$ *in a Banach space* X *satisfying the following conditions:*

i) $S(0) = I, S(t)S(s) = S(s)S(t) = S(t+s)$, *for all* $t, s \geq 0$;

ii) $S(t)u \to u$ *when* $t \to 0^+$ *for all* $u \in X$;

iii) $t \to S(t)u$ *is a real analytic function on* $0 < t < \infty$ *for each* $u \in X$.

The associated infinitesimal generator A *is defined as follows:* $Au = \lim_{t\to 0^+} \frac{1}{t}(S(t)u - u)$ *and its domain* $D(A) \subseteq X$ *consists of those elements* $u \in X$ *for which the limit exists in the space* X.

Definition 4. *[39] Let* $S_{a,\phi} = \{\lambda \in \mathbb{C} : \phi \leq \arg(\lambda - a) \leq 2\pi - \phi\}$ *be a sector of complex numbers. A closed densely defined linear operator* $A : D(A) \subset X \to X$ *is called a sectorial operator if there exists a constant* $M \geq 0$ *such that* $\|(A - \lambda)^{-1}\| \leq M/|\lambda - a|$ *for all* $\lambda \in S_{a,\phi} \subset \mathbb{C} \setminus \sigma(A)$.

Next, we briefly recall the construction and basic properties of Bessel potential spaces. It is worth noting that if A is a sectorial operator in a Banach space X, then $-A$ is a generator of an analytic semigroup $\{e^{-At}, t \geq 0\}$ acting on X (cf., [39, Chapter I]). For any $\gamma > 0$, we can introduce the operator $A^{-\gamma} : X \to X$ as follows: $A^{-\gamma} = \frac{1}{\Gamma(\gamma)} \int_0^\infty \xi^{\gamma-1} e^{-A\xi} \mathrm{d}\xi$. Then, the fractional power space $X^\gamma = D(A^\gamma)$ is the domain of the operator $A^\gamma = (A^{-\gamma})^{-1}$, i.e., $X^\gamma = \{u \in X : \exists \varphi \in X, u = A^{-\gamma}\varphi\}$. The norm is defined as follows: $\|u\|_{X^\gamma} = \|A^\gamma u\|_X = \|\varphi\|_X$. Furthermore, we have continuous embedding: $D(A) \equiv X^1 \hookrightarrow X^{\gamma_1} \hookrightarrow X^{\gamma_2} \hookrightarrow X^0 \equiv X$, for $0 \leq \gamma_2 \leq \gamma_1 \leq 1$.

Recall the convolution operator $(G * \varphi)(x) = \int_{\mathbb{R}^n} G(x-y)\varphi(y)\mathrm{d}y$. According to [39, Section 1.6], $A = -(\sigma^2/2)\Delta$ is a sectorial operator in the Lebesgue space $X = L^p(\mathbb{R}^n)$ for any $p \geq 1, n \geq 1$, and $D(A) \subset W^{2,p}(\mathbb{R}^n)$. It follows from [40, Chapter 5] that the space $X^\gamma, \gamma > 0$, can be identified with the Bessel potential space $\mathscr{L}^p_{2\gamma}(\mathbb{R}^n)$, where

$$\mathscr{L}^p_{2\gamma}(\mathbb{R}^n) := \{u \in X : \exists \varphi \in X, u = G_{2\gamma} * \varphi\}.$$

Here, $G_{2\gamma}$ is the Bessel potential function,

$$G_{2\gamma}(x) = \frac{1}{(4\pi)^{n/2}\Gamma(\gamma)} \int_0^\infty y^{-1+\gamma-n/2} e^{-(y+|x|^2/(4y))} \mathrm{d}y.$$

The norm of $u = G_{2\gamma} * \varphi$ is given by $\|u\|_{X^\gamma} = \|\varphi\|_{L^p}$. The space X^γ is continuously embedded in the fractional Sobolev–Slobodeckii space $W^{2\gamma,p}(\mathbb{R}^n)$ (cf., [39, Section 1.6]).

In what follows, we denote $C_0 > 0$ as a generic constant which is independent of the solution u; however, it may depend on the model parameters, e.g., $n \geq 1, p \geq 1, \gamma \in [0,1)$.

Proposition 1. *[3, Proposition 1] Let us define the mapping $Q(u,\xi)$ as follows:*

$$Q(u,\xi) = u(x+\xi(x)) - \xi(x) \cdot \nabla_x u(x), \quad x \in \mathbb{R}^n.$$

Then, there exists a constant $\hat{C} > 0$ such that, for any vector valued functions $\xi_1, \xi_2 \in (L^\infty(\mathbb{R}^n))^n$, and u such that $\nabla_x u \in (X^{\gamma-1/2})^n$, $1/2 \leq \gamma < 1$, the following estimate holds:

$$\|Q(u,\xi_1) - Q(u,\xi_2)\|_{L^p(\mathbb{R}^n)} \leq \hat{C}\|\xi_1-\xi_2\|_\infty^{2\gamma-1}(\|\xi_1\|_\infty + \|\xi_2\|_\infty)\|\nabla_x u\|_{X^{\gamma-1/2}}.$$

Proof. Let $u \in X$ be such that $\nabla_x u \in (X^{\gamma-1/2})^n$, i.e., $\partial_{x_i} u \in X^{\gamma-1/2}$ for each $i = 1, \cdots, n$. Then, $\nabla_x u = A^{-(2\gamma-1)/2}\varphi = G_{2\gamma-1} * \varphi$ for some $\varphi \in (L^p(\mathbb{R}^n))^n$, and $\|\nabla_x u\|_{X^{\gamma-1/2}} = \|A^{(2\gamma-1)/2}\nabla_x u\|_X = \|\varphi\|_{L^p}$. Here, $\varphi = (\varphi_1, \cdots, \varphi_n)$ and $\partial_{x_i} u = G_{2\gamma-1} * \varphi_i$. Let $x, \xi \in \mathbb{R}^n$. Then,

$$\nabla_x u(x+\xi) = G_{2\gamma-1}(x+\xi-\cdot) * \varphi(\cdot), \qquad \nabla_x u(x) = G_{2\gamma-1}(x-\cdot) * \varphi(\cdot).$$

Recall that the following inequality holds for convolution operator:

$$\|\psi * \varphi\|_{L^p(\mathbb{R}^n)} \leq \|\psi\|_{L^q(\mathbb{R}^n)}\|\varphi\|_{L^r(\mathbb{R}^n)},$$

where $p, q, r \geq 1$ and $1/p + 1 = 1/q + 1/r$ (see [39, Section 1.6]). In particular, for $q = 1$, we have $\|\psi * \varphi\|_{L^p} \leq \|\psi\|_{L^1}\|\varphi\|_{L^p}$. The following estimate holds for the modulus of continuity of the Bessel potential function $G_\alpha, \alpha \in (0,1)$:

$$\|G_\alpha(\cdot + h) - G_\alpha(\cdot)\|_{L^1} \leq C_0|h|^\alpha,$$

for any $h \in \mathbb{R}^n$ (cf., [40, Chapter 5.4, Proposition 7]). Let ξ_1, ξ_2 be bounded vector valued functions, i.e., $\xi_1, \xi_2 \in (L^\infty(\mathbb{R}^n))^n$. Then, for any $x \in \mathbb{R}^n$ and $\theta \in [0,1]$, we have

$$\begin{aligned}
& u(x+\xi_1(x)) - u(x+\xi_2(x)) - (\xi_1(x)-\xi_2(x))\cdot\nabla_x u(x) \\
= \; & u(x+\xi_1(x)) - \nabla_x u(x) - \xi_1(x)\cdot\nabla_x u(x) \\
& -[u(x+\xi_2(x)) - \nabla_x u(x) - \xi_2(x)\cdot\nabla_x u(x)] \\
= \; & (\xi_1(x)-\xi_2(x)) \int_0^1 \nabla_x u(x+\theta\xi_1(x)) - \nabla_x u(x) \mathrm{d}\theta \\
& + \int_0^1 \nabla_x u(x+\theta\xi_1(x)) - \nabla_x u(x+\theta\xi_2(x)) \mathrm{d}\theta \, .
\end{aligned}$$

Now,

$$\begin{aligned}
& \|Q(u,\xi_1) - Q(u,\xi_2)\|^p_{L^p(\mathbb{R}^n)} \\
& = \int_{\mathbb{R}^n} |u(x+\xi_1(x)) - u(x+\xi_2(x)) - (\xi_1(x)-\xi_2(x))\cdot\nabla_x u(x)|^p \mathrm{d}x \\
& \leq \int_{\mathbb{R}^n} \left|(\xi_1(x)-\xi_2(x)) \int_0^1 \nabla_x u(x+\theta\xi_1(x)) - \nabla_x u(x)\mathrm{d}\theta\right|^p \mathrm{d}x \\
& + \int_{\mathbb{R}^n} \left|\xi_2(x) \int_0^1 \nabla_x u(x+\theta\xi_1(x)) - \nabla_x u(x+\theta\xi_2(x))\mathrm{d}\theta\right|^p \mathrm{d}x \\
& \leq \|\xi_1-\xi_2\|^p_\infty \int_0^1 \int_{\mathbb{R}^n} |\nabla_x u(x+\theta\xi_1(x)) - \nabla_x u(x)|^p dx d\theta \\
& + \|\xi_2\|^p_\infty \int_0^1 \int_{\mathbb{R}^n} |\nabla_x u(x+\theta\xi_1(x)) - \nabla_x u(x+\theta\xi_2(x)|^p dx d\theta \\
& \leq \|\xi_1-\xi_2\|^p_\infty \int_0^1 \|\,(G_{2\gamma-1}(\cdot+\theta\xi_1) - G_{2\gamma-1}(\cdot)) * \varphi\|^p_{L^p} d\theta \\
& + \|\xi_2\|^p_\infty \int_0^1 \|\,(G_{2\gamma-1}(\cdot+\theta\xi_1) - G_{2\gamma-1}(\cdot+\theta\xi_2)) * \varphi\|^p_{L^p} d\theta \\
& \leq \|\xi_1-\xi_2\|^p_\infty \int_0^1 \|G_{2\gamma-1}(\cdot+\theta\xi_1) - G_{2\gamma-1}(\cdot)\|^p_{L^1} d\theta \|\varphi\|^p_{L^p} \\
& + \|\xi_2\|^p_\infty \int_0^1 \|G_{2\gamma-1}(\cdot+\theta\xi_1) - G_{2\gamma-1}(\cdot+\theta\xi_2)\|^p_{L^1} d\theta \|\varphi\|^p_{L^p} \\
& \leq \left(\|\xi_1-\xi_2\|^p_\infty \|\xi_1\|^{(2\gamma-1)p}_\infty + \|\xi_2\|^p_\infty \|\xi_1-\xi_2\|^{(2\gamma-1)p}_\infty\right) C_0^p \|\nabla_x u\|^p_{X^{\gamma-1/2}} \\
& \leq \|\xi_1-\xi_2\|^{(2\gamma-1)p}_\infty \left(\|\xi_1\|^p_\infty + \|\xi_2\|^{(2-2\gamma)p}_\infty \|\xi_1\|^{(2\gamma-1)p}_\infty + \|\xi_2\|^p_\infty\right) C_0^p \|\nabla_x u\|^p_{X^{\gamma-1/2}} \, .
\end{aligned}$$

By Young's inequality, we have $ab \leq \frac{a^\alpha}{\alpha} + \frac{b^\beta}{\beta}$ for any $a,b \geq 0$, and $\alpha,\beta > 1$ with $1/\alpha + 1/\beta = 1$ (cf., [41]). Set $\alpha = 1/(2-2\gamma), \beta = 1/(2\gamma-1)$. Then, $1/\alpha + 1/\beta = 1$, and we obtain $\|\xi_2\|^{(2-2\gamma)p}_\infty \|\xi_1\|^{(2\gamma-1)p}_\infty \leq (2-2\gamma)\|\xi_2\|^p_\infty + (2\gamma-1)\|\xi_1\|^p_\infty \leq 2\|\xi_2\|^p_\infty + \|\xi_1\|^p_\infty$.

Therefore,

$$\begin{aligned}
\|Q(u,\xi_1) - Q(u,\xi_2)\|^p_{L^p(\mathbb{R}^n)} & \leq 2\|\xi_1-\xi_2\|^{(2\gamma-1)p}_\infty \left(\|\xi_1\|^p_\infty + \|\xi_2\|^p_\infty\right) C_0^p \|\nabla_x u\|^p_{X^{\gamma-1/2}} \\
& \leq 2C_0^p \|\xi_1-\xi_2\|^{(2\gamma-1)p}_\infty (\|\xi_1\|_\infty + \|\xi_2\|_\infty)^p \|\nabla_x u\|^p_{X^{\gamma-1/2}} .
\end{aligned}$$

Therefore, the pointwise estimate holds with the constant $\hat{C} = 2^{1/p}C_0 > 0$. □

Applying Proposition 1 with $\xi_1 = \xi$ and $\xi_2 = 0$, we obtain the following corollary.

Corollary 1. *[3, Corollary 1] Let u be such that $\nabla_x u \in (X^{\gamma-1/2})^n$ where $1 > \gamma \geq 1/2$. Then, for any $\xi \in \mathbb{R}^n$, the following pointwise estimate holds:*

$$\|Q(u,\xi)\|_{L^p(\mathbb{R}^n)} \leq C_0|\xi|^{2\gamma}\|\nabla_x u\|_{X^{\gamma-1/2}}.$$

Next, we consider the case where the nonlocal integral term depends on the variables x and z. It is a generalization of the result [1, Lemma 3.4] due to Cruz and Ševčovič proven for the case where $\xi(x,z) \equiv z$.

Proposition 2. *[3, Proposition 2] Suppose that the shift mapping $\xi = \xi(x,z)$ satisfies $\sup_{x\in\mathbb{R}^n}|\xi(x,z)| \leq C_0|z|^\omega(1+e^{D_0|z|})$ for some constants $C_0 > 0, D_0 \geq 0, \omega > 0$ and any $z \in \mathbb{R}^n$. Assume ν is a Lévy measure with the shape parameters α, D, and either $\mu > 0, D \in \mathbb{R}$, or $\mu = 0$ and $D > D_0 \geq 0$. Assume $1/2 \leq \gamma < 1$, and $\gamma > (\alpha - n)/(2\omega)$. Then there exists a constant $C_0 > 0$ such that*

$$\|f(u)\|_{L^p} \leq C_0\|\nabla_x u\|_{X^{\gamma-1/2}},$$

provided that $\nabla_x u \in (X^{\gamma-1/2})^n$. If $u \in X^\gamma$ then $\|f(u)\|_{L^p} \leq C\|u\|_{X^\gamma}$, i.e., $f : X^\gamma \to X$ is a bounded linear operator.

Proof. The Lévy measure $\nu(\mathrm{d}z)$ is given by $\nu(\mathrm{d}z) = h(z)\mathrm{d}z$. Let us denote the auxiliary function $\tilde{h}(z) = |z|^\alpha h(z)$. Then, $0 \leq \tilde{h}(z) \leq C_0 e^{-D|z|-\mu|z|^2}$ since $h(z) = |z|^{-\alpha}\tilde{h}(z) = h_1(z)h_2(z)$, where $h_1(z) = |z|^{-\beta}\tilde{h}(z)^{\frac{1}{2}}$ and $h_2(z) = |z|^{\beta-\alpha}\tilde{h}(z)^{\frac{1}{2}}$. Applying Proposition 1 with $\xi_1 = \xi, \xi_2 = 0$, and using the Hölder inequality, we obtain

$$\begin{aligned}
\|f(u)\|^p_{L^p} &= \int_{\mathbb{R}^n}\left|\int_{\mathbb{R}^n}(u(x+\xi(x,z)) - u(x) - \xi(x,z)\cdot\nabla_x u(x))h(z)\mathrm{d}z\right|^p \mathrm{d}x \\
&\leq \int_{\mathbb{R}^n}\int_{\mathbb{R}^n}|u(x+\xi(x,z)) - u(x) - \xi(x,z)\cdot\nabla_x u(x)|^p h_1(z)^p\mathrm{d}z \\
&\qquad \times \left(\int_{\mathbb{R}^n} h_2(z)^q\mathrm{d}z\right)^{p/q}\mathrm{d}x \\
&= \int_{\mathbb{R}^n}\left(\int_{\mathbb{R}^n}|u(x+\xi(x,z)) - u(x) - \xi(x,z)\cdot\nabla_x u(x)|^p\,\mathrm{d}x\right)h_1(z)^p\mathrm{d}z \\
&\qquad \times \left(\int_{\mathbb{R}^n} h_2(z)^q\mathrm{d}z\right)^{p/q} \\
&\leq C_0^p\|\nabla_x u\|^p_{X^{\gamma-1/2}}\int_{\mathbb{R}^n}|\xi(x,z)|^{2\gamma p}|z|^{-\beta p}\tilde{h}(z)^{p/2}\mathrm{d}z\left(\int_{\mathbb{R}^n} h_2(z)^q\mathrm{d}z\right)^{p/q} \\
&\leq C_0^p\|\nabla_x u\|^p_{X^{\gamma-1/2}}\int_{\mathbb{R}^n}|z|^{(2\gamma\omega-\beta)p}\tilde{h}(z)^{p/2}\mathrm{d}z\left(\int_{\mathbb{R}^n} h_2(z)^q\mathrm{d}z\right)^{p/q}.
\end{aligned}$$

Assuming $p, q \geq 1, 1/p + 1/q = 1$ are such that

$$(2\gamma\omega - \beta)p > -n, \qquad (\beta - \alpha)q = (\beta - \alpha)\frac{p}{p-1} > -n,$$

then, the integrals $\int_{\mathbb{R}^n} |z|^{(2\gamma\omega-\beta)p}\tilde{h}(z)^{p/2}\mathrm{d}z$ and $\int_{\mathbb{R}^n} h_2(z)^q \mathrm{d}z = \int_{\mathbb{R}^n} |z|^{(\beta-\alpha)q}\tilde{h}(z)^{q/2}\mathrm{d}z$ are finite, provided that the shape parameters satisfy either $\mu > 0, D \in \mathbb{R}$, or $\mu = 0, D > D_0 \geq 0$. As $\gamma > (\alpha - n)/(2\omega)$, there exists $\beta > 1$ satisfying

$$\alpha - n + n/p < \beta < 2\gamma\omega + n/p.$$

Therefore, there exists $C_0 > 0$ such that $\|f(u)\|_{L^p} \leq C_0\|\nabla_x u\|_{X^{\gamma-1/2}}$. □

Let $C([0,T], X^\gamma)$ be the Banach space consisting of continuous functions from $[0,T]$ to X^γ with the maximum norm. The following proposition is due to Henry [39] (see also Cruz and Ševčovič [2]).

Proposition 3. *[39, Proposition 3.5] Suppose that the linear operator $-A$ is a generator of an analytic semigroup $\{e^{-At}, t \geq 0\}$ in a Banach space X. Assume the initial condition U_0 belongs to the space X^γ where $0 \leq \gamma < 1$. Suppose that the mappings $F : [0,T] \times X^\gamma \to X$ and $h : (0,T] \to X$ are Hölder continuous in the τ variable, $\int_0^T \|h(\tau)\|_X \mathrm{d}\tau < \infty$, and F is Lipschitz continuous in the U variable. Then, for any $T > 0$, there exists a unique solution to the abstract semilinear evolution equation: $\partial_\tau U + AU = F(\tau, U) + h(\tau)$ such that $U \in C([0,T], X^\gamma), U(0) = U_0, \partial_\tau U(\tau) \in X, U(\tau) \in D(A)$ for any $\tau \in (0,T)$. The function U is a solution in the mild (integral) sense, i.e., $U(\tau) = e^{-A\tau}U_0 + \int_0^\tau e^{-A(\tau-s)}(F(s,U(s)) + h(s))\mathrm{d}s$, $\tau \in [0,T]$.*

Applying Propositions 2 and 3, we can state the following result, which is a nontrivial generalization of the result shown by Ševčovič and Cruz [2] for $n = 1$.

Theorem 2. *[3, Theorem 1] Suppose that the shift mapping $\xi = \xi(x,z)$ satisfies $\sup_{x\in\mathbb{R}} |\xi(x,z)| \leq C_0|z|^\omega(1 + e^{D_0|z|}), z \in \mathbb{R}^n$, for some constants $C_0 > 0, D_0 \geq 0, \omega > 0$. Assume ν is an admissible activity Lévy measure with the shape parameters α, D, and, either $\mu > 0, D \in \mathbb{R}$, or $\mu = 0, D > D_0 \geq 0$. Assume $1/2 \leq \gamma < 1$ and $\gamma > (\alpha - n)/(2\omega)$, $n \geq 1$. Suppose that $g(\tau, x, u, \nabla_x u)$ is Hölder continuous in the τ variable and Lipschitz* *continuous in the remaining variables, respectively. Assume $u_0 \in X^\gamma$, and $T > 0$. Then, there exists a unique mild solution u to PIDE (22) that satisfies $u \in C([0,T], X^\gamma)$.*

4.2. Maximal Monotone Operator Technique for Solving Nonlinear Parabolic Equations

This section presents the existence and uniqueness results of a fully nonlinear parabolic equation using the monotone operator technique. We consider the HJB equation arising from portfolio optimization selection, where the goal is to maximize the conditional expected value of the terminal utility of the portfolio. Such a fully nonlinear HJB equation presented in an abstract setting can be viewed as a PIDE in some sense. First, we employ the so-called Riccati transformation method to transform the fully nonlinear HJB equation into a quasilinear parabolic equation, which can be viewed as the porous media type of equation with source term. Then, we showed that the underlying operator is maximally monotone in some Sobolev spaces. Next, we employed the Banach fixed point theorem and Fourier transform technique to obtain the existence and uniqueness of a solution to the general form of the transformed parabolic equation in an abstract setting in high-dimensional spaces. Furthermore, as a crucial requirement for solving the Cauchy problem, we find that the diffusion function to the quasilinear parabolic equation is globally Lipschitz continuous under some assumptions.

We consider the Cauchy problem for the nonlinear parabolic PDE of the following form:

$$\partial_\tau \varphi - \Delta\alpha(\tau, \varphi) = g_0(\tau, \varphi) + \nabla \cdot \boldsymbol{g}_1(\tau, \varphi), \tag{25}$$

$$\varphi(\cdot, 0) = \varphi_0, \tag{26}$$

where $\tau \in (0,T), x \in \mathbb{R}^d, d \geq 1$. The solution $\varphi = \varphi(x,\tau)$ to such a nonlinear parabolic equation is established in some Sobolev spaces in high-dimensional spaces (see [4]). To achieve such results, we assumed that the diffusion function $\alpha = \alpha(x,\tau,\varphi)$ is globally Lipschitz continuous and strictly increasing in the φ-variable. An example of such a Lipschitz continuous function $\alpha(x,\tau,\varphi)$ is the value function of the following parametric optimization problem:

$$\alpha(x,\tau,\varphi) = \min_{\boldsymbol{\theta}\in\triangle}\left(-\mu(x,t,\boldsymbol{\theta}) + \frac{\varphi}{2}\sigma(x,t,\boldsymbol{\theta})^2\right), \quad \tau\in(0,T), x\in\mathbb{R}^d, \varphi > \varphi_{min}, \tag{27}$$

where μ and σ^2 are given C^1 functions, and $\triangle \subset \mathbb{R}^n$ is a compact decision set. The properties of the value function depend on the structure of the decision set $\triangle$. It is smooth if $\triangle$ is a convex set; meanwhile, it can only be $C^{0,1}$ smooth if $\triangle$ is not connected.

4.2.1. Existence and Uniqueness of a Solution to the Cauchy Problem

First, we define our underlying function spaces. Let $V \hookrightarrow H \hookrightarrow V'$ be a Gelfand triple, where

$$H = L^2(\mathbb{R}^d) = \{f : \mathbb{R}^d \to \mathbb{R}, \quad \|f\|_{L^2}^2 = \int_{\mathbb{R}^d} |f(x)|^2 dx < \infty\}$$

is a Hilbert space endowed with the inner product $(f, g) = \int_{\mathbb{R}^d} f(x)g(x)dx$. The Banach spaces V and V' are defined as follows:

$$V = H^1(\mathbb{R}^d), \quad V' = H^{-1}(\mathbb{R}^d),$$

where the Sobolev spaces $H^s(\mathbb{R}^d)$ are defined by means of the Fourier transform

$$\hat{f}(\xi) = \frac{1}{(2\pi)^{d/2}}\int_{\mathbb{R}^d} e^{-ix\cdot\xi} f(x)dx, \quad \xi = (\xi_1, \xi_2, ..., \xi_d)^T \in \mathbb{R}^d,$$

$$H^s(\mathbb{R}^d) = \{f : \mathbb{R}^d \to \mathbb{R}, (1 + |\xi|^2)^{s/2}\hat{f}(\xi) \in L^2(\mathbb{R}^d)\}, \; s \in \mathbb{R}$$

endowed with the norm $\|f\|_{H^s}^2 = \int_{\mathbb{R}^d}(1 + |\xi|^2)^s|\hat{f}(\xi)|^2 d\xi$, and $|\xi| = (\xi_1^2 + \cdots + \xi_d^2)^{1/2}$. Let the linear operator $A : V \to V'$ be defined as follows:

$$A\psi = \psi - \Delta\psi.$$

It is worth nothing that the operator A is self-adjoint in the Hilbert space $H = L^2(\mathbb{R}^d)$ with the following Fourier transform representation:

$$\widehat{A\psi}(\xi) = (1 + |\xi|^2)\hat{\psi}(\xi).$$

The fractional power of A is defined by $\widehat{A^s\psi}(\xi) = (1 + |\xi|^2)^s\hat{\psi}(\xi),\ s \in \mathbb{R}$. In particular,

$$\widehat{A^{\pm 1/2}\psi}(\xi) = (1 + |\xi|^2)^{\pm 1/2}\hat{\psi}(\xi),$$

and $A^{-1/2}$ is a self-adjoint operator in the Hilbert space $H = L^2(\mathbb{R}^d)$. Moreover, $A^{-1} = A^{-1/2}A^{-1/2}$.

In the sequel, we denote the duality pairing between spaces V and V' by $\langle .,. \rangle$, i.e., the value of a functional $F \in V'$ at $u \in V$ is denoted by $\langle F, u \rangle$. We have the following definitions.

Definition 5. *[4, 42] An operator (in general, nonlinear) $B : V \to V'$ is said to be*

(i) monotone if

$$\langle B(u) - B(v), u - v \rangle \geq 0,\ \forall\, u, v \in V,$$

(ii) strongly monotone if there exists a constant $C > 0$ such that

$$\langle B(u) - B(v), u - v \rangle \geq C\|u - v\|_V^2,\ \forall\, u, v \in V,$$

(iii) hemicontinuous if for each $u, v \in V$, the real-valued function $t \mapsto B(u + tv)(v)$ is continuous.

Theorem 3. *[42, 43] Let V be a separable reflexive Banach space, dense, and continuous in a Hilbert space H, which is identified with its dual, so $V \hookrightarrow H \hookrightarrow V'$. Let $p \geq 2$ and set $\mathcal{V} = L^p((0,T); V)$. Assume a family of operators $\mathcal{A}(\tau, .) : V \to V', 0 \leq \tau < T$, is given such that*

(i) for each $\varphi \in V$, the function $\mathcal{A}(., \varphi) : [0, T] \to V'$ is measurable,

(ii) for a.e $\tau \in [0, T]$, the operator $\mathcal{A}(\tau, .) : V \to V'$ is monotone, hemicontinuous, and bounded by $\|\mathcal{A}(\tau, \varphi)\| \leq C(\|\varphi\|^{p-1} + k(\tau)), \varphi \in V, 0 \leq \tau < T$, where $k \in L^{p'}(0, T)$,

(iii) and there exists $\lambda > 0$ such that $\langle \mathcal{A}(\tau, \varphi), \varphi \rangle \geq \lambda\|\varphi\|^p - k(\tau), \varphi \in V, 0 \leq \tau < T$.

Then, for each $\hat{f} \in \mathcal{V}'$ and $\varphi_0 \in H$, there exists a unique solution $\varphi \in \mathcal{V}$ of the Cauchy problem

$$\partial_\tau\varphi(\tau) + \mathcal{A}(\tau, \varphi(\tau)) = \hat{f}(\tau) \text{ in } \mathcal{V}',\quad \varphi(0) = \varphi_0.$$

Consider the spaces $\mathcal{V} = L^2((0,T);V)$, $\mathcal{H} = L^2((0,T);H)$, and $\mathcal{V}' = L^2((0,T);V')$, i.e., $p = 2$. Thus, we have that these spaces satisfy the Gelfand triple, i.e., $\mathcal{V} \hookrightarrow \mathcal{H} \hookrightarrow \mathcal{V}'$, where $\mathcal{H}$ is a Hilbert space endowed with the norm

$$\|\varphi\|_{\mathcal{H}}^2 = \int_0^T \|\varphi(\tau)\|_H^2 d\tau, \ \forall \varphi \in \mathcal{H}.$$

For a given value φ_{min}, we denote $\mathcal{D} = \mathbb{R}^d \times (0,T) \times (\varphi_{min}, \infty)$.

Theorem 4. *[4, Theorem 2] Assume that the above settings on H and V hold. Let $g_0, g_{1j} : [0,T] \times H \to H, j = 1, \cdots, n$, be globally Lipschitz continuous functions. Suppose $\alpha \in C^{0,1}(\mathcal{D})$ is such that there exist constants $\omega, L, L_0 > 0$ such that $0 < \omega \le \alpha'_\varphi(x,\tau,\varphi) \le L$, $|\nabla_x \alpha(x,\tau,\varphi)| \le p(x,\tau) + L_0|\varphi|$, $\alpha(x,\tau,0) = h(x,\tau)$ for a.e. $(x,\tau,\varphi) \in \mathcal{D}$ and $p, h \in L^\infty((0,T);H)$. Then, for any $T > 0$ and $\varphi_0 \in H$, there exists a unique solution $\varphi \in \mathcal{V}$ of the Cauchy problem*

$$\partial_\tau \varphi + A\alpha(\cdot,\tau,\varphi) = g_0(\tau,\varphi) + \nabla \cdot \boldsymbol{g}_1(\tau,\varphi), \qquad \varphi(0) = \varphi_0. \tag{28}$$

We remark here that the above result and its proof are contained in our recent paper [4, Theorem 2].

Proof: Recall that $H = L^2(\mathbb{R}^d)$ and $V = H^1(\mathbb{R}^d)$, its dual space being $V' = H^{-1}(\mathbb{R}^d)$. Let the scalar products in V and V' be respectively defined by

$$(f,g)_V = (A^{1/2}f, A^{1/2}g)_H = (Af,g)_H, \ (f,g)_{V'} = (A^{-1/2}f, A^{-1/2}g)_H = (A^{-1}f,g)_H.$$

Let us define the operator $\mathcal{A}(\tau,\cdot) : V \to V'$ by

$$\langle \mathcal{A}(\tau,\varphi), \psi \rangle = (A^{-1}A\alpha(\cdot,\tau,\varphi), \psi)_H = (\alpha(\cdot,\tau,\varphi), \psi)_H.$$

Therefore, we conclude that the mapping $\varphi \mapsto \alpha(\cdot,\tau,\varphi)$ maps V into V under the assumption made on the function α. Indeed, if $\varphi \in V$ and $\eta = \alpha(\cdot,\tau,\varphi)$, then $\eta(x) = \alpha(x,\tau,\varphi(x)) - \alpha(x,\tau,0) + \alpha(x,\tau,0)$, and so

$$|\eta(x)| \le (\max_\varphi \alpha'_\varphi(x,\tau,\varphi))|\varphi(x)| + |h(x,\tau)| \le L|\varphi(x)| + |h(x,\tau)|.$$

Thus, $\int_{\mathbb{R}^d} |\eta(x)|^2 dx \le 2\int_{\mathbb{R}^d} L^2|\varphi(x)|^2 + |h(x,\tau)|^2 dx \le 2L^2\|\varphi\|_H^2 + 2\|h(\cdot,\tau)\|_H^2$. Since $\nabla\eta(x) = \nabla_x\alpha(x,\tau,\varphi(x)) + \alpha'_\varphi(x,\tau,\varphi(x))\nabla\varphi(x)$, we have

$$
\begin{aligned}
\|\eta\|_V^2 &= \int_{\mathbb{R}^d} |\eta(x)|^2 + |\nabla\eta(x)|^2 dx \\
&\leq 2\int_{\mathbb{R}^d} L^2|\varphi(x)|^2 + |h(x,\tau)|^2 dx + 2\int_{\mathbb{R}^d} |p(x,\tau)|^2 + L_0^2|\varphi(x)|^2 dx + \\
&+ 2\int_{\mathbb{R}^d} L^2|\nabla\varphi(x)|^2 dx \\
&\leq 2(L^2\|\varphi\|_V^2 + \|h(\cdot,\tau)\|_H^2 + \|p(\cdot,\tau)\|_H^2 + L_0^2\|\varphi\|_H^2) < \infty,
\end{aligned}
$$

because $p, h \in L^\infty((0,T);H)$. Consequently, $\eta \in V$, as claimed.

Next, we show that the operator $\mathcal{A}$ is monotone in the space V'. According to (77), we have $(\alpha(x,\tau,\varphi_1)-\alpha(x,\tau,\varphi_2))(\varphi_1-\varphi_2) \geq \omega(\varphi_1-\varphi_2)^2$, for any $\varphi_1, \varphi_2 \geq \varphi_{min}, x \in \mathbb{R}, \tau \in [0,T]$.

$$
\begin{aligned}
&\langle \mathcal{A}(\tau,\varphi_1) - \mathcal{A}(\tau,\varphi_2), \varphi_1 - \varphi_2\rangle = (\alpha(\cdot,\tau,\varphi_1) - \alpha(\cdot,\tau,\varphi_2), \varphi_1 - \varphi_2) \\
&= \int_{\mathbb{R}^d} (\alpha(x,\tau,\varphi_1(x)) - \alpha(x,\tau,\varphi_2(x)))(\varphi_1(x) - \varphi_2(x)) dx \\
&\geq \int_{\mathbb{R}^d} \omega|\varphi_1(x) - \varphi_2(x)|^2 dx = \omega\|\varphi_1 - \varphi_2\|_H^2.
\end{aligned}
$$

This implies that the operator $\mathcal{A}(\tau,\cdot)$ is strongly monotone.

For a given $\tilde{\varphi} \in \mathcal{H}$, we have $\hat{f} \in \mathcal{V}'$, where $\hat{f}(\tau) = g_0(\tau,\tilde{\varphi}(\cdot,\tau)) + \nabla\cdot \boldsymbol{g}_1(\tau,\tilde{\varphi}(\cdot,\tau))$, because $g_0, g_{1j} : [0,T]\times H \to H$ are globally Lipschitz continuous, $H \hookrightarrow V'$, and the operator ∇ maps H into V'. The hemicontinuity, boundedness, and coercivity of the operator $\mathcal{A}$ follow from the assumption that the function α is globally Lipschitz continuous and strictly increasing.

Applying Theorem 3, we deduce the existence of a solution $\varphi \in \mathcal{V}$ such that

$$
\partial_\tau\varphi + \mathcal{A}(\tau,\varphi) = \hat{f}(\tau), \qquad \varphi_0 \in H, \tag{29}
$$

where $\mathcal{A}(\tau,\varphi) = A\alpha(\cdot,\tau,\varphi)$. Next, we multiply (29) by A^{-1} to obtain

$$
\partial_\tau A^{-1}\varphi + \alpha(\cdot,\tau,\varphi) = f, \tag{30}
$$

where $f = f(\tau,\tilde{\varphi}) = A^{-1}\hat{f}(\tau)$. For $\tau \in [0,T]$, we denote $\tilde{f}(\tilde{\varphi}) = A^{-1/2}\hat{f}(\tau) = A^{-1/2}g_0(\tau,\tilde{\varphi}) + A^{-1/2}\sum_{j=1}^d \partial_{x_j} g_{1j}(\tau,\tilde{\varphi})$. The Fourier transform of $\tilde{f}$ is defined by

$$
\widehat{\tilde{f}(\tilde{\varphi})}(\xi) = \frac{1}{(1+|\xi|^2)^{1/2}}\widehat{g_0(\tau,\tilde{\varphi})}(\xi) + \sum_{j=1}^d \frac{(-i\xi_j)}{(1+|\xi|^2)^{1/2}}\widehat{g_{1j}(\tau,\tilde{\varphi})}(\xi).
$$

Let $\beta > 0$ be the Lipschitz constant of the mappings $g_0, g_{1j}, j = 1, \cdots, d$. Using Parseval's identity and Lipschitz continuity of g_0, g_{1j} in H, we obtain, for $\tilde{\varphi}_1, \tilde{\varphi}_2 \in \mathcal{H}$,

$$\|\tilde{f}(\tilde{\varphi}_1) - \tilde{f}(\tilde{\varphi}_2)\|_H^2 = \|\widehat{\tilde{f}(\tilde{\varphi}_1)} - \widehat{\tilde{f}(\tilde{\varphi}_2)}\|_H^2 = \int_{\mathbb{R}^d} |\widehat{\tilde{f}(\tilde{\varphi}_1)}(\xi) - \widehat{\tilde{f}(\tilde{\varphi}_2)}(\xi)|^2 d\xi$$

$$\leq 2\int_{\mathbb{R}^d} \frac{1}{1+|\xi|^2} |\widehat{g_0(\tau,\tilde{\varphi}_1)}(\xi) - \widehat{g_0(\tau,\tilde{\varphi}_2)}(\xi)|^2$$

$$+ \sum_{j=1}^{d} \frac{|\xi|^2}{1+|\xi|^2} |\widehat{g_{1j}(\tau,\tilde{\varphi}_1)}(\xi) - \widehat{g_{1j}(\tau,\tilde{\varphi}_2)}(\xi)|^2 d\xi$$

$$\leq 2\|\widehat{g_0(\tau,\tilde{\varphi}_1)} - \widehat{g_0(\tau,\varphi_2)}\|_H^2 + 2\sum_{j=1}^{d} \|\widehat{g_{1j}(\tau,\tilde{\varphi}_1)} - \widehat{g_{1j}(\tau,\varphi_2)}\|_H^2$$

$$= 2\|g_0(\tau,\tilde{\varphi}_1) - g_0(\tau,\tilde{\varphi}_2)\|_H^2 + 2\sum_{j=1}^{d} \|g_{1j}(\tau,\tilde{\varphi}_1) - g_{1j}(\tau,\tilde{\varphi}_2)\|_H^2$$

$$\leq \tilde{\beta}^2 \|\tilde{\varphi}_1 - \tilde{\varphi}_2\|_H^2,$$

where $\tilde{\beta}^2 = 2(1+d)\beta^2$. Hence, we obtain

$$\|\tilde{f}(\tilde{\varphi}_1) - \tilde{f}(\tilde{\varphi}_2)\|_H \leq \tilde{\beta}\|\tilde{\varphi}_1 - \tilde{\varphi}_2\|_H. \tag{31}$$

Suppose $\varphi_1, \varphi_2 \in \mathcal{H}$ are such that $\varphi_1 = F(\tilde{\varphi}_1)$ and $\varphi_2 = F(\tilde{\varphi}_2)$. Here, the map $F : \mathcal{H} \to \mathcal{H}$ is defined by $\varphi = F(\tilde{\varphi})$, where φ is a solution to the Cauchy problem

$$\partial_\tau A^{-1}\varphi + \alpha(\cdot,\tau,\varphi) - f(\tau,\tilde{\varphi}), \qquad \varphi(0) = \varphi_0$$

Letting $\varphi = \varphi_1 - \varphi_2 = F(\tilde{\varphi}_1) - F(\tilde{\varphi}_2)$, we obtain

$$\partial_\tau A^{-1}(\varphi_1 - \varphi_2) + \alpha(\cdot,\tau,\varphi_1) - \alpha(\cdot,\tau,\varphi_2) = f(\tilde{\varphi}_1) - f(\tilde{\varphi}_2). \tag{32}$$

Next, multiplying (32) by $\varphi_1 - \varphi_2$ and taking the scalar product in the space H, we obtain

$$\begin{aligned}(\partial_\tau A^{-1}(\varphi_1 - \varphi_2), \varphi_1 - \varphi_2) \quad &+ \quad (\alpha(\cdot,\tau,\varphi_1) - \alpha(\cdot,\tau,\varphi_2), \varphi_1 - \varphi_2) \\ &= \quad (f(\tau,\tilde{\varphi}_1) - f(\tau,\tilde{\varphi}_2), \varphi_1 - \varphi_2). \end{aligned} \tag{33}$$

Using (31) and the fact that $A^{-1/2}$ is self-adjoint in H, then (33) gives

$$\frac{1}{2}\frac{d}{d\tau}\|A^{-1/2}(\varphi_1-\varphi_2)\|_H^2+\omega\|\varphi_1-\varphi_2\|_H^2$$
$$\le \langle f(\tau,\tilde{\varphi}_1)-f(\tau,\tilde{\varphi}_2),\varphi_1-\varphi_2\rangle = \langle A^{1/2}(f(\tau,\tilde{\varphi}_1)-f(\tau,\tilde{\varphi}_2)),A^{-1/2}(\varphi_1-\varphi_2)\rangle$$
$$\le \|A^{1/2}(f(\tau,\tilde{\varphi}_1)-f(\tau,\tilde{\varphi}_2))\|_H\|\varphi_1-\varphi_2\|_{V'} = \|\tilde{f}(\tilde{\varphi}_1)-\tilde{f}(\tilde{\varphi}_2)\|_H\|\varphi_1-\varphi_2\|_{V'}$$
$$\le \tilde{\beta}\|\tilde{\varphi_1}-\tilde{\varphi_2}\|_H\|\varphi_1-\varphi_2\|_{V'}.$$

This implies

$$\frac{1}{2}\frac{d}{d\tau}\|\varphi_1-\varphi_2\|_{V'}^2+\omega\|\varphi_1-\varphi_2\|_H^2 \le \tilde{\beta}\|\tilde{\varphi_1}-\tilde{\varphi_2}\|_H\|\varphi_1-\varphi_2\|_{V'}.$$

Then, integrating on a small time interval $[0,T]$ from 0 to t and noting that $\varphi_1(0)=\varphi_2(0)=\varphi_0$, we obtain

$$\frac{1}{2}\|\varphi_1(\tau)-\varphi_2(\tau)\|_{V'}^2+\omega\int_0^{\tau}\|\varphi_1(s)-\varphi_2(s)\|_H^2 ds$$
$$\le \tilde{\beta}\int_0^{\tau}\|\tilde{\varphi}_1(s)-\tilde{\varphi}_2(s)\|_H\|\varphi_1(s)-\varphi_2(s)\|_{V'}ds$$
$$\le \tilde{\beta}\max_{\tau\in[0,T]}\|\varphi_1(\tau)-\varphi_2(\tau)\|_{V'}\int_0^T\|\tilde{\varphi}_1(\tau)-\tilde{\varphi}_2(\tau)\|_H d\tau.$$

Taking the maximum over $\tau\in[0,T]$ and using the fact that for any $a,b\in\mathbb{R}$, $ab\le\frac{1}{2}a^2+\frac{1}{2}b^2$, we obtain

$$\frac{1}{2}(\max_{\tau\in[0,T]}\|\varphi_1(\tau)-\varphi_2(\tau)\|_{V'})^2+\omega\int_0^T\|\varphi_1(\tau)-\varphi_2(\tau)\|_H^2 d\tau$$
$$\le \tilde{\beta}\max_{\tau\in[0,T]}\|\varphi_1(\tau)-\varphi_2(\tau)\|_{V'}\int_0^T\|\tilde{\varphi}_1(\tau)-\tilde{\varphi}_2(\tau)\|_H d\tau$$
$$\le \frac{1}{2}(\max_{\tau\in[0,T]}\|\varphi_1(\tau)-\varphi_2(\tau)\|_{V'})^2+\frac{\tilde{\beta}^2}{2}(\int_0^T\|\tilde{\varphi}_1(\tau)-\tilde{\varphi}_2(\tau)\|_H d\tau)^2.$$

Using the Cauchy–Schwartz inequality, we obtain $\omega\int_0^T\|\varphi_1(\tau)-\varphi_2(\tau)\|_H^2 d\tau \le \frac{\tilde{\beta}^2}{2}\int_0^T d\tau\int_0^T\|\tilde{\varphi}_1(\tau)-\tilde{\varphi}_2(\tau)\|_H^2 d\tau = \frac{\tilde{\beta}^2 T}{2}\int_0^T\|\tilde{\varphi}_1(\tau)-\tilde{\varphi}_2(\tau)\|_H^2 d\tau$. This implies that

$$\|F(\tilde{\varphi_1})-F(\tilde{\varphi_2})\|_{\mathcal{H}}^2 \le \frac{\tilde{\beta}^2 T}{2\omega}\|\tilde{\varphi_1}-\tilde{\varphi_2}\|_{\mathcal{H}}^2.$$

Thus, for sufficiently small value of T such that $\frac{\tilde{\beta}^2 T}{2\omega} < 1$, the operator F is a contraction on the space $\mathcal{H}$. Therefore, by the Banach fixed point theorem, F has a unique fixed point in $\mathcal{H}$. It is worth noting that $\tilde{\beta}$ and ω are given such that they are independent of T. If $T > 0$ is arbitrary, then we can apply a simple continuation argument. In other words, if the solution exists in $(0, T_0)$ interval with $\frac{\tilde{\beta}^2 T_0}{2\omega} < 1$, then starting from the initial condition $\varphi_0 = \varphi(T_0/2)$, we can continue the solution φ from the interval $(0, T_0)$ over the interval $(0, T_0) \cup (T_0/2, T_0/2 + T_0) \equiv (0, 3T_0/2)$. Continuing in this manner, we obtain the existence and uniqueness of a solution $\varphi \in \mathcal{H}$ defined on the time interval $(0, T)$.

Finally, the solution belongs to the space $\mathcal{V}$ because the right-hand side, i.e., the function $\hat{f}(\tau) = g_0(\tau, \varphi(\cdot, \tau)) + \nabla \cdot \boldsymbol{g}_1(\tau, \varphi(\cdot, \tau))$ belongs to $\mathcal{V}'$. Applying Theorem 3 we conclude $\varphi \in \mathcal{V}$, as claimed. ◇

The following result demonstrates the absolute continuity and a-priori energy estimate property of the solution. Based on the assumption of the previous theorem, we have $\alpha(\cdot, 0), g_0(\cdot, 0), g_{1j}(\cdot, 0) \in \mathcal{H}$. Here, the space $\mathcal{X} = L^\infty((0,T); V')$ is endowed with the norm

$$\|\varphi\|_{\mathcal{X}}^2 = \sup_{\tau \in [0,T]} \|\varphi(\tau)\|_{V'}^2, \ \forall \varphi \in \mathcal{X}.$$

Again, the following result and its proof are contained in our recent paper [4, Theorem 3].

Theorem 5. *[4, Theorem 3] Suppose that the functions α, g_0, g_{1j} satisfy the assumptions of Theorem 4. Then, the unique solution $\varphi \in \mathcal{V}$ to the Cauchy problem (29) is absolutely continuous, i.e., $\varphi \in C([0,T]; H)$. Moreover, there exists a constant $\tilde{C} > 0$, such that the unique solution satisfies the following inequality:*

$$\|\varphi\|_{\mathcal{X}}^2 + \|\varphi\|_{\mathcal{H}}^2 \leq \tilde{C}\big(\|\varphi_0\|_{V'}^2 + \|\alpha(\cdot,0)\|_{\mathcal{H}}^2 + \|g_0(\cdot,0)\|_{\mathcal{H}}^2 + \sum_{j=1}^{d} \|g_{1j}(\cdot,0)\|_{\mathcal{H}}^2\big). \tag{34}$$

Proof. Since $\hat{f} \in \mathcal{V}'$, where $\hat{f} = g_0 + \nabla \cdot \boldsymbol{g}_1$ and $\mathcal{A}(\tau, \varphi) \in \mathcal{V}'$, then $\partial_\tau \varphi \in \mathcal{V}'$. Therefore, for each $\varphi_0 \in H$, we have $\varphi \in W$, where W is the Banach space $W = \{\varphi, \varphi \in \mathcal{V}, \partial_\tau \varphi \in \mathcal{V}'\}$. According to [43, Proposition

1.2], we have $W \hookrightarrow C([0,T];H)$. Hence, the unique solution φ to the Cauchy problem (4) belongs to the space $C([0,T];H)$, as claimed.

Next, we show that the unique solution satisfies a-priori energy estimate (34). Let φ be a unique solution to the Cauchy problem (28). Multiply (30) by φ and take the scalar product in H to obtain

$$(\partial_\tau A^{-1}\varphi,\varphi)_H + (\alpha(\cdot,\tau,\varphi),\varphi)_H = (A^{-1}g_0(\tau,\varphi) + A^{-1}\nabla\cdot\boldsymbol{g}_1(\tau,\varphi),\varphi). \tag{35}$$

Using the Lipschitz continuity of $g_0, \boldsymbol{g}_1$, and strong monotonicity of α, we obtain

$$\begin{aligned}
\frac{1}{2}\frac{d}{d\tau}\|\varphi\|_{V'}^2 + \omega\|\varphi\|_H^2 &= (\partial_\tau A^{-1}\varphi,\varphi) + \omega\|\varphi\|_H^2\\
&\le (\partial_\tau A^{-1}\varphi,\varphi) + (\alpha(\cdot,\varphi)-\alpha(\cdot,0),\varphi)\\
&= (A^{-1}(g_0(\cdot,\varphi) + \nabla\cdot\boldsymbol{g}_1(\tau,\varphi)) - \alpha(\cdot,0),\varphi)\\
&= (A^{-1}(g_0(\cdot,\varphi) - g_0(\cdot,0) + \nabla\cdot\boldsymbol{g}_1(\cdot,\varphi) - \nabla\cdot\boldsymbol{g}_1(\cdot,0)),\varphi)\\
&\quad +(A^{-1}(g_0(\cdot,0) + \nabla\cdot\boldsymbol{g}_1(\cdot,0)),\varphi) - (\alpha(\cdot,0),\varphi)\\
&= (A^{-1/2}(g_0(\cdot,\varphi) - g_0(\cdot,0) + \nabla\cdot\boldsymbol{g}_1(\cdot,\varphi) - \nabla\cdot\boldsymbol{g}_1(\cdot,0)), A^{-1/2}\varphi)\\
&\quad +(A^{-1/2}(g_0(\cdot,0) + \nabla\cdot\boldsymbol{g}_1(\cdot,0)), A^{-1/2}\varphi) - (\alpha(\cdot,0),\varphi)\\
&\le \beta(1+d)\|\varphi\|_H\|\varphi\|_{V'} + \|A^{-1/2}(g_0(\cdot,0) + \nabla\cdot\boldsymbol{g}_1(\cdot,0))\|_H\|\varphi\|_{V'}\\
&\quad +\|\alpha(\cdot,0)\|_H\|\varphi\|_H\\
&\le \frac{\omega}{4}\|\varphi\|_H^2 + \frac{\beta^2(1+d)^2}{\omega}\|\varphi\|_{V'}^2 + \frac{1}{2}\|A^{-1/2}(g_0(\cdot,0) + \nabla\cdot\boldsymbol{g}_1(\cdot,0))\|_H^2\\
&\quad +\frac{1}{2}\|\varphi\|_{V'}^2 + \frac{1}{\omega}\|\alpha(\cdot,0)\|_H^2 + \frac{\omega}{4}\|\varphi\|_H^2.
\end{aligned}$$

Hence, there exist constants $C_0, C_1 > 0$ such that

$$\frac{d}{d\tau}\|\varphi\|_{V'}^2 + \omega\|\varphi\|_H^2 \quad\le\quad C_1\|\varphi\|_{V'}^2 + C_0\big(\|g_0(\cdot,0)\|_H^2 + \sum_{j=1}^d \|g_{1j}(\cdot,0)\|_H^2 + \|\alpha(\cdot,0)\|_H^2\big).$$

Solving the differential inequality $y'(\tau) \le C_1 y(\tau) + r(\tau)$, where $y(\tau) = \|\varphi(\cdot,\tau)\|_{V'}^2$ and $r(\tau) = C_0\big(\|g_0(\cdot,\tau,0)\|_H^2 + \sum_{j=1}^d \|g_{1j}(\cdot,\tau,0)\|_H^2 + \|\alpha(\cdot,\tau,0)\|_H^2\big)$, we obtain

$$y(\tau) \le e^{C_1 T}\big(y(0) + \int_0^T r(s)ds\big),$$

and the proof of the theorem follows. □

5. Applications to Option Pricing

The classical linear Black–Scholes model and its multidimensional generalizations have been widely used in the financial market analysis. It is well known that the price $V = V(t, S)$ of an option on an underlying asset price S at time $t \in [0, T]$ can be obtained as a solution to the linear Black–Scholes parabolic equation of the form (1). Generally, the underlying asset price is assumed to follow the geometric Brownian motion $dS/S = \mu dt + \sigma dW$. Here, $\{W_t, t \geq 0\}$ is the standard Wiener process. The terminal condition $\Phi(S)$ represents the payoff diagram at maturity $t = T$, $\Phi(S) = (S - K)^+$ (call option case) or $\Phi(S) = (K - S)^+$ (put option case).

For the multidimensional case, where the option price $V(t, S_1, \cdots, S_n)$ depends on the vector of n underlying stochastic assets $S = (S_1, \cdots, S_n)$ with the volatilities σ_i and mutual correlations $\varrho_{ij}, i, j = 1, \cdots, n$, the Black–Scholes pricing equation can be expressed as follows:

$$\frac{\partial V}{\partial t} + \frac{1}{2}\sum_{i=1}^{n}\sum_{j=1}^{n} \rho_{ij}\sigma_i\sigma_j S_i S_j \frac{\partial^2 V}{\partial S_i \partial S_j} + r\sum_{i=1}^{n} S_i \frac{\partial V}{\partial S_i} - rV = 0, \quad V(T, S) = \Phi(S). \tag{36}$$

Equations (1) and (36) can be transformed into equation (21) defined on the whole space $\mathbb{R}^n$ (cf., Ševčovič, Stehlíková, Mikula [44, Chapter 4, Section 5]).

According to stock market observations, the models (1) and (36) were derived under some restrictive assumptions, e.g., completeness and frictionless of the financial market, perfect replication of a portfolio and its liquidity, and absence of transaction costs. However, these assumptions are often violated in financial markets. In recent decades, several attempts have been made to investigate the effects of nontrivial transaction costs [7, 6, 45, 46]. For example, Schönbucher and Willmott [8], Frey and Patie [9], Frey and Stremme [10] investigated the feedback and illiquid market effects due to large traders choosing given stock-trading strategies. Jandačka and Ševčovič recently investigated the effects of risk arising from an unprotected portfolio. Barles and Soner [47] analyzed option pricing models based on utility maximization. The common feature of these generalizations of the linear Black–Scholes equation (1) is that the constant volatility σ is replaced by a nonlinear function depending on the second derivative $\partial_S^2 V$ of the option price V. Among these generalizations, Frey and Stremme [11] derived a nonlinear Black–Scholes model by assuming that the underlying asset dynamics takes into account the presence of feedback effects due to the influence of a large trader choosing a particular stock trading strategy (see also [8, 9, 12]).

Recently, Cruz and Ševčovič [1] generalized the Black–Scholes equation in two important directions. First, they used the ideas of Frey and Stremme [11] to incorporate the effect of a large trader into the model. Second, they relaxed the assumption on liquidity of market by assuming that the underlying asset price follows a Lévy stochastic process with jumps to obtain the following nonlinear PIDE:

$$
\begin{aligned}
0 &= \frac{\partial V}{\partial t} + \frac{1}{2}\frac{\sigma^2 S^2}{(1-\varrho S\partial_S\phi)^2}\frac{\partial^2 V}{\partial S^2} + rS\frac{\partial V}{\partial S} - rV \\
&+ \int_{\mathbb{R}} \left(V(t, S+H) - V(t,S) - H\frac{\partial V}{\partial S} \right) \nu(\mathrm{d}z),
\end{aligned} \tag{37}
$$

where the shift function $H = H(\phi, S, z)$ depends on the large investor stock-trading strategy function $\phi = \phi(t, S)$. Moreover, this shift function is a solution to the following implicit algebraic equation:

$$H = \rho S(\phi(t, S+H) - \phi(t,S)) + S(e^z - 1). \tag{38}$$

The large trader strategy function ϕ may depend on the derivative $\partial_S V$ of the option price V, e.g., $\phi(t,S) = \partial_S V(t,S)$. However, in our application, we assume the trading strategy function $\phi(t,S)$ is prescribed and globally Hölder continuous. Next, we present the analysis of this equation depending the behavior of the parameter $\rho = 0$.

If $\rho = 0$, then $H = S(e^z - 1)$. Thus, the equation (37) can be reduced to a linear PIDE of the form (21) in the one-dimensional space ($n = 1$). This is obtained using the standard transformation $\tau = T - t, x = \ln(\frac{S}{K})$ and setting $V(t,S) = e^{-r\tau}u(\tau, x)$.

However, if $\rho > 0$, then (37) can be transformed into a nonlinear parabolic PIDE. Indeed, suppose that the transformed large trader stock-trading strategy $\psi(\tau, x) = \phi(t, S)$. Then, $V(t,S)$ solves equation (37) if and only if the transformed function $u(\tau, x)$ is a solution to the following nonlinear parabolic equation:

$$
\begin{aligned}
&\frac{\partial u}{\partial \tau} = \frac{\sigma^2}{2}\frac{1}{(1-\rho\partial_x\psi)^2}\frac{\partial^2 u}{\partial^2 x} + \left(r - \frac{\sigma^2}{2}\frac{1}{(1-\rho\partial_x\psi)^2} - \delta(\tau,x) \right)\frac{\partial u}{\partial x} \\
&+ \int_{\mathbb{R}} \left(u(\tau, x+\xi) - u(\tau,x) - \xi\frac{\partial u}{\partial x}(\tau,x) \right)\nu(\mathrm{d}z), \quad u(0,x) = \Phi(Ke^x)
\end{aligned} \tag{39}
$$

$\tau \in [0, T], x \in \mathbb{R}$. The shift function $\xi(\tau, x, z)$ is a solution to the following algebraic equation:

$$e^{\xi} = e^{z} + \rho(\psi(\tau, x+\xi) - \psi(\tau, x)), \tag{40}$$

and $\delta(\tau, x) = \int_{\mathbb{R}}(e^{\xi} - 1 - \xi)\nu(\mathrm{d}z) = \int_{\mathbb{R}}(e^{z} - 1 - \xi + \rho(\psi(\tau, x+\xi) - \psi(\tau, x)))\nu(\mathrm{d}z)$.

For small values of $0 < \rho \ll 1$, we can construct the first order asymptotic expansion $\xi(\tau, x, z) = \xi_0(\tau, x, z) + \rho\xi_1(\tau, x, z)$. For $\rho = 0$, we obtain $\xi_0(\tau, x, z) = z$. Hence,

$$e^{z+\rho\xi_1} = e^{z} + \rho(\psi(\tau, x+z+\rho\xi_1) - \psi(\tau, x)).$$

Taking the first derivative of the above implicit equation with respect to ρ and evaluating it at the origin $\rho = 0$, we obtain $\xi_1 = e^{-z}(\psi(\tau, x+z) - \psi(\tau, x))$, i.e.,

$$\xi(\tau, x, z) = z + \rho e^{-z}(\psi(\tau, x+z) - \psi(\tau, x)). \tag{41}$$

Consequently, we obtain the following lemma.

Lemma 6. *[3, Lemma 1] Assume that the stock-trading strategy $\phi = \phi(t, S)$ is a globally ω-Hölder continuous function, $0 < \omega \leq 1$. Then, the transformed function $\psi(\tau, x) = \phi(t, S)$ is ω-Hölder continuous, and the first order asymptotic expansion $\xi(\tau, x, z)$ of the nonlinear algebraic equation (40) is ω-Hölder continuous in all variables. Furthermore, there exists a constant $C_0 > 0$ such that $\sup_{\tau,x} |\xi(\tau, x, z)| \leq C_0|z|^{\omega}(1 + e^{|z|})$ for any $z \in \mathbb{R}$.*

5.1. Linearization of PIDE

In what follows, we consider a simplified linear approximation of (37) by setting $\rho = 0$ in the diffusion function, but we keep the shift function H depending on the parameter ρ. Then, the transformed Cauchy problem for the solution u with the first order approximation of the shift function ξ is given as follows:

$$\begin{aligned}\frac{\partial u}{\partial \tau} &= \frac{\sigma^2}{2}\frac{\partial^2 u}{\partial^2 x} + \left(r - \frac{\sigma^2}{2} + \delta(\tau, x)\right)\frac{\partial u}{\partial x} \\ &\quad + \int_{\mathbb{R}} \left(u(\tau, x+\xi) - u(\tau, x) - \xi\frac{\partial u}{\partial x}(\tau, x)\right)\nu(\mathrm{d}z), \end{aligned} \tag{42}$$

$\tau \in [0, T], x \in \mathbb{R}$, where $\xi(\tau, x, z) = z + \rho(\psi(\tau, x+z)) - \psi(\tau, x))$.

Note that the call/put option payoff functions $\Phi(S) = \Phi(Ke^{x}) = (S-K)^{+} = K(e^{x}-1)^{+}$ / $\Phi(S) = \Phi(Ke^{x}) = (K-S)^{+} = K(1-e^{x})^{+}$ do not belong to the Banach space X^{γ}. According to [2], the procedure on how to overcome this problem and formulate the existence and uniqueness of a solution to the PIDE (42) is based on shifting the solution u by u^{BS}. Here, $u^{BS}(\tau, x) = e^{r\tau}V^{BS}(T-\tau, Ke^{x})$

is an explicitly given solution to the linear Black–Scholes equation without the PIDE part. In other words, u^{BS} solves the following linear parabolic equation:

$$\frac{\partial u^{BS}}{\partial \tau} - \frac{\sigma^2}{2}\frac{\partial^2 u^{BS}}{\partial x^2} - \left(r - \frac{\sigma^2}{2}\right)\frac{\partial u^{BS}}{\partial x} = 0, \quad u^{BS}(0,x) = \Phi(Ke^x),\ \tau \in (0,T), x \in \mathbb{R}. \tag{43}$$

Recall that $u^{BS}(\tau,x) = Ke^{x+r\tau}N(d_1) - KN(d_2)$ (call option case), where $d_{1,2} = (x+(r\pm\sigma^2/2)\tau)/(\sigma\sqrt{\tau})$ (cf., [6, 44]). Here, $N(d) = \frac{1}{\sqrt{2\pi}}\int_{-\infty}^{d} e^{-\xi^2/2}\mathrm{d}\xi$ is the cumulative density function of the normal distribution. The next result and its proof are based on the recent paper by Ševčovič and Udeani [3].

Theorem 7. *[3, Theorem 2] Assume the transformed stock-trading strategy function $\psi(\tau,x)$ is globally ω-Hölder continuous in both variables. Suppose that ν is a Lévy measure with the shape parameters $\alpha < 3, D \in \mathbb{R}$, where $\mu > 0$, or $\mu = 0$ and $D > 1$. Let X^γ be the space of Bessel potentials space $\mathscr{L}^p_{2\gamma}(\mathbb{R})$, where $\frac{\alpha-1}{2\omega} < \gamma < \frac{p+1}{2p}$ and $\frac{1}{2} \le \gamma < 1$. Let $T > 0$. Then, the linear PIDE (42) has a unique mild solution u with the property that the difference $U = u - u^{BS}$ belongs to the space $C([0,T],X^\gamma)$.*

Proof. We first outline the idea of the proof. The initial condition $u(0,\cdot) \notin X^\gamma$ for two reasons. It is not smooth for $x = 0$ and grows exponentially for $x \to \infty$ (call option) or $x \to -\infty$ (put option). The shift function $U = u - u^{BS}$ satisfies $U(0,\cdot) \equiv 0$, and so the initial condition $U(0,\cdot)$ belongs to X^γ. However, the shift function u^{BS} enters the governing PIDE as it includes the term $f(u^{BS}(\tau,\cdot))$ in the right-hand side. Since $u^{BS}(0,x)$ is not sufficiently smooth for $x = 0$, the shift term $f(u^{BS}(\tau,\cdot))$ is singular for $\tau \to 0^+$. Following the ideas of [2], for the shift term $f(u^{BS}(\tau,\cdot))$, we can provide Hölder estimates, which are sufficient to prove the main result of this theorem (cf., [2, Lemma 4.1]). Furthermore, the exponential growth of the function u^{BS} will be overcome since $\tilde{f}(e^x) = 0$, where $\tilde{f}(u) = f(u) - \delta(\tau,\cdot)\partial_x u$, i.e.,

$$\tilde{f}(u)(x) = \int_{\mathbb{R}} \left(u(x+\xi) - u(x) - (e^\xi - 1)\partial_x u(x)\right)\nu(\mathrm{d}z).$$

Next, we present more details of the proof. The function u^{BS} solves the linear PDE (43). Thus, the difference $U = u - u^{BS}$ of the solution u to (42) and u^{BS} satisfies the PIDE with the right-hand side:

$$\begin{aligned}\frac{\partial U}{\partial \tau} &= \frac{\sigma^2}{2}\frac{\partial^2 U}{\partial x^2} + \left(r - \frac{\sigma^2}{2} - \delta(\tau,x)\right)\frac{\partial U}{\partial x} + f(U) + f(u^{BS}) - \delta(\tau,x)\frac{\partial u^{BS}}{\partial x}\\ &= \frac{\sigma^2}{2}\frac{\partial^2 U}{\partial x^2} + f(U) + g(\tau,x,\partial_x U) + h(\tau,\cdot),\end{aligned}$$

$U(0,x) = 0,\ x \in \mathbb{R}, \tau \in (0,T)$. Here $g(\tau, x, \partial_x U) = (r - \sigma^2/2 - \delta(\tau, x))\partial_x U$, and $h(\tau, \cdot) = \tilde{f}(u^{BS}(\tau, \cdot))$. According to Proposition 2, $f : X^\gamma \to X$ is a bounded linear mapping. Consequently, it is Lipschitz continuous, provided that $1/2 \le \gamma < 1$ and $\gamma > (\alpha - 1)/(2\omega)$. Clearly, $\tilde{f}(e^x) = 0$. Hence,

$$\tilde{f}(u^{BS}) = \tilde{f}(u^{BS} - Ke^{r\tau + x}), \quad \text{and } \partial_\tau \tilde{f}(u^{BS}) = \tilde{f}(\partial_\tau(u^{BS} - Ke^{r\tau+x})).$$

Now, it follows from [2, Lemma 4.1] that the following estimate holds true:

$$\|h(\tau_1, \cdot) - h(\tau_2, \cdot)\|_{L^p} = \|\tilde{f}(u^{BS}(\tau_1, \cdot)) - \tilde{f}(u^{BS}(\tau_2, \cdot))\|_{L^p} \le C_0 |\tau_1 - \tau_2|^{-\gamma + \frac{p+1}{2p}},$$

$$\|h(\tau, \cdot)\|_{L^p} = \|\tilde{f}(u^{BS})(\tau, \cdot))\|_{L^p} \le C_0 |\tau^{-(2\gamma-1)(\frac{1}{2} - \frac{1}{2p})},$$

for any $0 < \tau_1, \tau_2, \tau \le T$. The function $h : [0,T] \to X \equiv L^p(\mathbb{R})$ is $((p+1)/(2p) - \gamma)$-Hölder continuous because $\gamma < \frac{p+1}{2p}$. Moreover,

$$\int_0^T \|h(\tau, \cdot)\|_{L^p} d\tau = \int_0^T \|\tilde{f}(u^{BS}(\tau, \cdot))\|_{L^p} d\tau \le C_0 \int_0^T \tau^{-(2\gamma-1)(\frac{1}{2} - \frac{1}{2p})} d\tau < \infty,$$

because $(2\gamma - 1)\left(\frac{1}{2} - \frac{1}{2p}\right) < 1$. Recall that the crucial part of the proof of [2, Lemma 4.1] was based on the estimates:

$$\|\tilde{f}(u^{BS}(\tau, \cdot))\|_{L^p} \le C_0 \|v(\tau, \cdot)\|_{X^{\gamma - 1/2}}, \text{and } \|\partial_\tau \tilde{f}(u^{BS}(\tau, \cdot))\|_{L^p} \le C_0 \|\partial_\tau v(\tau, \cdot)\|_{X^{\gamma-1/2}},$$

where $v(\tau, x) = \partial_x \left(u^{BS}(\tau, x) - Ke^{r\tau + x}\right) = Ke^{r\tau + x}(N(d_1(\tau, x)) - 1)$. This estimate is satisfied because of Proposition 2 under the assumptions made on γ. The proof for the case of a put option is similar. The final estimate on the Hölder continuity of the mapping h follows from careful estimates of the solution u^{BS} derived in the proof of [2, Lemma 4.1]. Now, the proof follows from Theorem 2 and Proposition 3. □

6. Feedback Effects under Jump-Diffusion Asset Price Dynamics

Suppose that a large trader uses a stock-holding strategy α_t and S_t is a cad-lag process (right continuous with limits to the left). In what follows, we will identify S_t with S_{t-}. We assume S_t has the following dynamics:

$$\mathrm{d}S_t = \mu S_t \mathrm{d}t + \sigma S_t \mathrm{d}W_t + \rho S_t \mathrm{d}\alpha_t + \int_{\mathbb{R}} S_t(e^x - 1) J_X(\mathrm{d}t, \mathrm{d}x), \quad (44)$$

which can seen as a perturbation of the classical jump-diffusion model. For instance, if a large trader does not trade, then $\alpha_t = 0$ or the market liquidity parameter ρ is set to zero, then the stock price S_t follows the classical jump-diffusion model.

We will assume the following structural hypothesis in this chapter:

Assumption 1. *[1, Assumption 1] Assume the trading strategy $\alpha_t = \phi(t, S_t)$ and the parameter $\rho \geq 0$ satisfy $\rho L < 1$, where $L = \sup_{S>0} |S \frac{\partial \phi}{\partial S}|$.*

Next, we show an explicit formula for the dynamics of S_t satisfying (44) under certain regularity assumptions made on the stock-holding function $\phi(t, S)$.

Proposition 4. *Suppose that the stock-holding strategy $\alpha_t = \phi(t, S_t)$ satisfies Assumption 1, where $\phi \in C^{1,2}([0, T] \times \mathbb{R}^+)$. If the process $S_t, t \geq 0$, satisfies the implicit stochastic equation (44), then the process S_t is driven by the following stochastic differential equation (SDE):*

$$\mathrm{d}S_t = b(t, S_t) S_t \mathrm{d}t + v(t, S_t) S_t \mathrm{d}W_t + \int_{\mathbb{R}} H(t, x, S_t) J_X(\mathrm{d}t, \mathrm{d}x), \tag{45}$$

where

$$b(t, S) = \frac{1}{1 - \rho S \frac{\partial \phi}{\partial S}(t, S)} \left(\mu + \rho \left(\frac{\partial \phi}{\partial t} + \frac{1}{2} v(t, S)^2 S^2 \frac{\partial^2 \phi}{\partial S^2} \right) \right), \tag{46}$$

$$v(t, S) = \frac{\sigma}{1 - \rho S \frac{\partial \phi}{\partial S}(t, S)}, \tag{47}$$

$$H(t, x, S) = S(e^x - 1) + \rho S \left[\phi(t, S + H(t, x, S)) - \phi(t, S) \right]. \tag{48}$$

Proof. First, the SDE (45) can be expressed for S_t as follows:

$$\begin{aligned} \mathrm{d}S_t &= \left(b(t, S_t) S_t + \int_{|x|<1} H(t, x, S_t) \nu(\mathrm{d}x) \right) \mathrm{d}t + v(t, S_t) S_t \mathrm{d}W_t \\ &\quad + \int_{|x|\geq 1} H(t, x, S_t) J_X(\mathrm{d}t, \mathrm{d}x) + \int_{|x|<1} H(t, x, S_t) \tilde{J}_X(\mathrm{d}t, \mathrm{d}x). \end{aligned}$$

Since the function $\phi(t, S)$ is smooth, by applying Itô formula (12) to the process $\phi(t, S_t)$, we obtain

$$\begin{aligned} \mathrm{d}\alpha_t &= \left(\tfrac{\partial \phi}{\partial t} + \tfrac{1}{2} v(t, S_t)^2 S_t^2 \tfrac{\partial^2 \phi}{\partial S^2} \right) \mathrm{d}t + \tfrac{\partial \phi}{\partial S} \mathrm{d}S_t \\ &+ \textstyle\int_{\mathbb{R}} \phi(t, S_t + H(t, x, S_t)) - \phi(t, S_t) - H(t, x, S_t) \tfrac{\partial \phi}{\partial S}(t, S_t) J_X(\mathrm{d}t, \mathrm{d}x). \end{aligned} \tag{49}$$

Substituting the differential $\mathrm{d}\alpha_t$ into (44), we obtain

$$\begin{aligned}\mathrm{d}S_t &= \mu S_t \mathrm{d}t + \sigma S_t \mathrm{d}W_t + \textstyle\int_{\mathbb{R}} S_t(e^x - 1) J_X(\mathrm{d}t, \mathrm{d}x) + \rho S_t \frac{\partial \phi}{\partial S} \mathrm{d}S_t \\ &\quad + \rho S_t \left(\frac{\partial \phi}{\partial t} + \frac{1}{2} v(t, S_t)^2 S_t^2 \frac{\partial^2 \phi}{\partial S^2} \right) \mathrm{d}t \qquad (50) \\ &+ \rho S_t \textstyle\int_{\mathbb{R}} \phi(t, S_t + H(t, x, S_t)) - \phi(t, S_t) - H(t, x, S_t) \frac{\partial \phi}{\partial S}(t, S_t) J_X(\mathrm{d}t, \mathrm{d}x).\end{aligned}$$

Next, rearrange terms in (50) to obtain

$$\begin{aligned}(1 - \rho S_t \frac{\partial \phi}{\partial S}(t, S_t))\mathrm{d}S_t &= (\mu S_t + \rho S_t(\frac{\partial \phi}{\partial t} + \frac{1}{2} v(t, S_t)^2 S_t^2 \frac{\partial^2 \phi}{\partial S^2}))\mathrm{d}t \\ &+ \sigma S_t \mathrm{d}W_t - \rho S_t \int_{\mathbb{R}} H(t, x, S_t) \frac{\partial \phi}{\partial S}(t, S_t) J_X(\mathrm{d}t, \mathrm{d}x) \qquad (51) \\ &+ \int_{\mathbb{R}} S_t(e^x - 1) + \rho S_t \left(\phi(t, S_t + H(t, x, S_t)) - \phi(t, S_t) \right) J_X(\mathrm{d}t, \mathrm{d}x).\end{aligned}$$

By comparing terms in (45) and (51), we obtain the expressions (46), (47), and the implicit equation for the function H:

$$\begin{aligned}H(t, x, S) &= \frac{1}{1 - \rho S \frac{\partial \phi}{\partial S}(t, S)} \left(S(e^x - 1) + \rho S \left(\phi(t, S + H(t, x, S)) - \phi(t, S) \right) \right) \\ &\quad - \frac{1}{1 - \rho S \frac{\partial \phi}{\partial S}(t, S)} \rho S \frac{\partial \phi}{\partial S}(t, S) H(t, x, S). \qquad (52)\end{aligned}$$

Simplifying this expression for H, we establish (48), as claimed. □

Since the function H is implicitly given by equation (48), we can expand its solution in terms of a small parameter ρ as follows:

$$H(t, x, S) = H^0(t, x, S) + \rho H^1(t, x, S) + O(\rho^2) \ as \ \rho \to 0,$$

to deduce the following proposition:

Proposition 5. *Assume ρ is sufficiently small. Then, the first-order approximation of the function $H(t, x, S)$ is given as follows:*

$$H(t, x, S) = S(e^x - 1) + \rho S \left(\phi(t, Se^x) - \phi(t, S) \right) + O(\rho^2) \quad as \ \rho \to 0. \quad (53)$$

The next proposition and its proof is based on the recent paper by Cruz and Ševčovič [1].

Proposition 6. *[1, Proposition 3.4]*

Assume that the asset price process $S_t = e^{X_t + rt}$ *satisfies SDE* (45)*, where the Lévy measure* ν *is such that* $\int_{|x|\geq 1} e^{2x}\nu(\mathrm{d}x) < \infty$*. Let the price of a derivative security* $V(t,S)$ *be define by*

$$V(t,S) = \mathbb{E}\left[e^{-r(T-t)}\Phi(S_T)|S_t = S\right] = e^{-r(T-t)}\mathbb{E}\left[\Phi(Se^{r(T-t)+X_{T-t}})\right]. \quad (54)$$

Suppose that the payoff function Φ *is Lipschitz continuous and the function* ϕ *has a bounded derivative. Then,* $V(t,S)$ *is a solution to the PIDE of the form:*

$$\begin{aligned} 0 &= \tfrac{\partial V}{\partial t} + \tfrac{1}{2}v(t,S)^2 S^2 \tfrac{\partial^2 V}{\partial S^2} + rS\tfrac{\partial V}{\partial S} - rV \\ &+ \textstyle\int_{\mathbb{R}} V(t, S + H(t,x,S)) - V(t,S) - H(t,x,S)\tfrac{\partial V}{\partial S}(t,S)\nu(\mathrm{d}x), \end{aligned} \quad (55)$$

where $v(t,S)$ *and* $H(t,x,S)$ *are given by* (47) *and* (48)*, respectively.*

Proof. Recall that the asset price dynamics of S_t under the measure $\mathbb{Q}$ is given by

$$\mathrm{d}S_t = rS_t\mathrm{d}t + v(t,S_t)S_t\mathrm{d}W_t + \int_{\mathbb{R}} H(t,x,S_t)\tilde{J}_X(\mathrm{d}t,\mathrm{d}x). \quad (56)$$

Applying Itô's lemma to $V(t,S_t)$, we obtain $\mathrm{d}(V(t,S_t)e^{-rt}) = a(t)\mathrm{d}t + \mathrm{d}M_t$, where

$$\begin{aligned} a(t) &= \tfrac{\partial V}{\partial t} + \tfrac{1}{2}v(t,S_t)^2 S_t^2 \tfrac{\partial^2 V}{\partial S^2} + rS_t\tfrac{\partial V}{\partial S} - rV \\ &+ \textstyle\int_{\mathbb{R}} V(t, S_t + H(t,x,S_t)) - V(t,S_t) - H(t,x,S_t)\tfrac{\partial V}{\partial S}(t,S_t)\nu(\mathrm{d}x), \end{aligned}$$

and

$$\mathrm{d}M_t = e^{-rt}S_t v(t,S_t)\frac{\partial V}{\partial S}\mathrm{d}W_t + e^{-rt}\int_{\mathbb{R}} V(t,S_t + H(t,x,S_t)) - V(t,S_t)\tilde{J}_X(\mathrm{d}t,\mathrm{d}x).$$

Our aim is to show that M_t is a martingale. Consequently, we have $a \equiv 0$ a.s., and V is a solution to (55) ([37, Proposition 8.9]). To show that the term $\int_0^T e^{-rt}\int_{\mathbb{R}} V(t,S_t + H(t,x,S_t)) - V(t,S_t)\tilde{J}_S(\mathrm{d}t,\mathrm{d}y)$ is a martingale, it suffices to show that

$$\mathbb{E}\left[\int_0^T e^{-2rt}\left(\int_{\mathbb{R}} V(t,S_t + H(t,x,S_t)) - V(t,S_t)\nu(\mathrm{d}x)\right)^2 \mathrm{d}t\right] < \infty. \quad (57)$$

Since $\sup_{0\leq t\leq T}\mathbb{E}\left[e^{X_{T-t}}\right] < \infty$, and the payoff function Φ is Lipschitz continuous, $V(t,S)$ is Lipschitz continuous with the Lipschitz constant $C > 0$. Furthermore, since the function $\phi(t,S)$ has bounded derivatives, we have

$$S\left|\phi(t,S + H(t,x,S)) - \phi(t,S)\right| \leq S\left|\frac{\partial\phi}{\partial S}\right| |H(t,x,S)| \leq L|H(t,x,S)|.$$

(see Assumption 1). Since $H(t,x,S) = S(e^x - 1) + \rho S(\phi(t, S + H(t,x,S)) - \phi(t,S))$, we obtain $|H(t,x,S)|^2 \leq S^2(e^x-1)^2/(1-\rho L)^2$. Again, since V is Lipschitz continuous with the Lipschitz constant $C > 0$, we have

$$\mathbb{E}\left[\int_0^T e^{-2rt}\left(\int_{\mathbb{R}} V(t, S_t + H(t,x,S_t)) - V(t,S_t)\nu(\mathrm{d}x)\right)^2 \mathrm{d}t\right]$$
$$\leq \frac{C^2}{(1-\rho L)^2}\mathbb{E}\left[\int_0^T \int_{\mathbb{R}} e^{-2rt}|S_t|^2(e^x-1)^2\nu(\mathrm{d}x)\mathrm{d}t\right] < \infty,$$

since $\sup_{t\in[0,T]}\mathbb{E}\left[S_t^2\right] < \infty$. Here, $C_0 = \int_{\mathbb{R}}(e^x-1)^2\nu(\mathrm{d}x) < \infty$ due to the assumptions made on the measure ν. Next, we show that $\int_0^T e^{-rt}S_t v(t,S_t)\frac{\partial V}{\partial S}(t,S_t)\mathrm{d}W_t$ is a martingale. Since $S\frac{\partial \phi}{\partial S}(t,S)$ is bounded, we have

$$0 < v(t,S) = \frac{\sigma}{1-\rho S\frac{\partial\phi}{\partial S}(t,S)} \leq \frac{\sigma}{1-\rho L} \equiv C_1 < \infty.$$

Therefore, $\mathbb{E}[\int_0^T e^{-2rt}(\frac{\partial V}{\partial S}(t,S_t)v(t,S_t)S_t)^2\mathrm{d}t] \leq C^2C_1^2\int_0^T e^{-2rt}\mathbb{E}[S_t^2]\mathrm{d}t < \infty$ because S_t is a martingale. Thus, M_t is also a martingale, and $a \equiv 0$. Hence, V is a solution to PIDE (55), as claimed. □

Remark 3. *If $\rho = 0$, then $H(t,x,S) = S(e^x - 1)$. Moreover, equation (55) becomes*

$$\frac{\partial V}{\partial t} + \frac{\sigma^2}{2}S^2\frac{\partial^2 V}{\partial S^2} + rS\frac{\partial V}{\partial S} - rV + \int_{\mathbb{R}} V(t,Se^x) - V(t,S) - S(e^x-1)\frac{\partial V}{\partial S}(t,S)\nu(\mathrm{d}x) = 0, \tag{58}$$

which is the well-known classical PIDE. If there are no jumps ($\nu = 0$) and a trader follows the delta hedging strategy, i.e., $\phi(t,S) = \partial_S V(t,S)$, then equation (55) reduces to the Frey–Stremme option pricing model:

$$\frac{\partial V}{\partial t} + \frac{1}{2}\frac{\sigma^2}{\left(1-\varrho S\partial_S^2 V\right)^2}S^2\frac{\partial^2 V}{\partial S^2} + rS\frac{\partial V}{\partial S} - rV = 0 \tag{59}$$

(cf. [12]). Finally, if $\rho = 0$ and $\nu = 0$, equation (55) reduces to the classical linear Black–Scholes equation.

For simplicity, we assume that the interest rate is zero, i.e., $r = 0$. Then, the function $V(t,S)$ is a solution to the following PIDE:

$$\begin{aligned} &\tfrac{\partial V}{\partial t} + \tfrac{1}{2}v(t,S)^2S^2\tfrac{\partial^2 V}{\partial S^2} \\ &+ \textstyle\int_{\mathbb{R}} V(t, S + H(t,x,S)) - V(t,S) - H(t,x,S)\frac{\partial V}{\partial S}(t,S)\nu(\mathrm{d}x) = 0. \end{aligned} \tag{60}$$

Next, let the tracking error of a trading strategy $\alpha_t = \phi(t, S_t)$ be $e_T^M := \Phi(S_T) - V_0 = V(T, S_T) - V_0 - \int_0^T \alpha_t \mathrm{d}S_t$. By applying Itô's formula to $V(t, S_t)$ and using (60), we obtain

$$
\begin{aligned}
V(T,S_T) - V_0 &= V(T,S_T) - V(0,S_0) = \textstyle\int_0^T \mathrm{d}V(t,S_t) \\
&= \textstyle\int_0^T \frac{\partial V}{\partial S}\mathrm{d}S_t + \int_0^T \frac{\partial V}{\partial t} + \frac{1}{2}v(t,S_t)^2 S_t^2 \frac{\partial^2 V}{\partial^2 S}\mathrm{d}t \\
&\quad + \textstyle\int_0^T \int_{\mathbb{R}} V(t, S_t + H(t,x,S_t)) - V(t,S_t) - H(t,x,S_t)\frac{\partial V}{\partial S} J_X(\mathrm{d}t, \mathrm{d}x) \\
&= \textstyle\int_0^T \frac{\partial V}{\partial S}\mathrm{d}S_t - \int_0^T \int_{\mathbb{R}} V(t, S_t + H(t,x,S_t)) - V(t,S_t) - H(t,x,S_t)\frac{\partial V}{\partial S}\nu(\mathrm{d}x)\mathrm{d}t \\
&\quad + \textstyle\int_0^T \int_{\mathbb{R}} V(t, S_t + H(t,x,S_t)) - V(t,S_t) - H(t,x,S_t)\frac{\partial V}{\partial S} J_X(\mathrm{d}t, \mathrm{d}x) \\
&= \textstyle\int_0^T \frac{\partial V}{\partial S}\mathrm{d}S_t + \int_0^T \int_{\mathbb{R}} V(t, S_t + H(t,x,S_t)) - V(t,S_t) - H(t,x,S_t)\frac{\partial V}{\partial S}\tilde{J}_X(\mathrm{d}t, \mathrm{d}x).
\end{aligned}
$$

Using expression (56) for the dynamics of the asset price S_t (with $r = 0$), the tracking error e_T^M can be expressed as follows:

$$
\begin{aligned}
e_T^M &= V(T,S_T) - V_0 - \textstyle\int_0^T \alpha_t \mathrm{d}S_t = \int_0^T \left(\frac{\partial V}{\partial S}(t,S_t) - \alpha_t\right)\mathrm{d}S_t \\
&\quad + \textstyle\int_0^T \int_{\mathbb{R}} V(t, S_t + H(t,x,S_t)) - V(t,S_t) - H(t,x,S_t)\frac{\partial V}{\partial S}\tilde{J}_X(\mathrm{d}t, \mathrm{d}x) \\
&= \textstyle\int_0^T v(t,S_t)S_t \left(\frac{\partial V}{\partial S} - \alpha_t\right)\mathrm{d}W_t \\
&\quad + \textstyle\int_0^T \int_{\mathbb{R}} V(t, S_t + H(t,x,S_t)) - V(t,S_t) - \alpha_t H(t,x,S_t)\tilde{J}_X(\mathrm{d}t, \mathrm{d}x).
\end{aligned}
\tag{61}
$$

Remark 4. *For the delta hedging strategy $\alpha_t = \phi(t, S_t) = \frac{\partial V}{\partial S}(t, S_t)$, the tracking error function e_T^M can be expressed as follows:*

$$
e_T^M = \int_0^T \int_{\mathbb{R}} V(t, S_t + H(t,x,S_t)) - V(t,S_t) - H(t,x,S_t)\frac{\partial V}{\partial S}(t,S_t)\tilde{J}_X(\mathrm{d}t, \mathrm{d}x).
$$

It is obvious that the tracking error for the delta hedging strategy need not be zero for $\nu \not\equiv 0$.

The next proposition presents a criterion that can be used to find the optimal hedging strategy. The proposition and its proof are based on a recent paper by Cruz and Ševčovič [1].

Proposition 7. *[1, Proposition 3.5] The trading strategy $\alpha_t = \phi(t, S_t)$ of a large trader minimizing the variance $\mathbb{E}\left[(\epsilon_T^M)^2\right]$ of the tracking error is given by the implicit equation:*

$$
\begin{aligned}
\phi(t,S_t) &= \beta^\rho(t,S_t)\big[v(t,S_t)^2 S_t^2 \tfrac{\partial V}{\partial S}(t,S_t) \\
&\quad + \textstyle\int_{\mathbb{R}} (V(t, S_t + H(t,x,S_t)) - V(t,S_t))H(t,x,S_t)\,\nu(\mathrm{d}x)\big],
\end{aligned}
\tag{62}
$$

where $\beta^\rho(t, S_t) = 1/[v(t, S_t)^2 S_t^2 + \int_{\mathbb{R}} H(t, x, S_t)^2 \nu(\mathrm{d}x)]$ *and* $H(t, x, S) = S(e^x - 1) + \rho S[\phi(t, S + H(t, x, S)) - \phi(t, S)]$.

Proof. Using equation (61) for the tracking error ϵ_T^M and Itô's isometry, we have

$$\mathbb{E}\left[(\epsilon_T^M)^2\right] = \mathbb{E}\left[\int_0^T v(t, S_t)^2 S_t^2 \left(\tfrac{\partial V}{\partial S}(t, S_t) - \alpha_t\right)^2 \mathrm{d}t\right]$$
$$+\mathbb{E}\left[\int_0^T \int_{\mathbb{R}} \left(V(t, S_t + H(t, x, S_t)) - V(t, S_t) - \alpha_t H(t, x, S_t)\right)^2 \nu(\mathrm{d}x)\mathrm{d}t\right].$$

The minimizer α_t of the above convex quadratic minimization problem satisfies the first-order necessary conditions $\mathrm{d}(\mathbb{E}\left[\epsilon_T^2\right], \alpha_t) = 0$, that is,

$$0 = -2\mathbb{E}\left[\int_0^T \left(v(t, S_t)^2 S_t^2 \left(\tfrac{\partial V}{\partial S}(t, S_t) - \alpha_t\right)\right.\right.$$
$$\left.\left. + \int_{\mathbb{R}} H(t, x, S_t)\big(V(t, S_t + H(t, x, S_t)) - V(t, S_t) - \alpha_t H(t, x, S_t)\big)\nu(\mathrm{d}x)\right)\omega_t \mathrm{d}t\right]$$

for any variation ω_t. Thus, the tracking error minimizing strategy α_t is given by (62). □

Remark 5. *It is worth noting that the optimal trading strategy minimizing the variance of the tracking error need not satisfy the structural Assumption 1. For instance, if* $\nu = 0$*, then the tracking error minimizer is the delta hedging strategy* $\phi = \partial_S V$*. For a call or put option, its gamma, i.e.,* $\partial_S^2 V(t, S)$*, becomes infinite as* $t \to T$ *and* $S = K$*. However, given a level* $L > 0$*, we can minimize the tracking error* $\mathbb{E}\left[\epsilon_T^2\right]$ *under the additional constraint* $\sup_{S>0} |S \frac{\partial \phi}{\partial S}(t, S)| \le L$*. In other words, we can solve the following convex constrained nonlinear optimization problem*

$$\min_{\phi} \ \mathbb{E}\left[\epsilon_T^2\right] \quad s.t. \ |S\partial_S \phi| \le L$$

instead of the unconstrained minimization problem proposed in Proposition 7.

Remark 6. *Note that if* $\nu = 0$ *and* $\rho \ge 0$*, the trading strategy* α_t *reduces to the Black–Scholes delta hedging strategy, i.e.,* $\alpha_t = \frac{\partial V}{\partial S}(t, S_t)$*. Meanwhile, if* $\nu \not\equiv 0$ *and* $\rho = 0$*, then the optimal trading strategy becomes* $\alpha_t = \phi^0(t, S_t)$*, where*

$$\phi^0(t, S_t) = \beta^0(t, S_t)\left(\sigma^2 S_t^2 \frac{\partial V}{\partial S}(t, S_t) + \int_{\mathbb{R}} S_t(e^x - 1)\left(V(t, S_t e^x) - V(t, S_t)\right)\nu(\mathrm{d}x)\right),$$

where $\beta^0(t, S_t) = 1/[\sigma^2 S_t^2 + \int_{\mathbb{R}} S_t^2 (e^x - 1)^2 \nu(\mathrm{d}x)]$.

We conclude this section with the following proposition providing the first-order approximation of the tracking error minimizing trading strategy for small values of the parameter $\rho \ll 1$. In what follows, we derive the first-order approximation of $\phi^\rho(t, S_t)$ as $\phi^\rho(t, S_t) = \phi^0(t, S_t) + \rho\phi^1(t, S_t) + O(\rho^2)$ as $\rho \to 0$.

Clearly, the first-order Taylor expansion of the volatility function $v(t, S)$ is given by

$$v(t,S)^2 = \frac{\sigma^2}{(1-\rho S\partial_S\phi)^2} = \sigma^2 + 2\rho\sigma^2 S\frac{\partial\phi^0}{\partial S}(t,S) + O(\rho^2), \text{ as } \rho \to 0.$$

According to Proposition 5 (see (53)), we have $H(t,x,S) = H^0(t,x,S) + \rho H^1(t,x,S) + O(\rho^2)$, where

$$H^0(t,x,S) = S(e^x - 1), \qquad H^1(t,x,S) = S[\phi^0(t,Se^x) - \phi^0(t,S)]. \tag{63}$$

The function β^ρ can be expanded as follows: $\beta^\rho(t,S) = \beta^0(t,S) + \rho\beta^{(1)}(t,S) + O(\rho^2)$,

$$\beta^0(t,S) = 1/[\sigma^2 S^2 + S^2 \textstyle\int_{\mathbb{R}} (e^x - 1)^2 \nu(\mathrm{d}x)], \tag{64}$$
$$\beta^{(1)}(t,S) = -(\beta^0(t,S))^2 \left[2\sigma^2 S^3 \tfrac{\partial\phi^0}{\partial S}(t,S) + 2S^2 \textstyle\int_{\mathbb{R}} (e^x - 1)[\phi^0(t,Se^x) - \phi^0(t,S)]\nu(\mathrm{d}x)\right].$$

Using the first-order expansions of the functions v^2, β^ρ, and H, we obtain the following results.

Proposition 8. *[1, Proposition 3.6] For small values of the parameter $\rho \ll 1$, the tracking error variance minimizing strategy $\alpha_t = \phi^\rho(t, S_t)$ is given by*

$$\phi^\rho(t,S_t) = \phi^0(t,S_t) + \rho\phi^{(1)}(t,S_t) + O(\rho^2), \text{ as } \rho \to 0, \tag{65}$$

where

$$\phi^{(1)}(t,S) = \beta^0(t,S)\left[2\sigma^2 S^3 \tfrac{\partial V}{\partial S}(t,S)\tfrac{\partial\phi^0}{\partial S}(t,S)\right.$$
$$\left. + \textstyle\int_{\mathbb{R}} \left(V(t,Se^x) - V(t,S) + \tfrac{\partial V}{\partial S}(t,Se^x)H^0(t,x,S)\right) H^1(t,x,S)\nu(\mathrm{d}x)\right]$$
$$+\beta^{(1)}(t,S)\left[\sigma^2 S^2 \tfrac{\partial V}{\partial S}(t,S) + \textstyle\int_{\mathbb{R}} (V(t,Se^x) - V(t,S)) H^0(t,x,S)\nu(\mathrm{d}x)\right],$$

and the functions H^0, H^1, β^0, and $\beta^{(1)}$ are defined as in (63) *and* (64).

6.1. Numerical Simulation for the Underlying PIDE

In this section, we illustrate the behavior of the solutions to the linear PIDE with various Lévy measures. Specifically, we consider European put options, i.e., $\Phi(S) = (K - S)^+$. The goal of the numerical simulation is to compare the solution to the linear Black–Scholes equation with solutions to the Merton and variance gamma PIDE models. The common model parameters were chosen as follows: $\sigma = 0.23, K = 100, T = 1$, and $r \in \{0, 0.1\}$. For the underlying Lévy process, we consider the variance gamma process with parameters $\theta = -0.43, \kappa = 0.27$ and the Merton processes with parameters $\lambda = 0.1$,

$m = -0.2$, and $\delta = 0.15$. First, we employ the finite difference discretization scheme proposed and analyzed by Cruz and Ševčovič [1] to calculate the numerical solution to the equation. Their scheme is based on a uniform spatial finite difference discretization with a spatial step $\Delta x = 0.01$, and implicit time discretization with a step $\Delta t = 0.005$. Then, we set the total number of spatial discretization steps as $N = 400$ and the number of time discretization steps as $M = 200$. We restricted the spatial computational domain to $x \in [-L, L]$, where $L = 4$. For more details on the discretization scheme, see the recent paper by Cruz and Ševčovič [1].

Figure 1 shows the comparison of European put option prices between PIDE and linear Black–Scholes models. Figures 1 (a) and (b) show the graphs of the put option prices $V(0, S)$ for $S \in [80, 125]$ for the interest rates $r = 0$ and $r = 0.1$, respectively. Table 1 summarizes the numerical values of option prices for two different values of the interest rate $r = 0.1$ and $r = 0$. The option price for the Merton and variance gamma models are higher than those of the classical Black-Scholes model. This is based on the idea that prices of put or call options should be higher on underlying assets following stochastic processes with jumps than those following a continuous geometric Brownian motion.

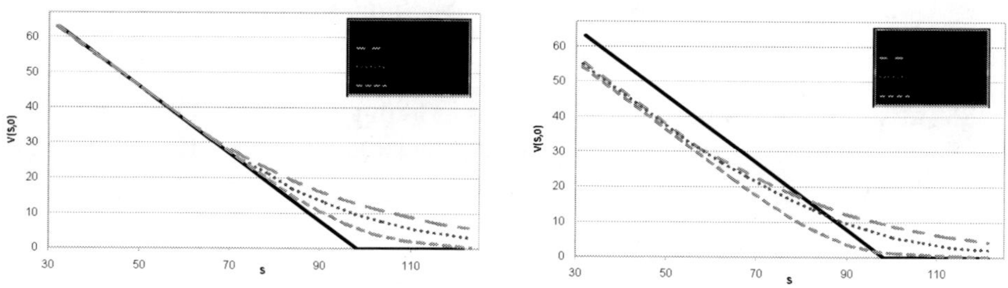

a) b)

Figure 1. Comparison of European put option prices for the Black–Scholes model and PIDE models under variance gamma (VarGamma) and Merton PIDE processes. Source: own computations based on the paper [1] by J. Cruz and D. Ševčovič.

Table 1. European put option prices $V(0,S)$ for the Black-Scholes and PIDE models under Variance Gamma and Merton processes for $r=0$ and $r=0.1$. Source: J. Cruz and D. Ševčovič, [1, 5].

	BS		PIDE-VG		PIDE-Merton		Payoff
S	$r=0$	$r=0.1$	$r=0$	$r=0.1$	$r=0$	$r=0.1$	
85.2144	15.2547	7.35166	19.2687	14.9855	17.1692	12.9056	14.7856
88.692	12.2484	5.24145	17.2948	13.3899	14.8335	10.9901	11.308
92.3116	9.42895	3.51944	15.428	11.8822	12.6423	9.21922	7.68837
96.0789	6.90902	2.21106	13.674	10.4691	10.6201	7.61307	3.92106
100.	4.78444	1.29196	12.0372	9.15576	8.78655	6.18483	0.
104.081	3.1099	0.69843	10.52	7.94499	7.155	4.94044	0.
108.329	1.88555	0.34773	9.12343	6.83762	5.73137	3.87864	0.
112.75	1.0604	0.15881	7.84623	4.51403	5.83246	2.99166	0.

7. Hamilton-Jacobi-Bellman Equation

In this section, we present the motivation for studying the fully nonlinear parabolic equation (25), which can be viewed as a parabolic PIDE in some sense. We also present the relationship between the nonlinear parabolic equation and the transformed equation using the so-called Ricatti transform.

The motivation for studying the nonlinear parabolic equation of the form (25) arises from dynamic stochastic programming for $d=1$. The fully nonlinear HJB equation describing optimal portfolio selection strategy is represented by the following nonlinear parabolic equation:

$$\partial_t V + \max_{\boldsymbol{\theta}\in\triangle}\left(\mu(x,t,\boldsymbol{\theta})\,\partial_x V + \frac{1}{2}\sigma(x,t,\boldsymbol{\theta})^2\,\partial_x^2 V\right) = 0\,, \tag{66}$$

$$V(x,T) = u(x), \tag{67}$$

where $x\in\mathbb{R}, t\in[0,T)$. A solution $V=V(x,t)$ to the parabolic equation (66) is subject to the terminal condition $V(x,T)=u(x)$. According to Kilianová and Ševčovič [48, 49, 50], such HJB equation of the form (66) arises from the dynamic stochastic programming, where the goal of an investor is to maximize the conditional expected value of the terminal utility of the portfolio:

$$\max_{\boldsymbol{\theta}|_{[0,T)}} \mathbb{E}\left[u(x_T^{\boldsymbol{\theta}})\,\middle|\,x_0^{\boldsymbol{\theta}}=x_0\right], \tag{68}$$

on a finite time horizon $[0,T]$. Here, $u:\mathbb{R}\to\mathbb{R}$ is an increasing terminal utility function, and x_0 is a given initial state condition of the process $\{x_t^{\boldsymbol{\theta}}\}$ at $t=0$. The underlying stochastic process $\{x_t^{\boldsymbol{\theta}}\}$ with a drift $\mu(x,t,\boldsymbol{\theta})$ and volatility $\sigma(x,t,\boldsymbol{\theta})$ is assumed to satisfy the following Itô's SDE:

$$dx_t^{\boldsymbol{\theta}} = \mu(x_t^{\boldsymbol{\theta}},t,\boldsymbol{\theta}_t)dt + \sigma(x_t^{\boldsymbol{\theta}},t,\boldsymbol{\theta}_t)dW_t\,, \tag{69}$$

where the control process $\{\boldsymbol{\theta}_t\}$ is adapted to the process $\{x_t\}$, and $\{W_t\}$ is the standard one-dimensional Wiener process. The control parameter $\boldsymbol{\theta}$ belongs to a given compact subset $\triangle$ in $\mathbb{R}^n$. An example of such a subset is the compact convex simplex $\triangle \equiv \mathcal{S}^n = \{\boldsymbol{\theta} \in \mathbb{R}^n \mid \boldsymbol{\theta} \geq \mathbf{0}, \mathbf{1}^T\boldsymbol{\theta} = 1\} \subset \mathbb{R}^n$, where $\mathbf{1} = (1, \cdots, 1)^T \in \mathbb{R}^n$.

Consider the value function

$$V(x,t) := \sup_{\boldsymbol{\theta}|_{[t,T)}} \mathbb{E}\left[u(x_T^{\boldsymbol{\theta}})|x_t^{\boldsymbol{\theta}} = x\right]. \tag{70}$$

Then, according to Bertsekas [51], the value function $V = V(x,t)$ solves the fully nonlinear HJB parabolic equation (66) and $V(x,T) := u(x)$.

Several attempts have been made for solving the HJB equation (66). Macová and Ševčovič [52] analyzed solutions to a fully nonlinear parabolic equation that models the problem of optimal portfolio construction. Consequently, they formulated the problem of optimal stock to bond proportion in the management of a pension fund portfolio could be formulated in terms of the solutions to the HJB equation. Federicol et al. [53] investigated the utility maximization problem for an investment-consumption portfolio when the current utility depends on the wealth process and the regularity of solutions to the HJB equation. They defined a dual problem and treated it by means of dynamic programming to show that the viscosity solutions of the associated HJB equation belong to a class of smooth function. Recently, Ishimura and Ševčovič [54] constructed and analyzed solutions to the class of HJB equation (66) with range bounds on the optimal response variable. They constructed monotone traveling wave solutions and identified parametric regions for which the traveling wave solutions have a positive or negative wave speed. Abe and Ishimura [55] employed the Riccati transformation method for solving the full nonlinear HJB equations, which was later studied and generalized by Kilianová and Ševčovič [49]. They investigated solutions of a fully nonlinear HJB equation for a constrained dynamic stochastic optimal allocation problem. However, no attempt has been made in solving the fully nonlinear HJB parabolic equation arising from portfolio optimization in high-dimensional space using the monotone operator technique. The monotone operator method is essential because it not only gives constructive proof of existence theorems, but also leads to various comparison results, which are effective tools for studying qualitative properties of solutions. In this chapter, we consider the case when the utility function u increases, as a consequence, $\partial_x V(x,\tau) > 0$.

7.1. Static Markowitz Model for Portfolio Optimization

This subsection presents the motivation for studying such nonlinear parabolic equation (66). It describes the mathematical formulation of the classical Markowitz model for portfolio optimization. In static portfolio optimization, this model aims to maximize the mean return of the set of stochastic returns $X^i, i = 1, \ldots, n$, under the constraint that variance of the portfolio is bounded by a given constant $\sigma_0^2/2$. Given a vector $\boldsymbol{\theta} = (\theta^1, \ldots, \theta^n)^T$ of weights, we construct a portfolio $X = \sum_{i=1}^n \theta^i X^i$. Let $\boldsymbol{\mu} \in \mathbb{R}^n, \mu^i = \mathbb{E}(X^i)$ be the vector of mean returns of stochastic asset returns and $\boldsymbol{\Sigma}$ be their covariance matrix, $\boldsymbol{\Sigma}_{ij} = cov(X^i X^j)$, then $\mathbb{E}(X) = \boldsymbol{\mu}^T\boldsymbol{\theta}$, and the variance $\mathbb{D}(X) = \boldsymbol{\theta}^T\boldsymbol{\Sigma}\boldsymbol{\theta}$. The Markowitz optimal portfolio optimization problem can be formulated as the following convex optimization problem:

$$
\begin{aligned}
&\max_{\boldsymbol{\theta}\in\mathbb{R}^n} \boldsymbol{\mu}^T\boldsymbol{\theta} && \text{– maximize the mean return,}\\
&s.t.\ \frac{1}{2}\boldsymbol{\theta}^T\boldsymbol{\Sigma}\boldsymbol{\theta} \le \frac{1}{2}\sigma_0^2 && \text{– the variance is prescribed,}\\
&\sum_{i=1}^n \theta^i = 1 && \text{– weights sum up to 100\%,}\\
&\boldsymbol{\theta} \ge 0 && \text{– no short positions allowed.}
\end{aligned}
$$

The corresponding Lagrange function for the minimization of $-\boldsymbol{\mu}^T\boldsymbol{\theta}$ has the following form:

$$
\mathcal{L}(\boldsymbol{\theta}, \varphi, \lambda, \xi) = -\boldsymbol{\mu}^T\boldsymbol{\theta} + \varphi\frac{1}{2}\boldsymbol{\theta}^T\boldsymbol{\Sigma}\boldsymbol{\theta} + \lambda 1^T\boldsymbol{\theta} + \xi^T\boldsymbol{\theta}, \tag{71}
$$

where $\varphi \in \mathbb{R}, \lambda \in \mathbb{R}, \xi \in \mathbb{R}^n$, and $\xi \ge 0$ are Lagrange multipliers. It is worth noting that the same Lagrange function corresponds to the minimization problem:

$$
\alpha(\varphi) := \min_{\boldsymbol{\theta}\in\triangle} -\boldsymbol{\mu}^T\boldsymbol{\theta} + \varphi\frac{1}{2}\boldsymbol{\theta}^T\boldsymbol{\Sigma}\boldsymbol{\theta}
$$

$$
\text{where } \triangle \equiv \{\boldsymbol{\theta} \in \mathbb{R}^n, \ \sum_{i=1}^n \theta^i = 1, \ \ \boldsymbol{\theta} \ge 0\},
$$

provided the Lagrange multiplier $\varphi > 0$ is known and fixed. Figure 2 shows the optimal asset selection for the German DAX30 stock index for various values of $\varphi > 0$. The optimal value is denoted by $\alpha(\varphi)$. The value of the Lagrange multiplier φ can be viewed as a measure of investor's risk-aversion (see Fig. 2). Therefore, the higher the value of risk aversion, the more portfolio is diversified among less risky assets with smaller mean returns.

7.2. Riccati Transformation of the HJB Equation and Application to Optimal Portfolio Selection Problem

7.2.1. Riccati Transformation

This section presents how the HJB equation (66) can be transformed into a quasilinear PDE using the so-called Ricatti transformation techniques. Such

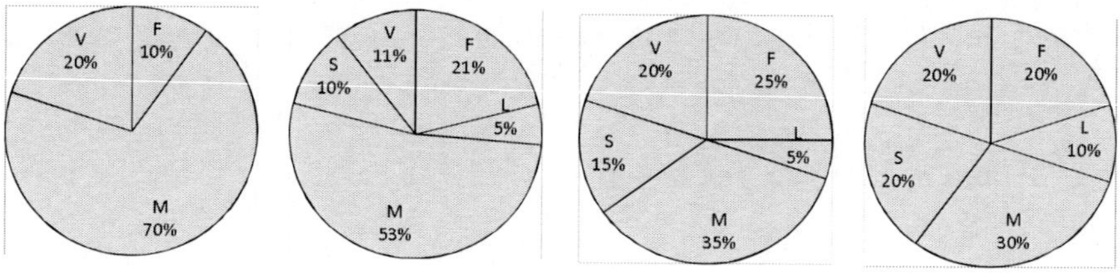

$\varphi \approx 1$ $\varphi \approx 4$ $\varphi \approx 6$ $\varphi \approx 8$

Figure 2. Optimal asset selection for German DAX30 stock index for various values of $\varphi > 0$. Composition of stocks: Fressenius (F), Linde (L), Merck (M), SAP (S), Volkswagen (V).

transformed parabolic equation corresponds to the Cauchy problem (25), which is obtained after some perturbation in the main operator.

The Riccati transformation φ of the value function V can be defined based on the approach introduced by Abe and Ishimura [55], Ishimura and Ševčovič [54], Ševčovič and Macová [52], and Kilianová and Ševčovič [49] as follows:

$$\varphi(x,\tau) = -\frac{\partial_x^2 V(x,t)}{\partial_x V(x,t)}, \quad \text{where } \tau = T - t. \tag{72}$$

Suppose the value function $V(x,t)$ is increasing in the x-variable. In other words, assume that the terminal utility function $u(x)$ is increasing. Then, the HJB equation (66) can be rewritten as follows:

$$\partial_t V - \alpha(\cdot,\varphi)\partial_x V = 0, \qquad V(\cdot,T) = u(\cdot), \tag{73}$$

where $\alpha(x,\tau,\varphi)$ is the value function of the following parametric optimization problem:

$$\alpha(x,\tau,\varphi) = \min_{\boldsymbol{\theta}\in\triangle} \left(-\mu(x,t,\boldsymbol{\theta}) + \frac{\varphi}{2}\sigma(x,t,\boldsymbol{\theta})^2 \right), \quad \tau = T - t. \tag{74}$$

Remark 7. *It is worth noting that the optimization problem (74) is related to the classical Markowitz model on optimal portfolio selection problem formulated as maximization of the mean return* $\mu(\boldsymbol{\theta}) \equiv \boldsymbol{\mu}^T\boldsymbol{\theta}$ *under the volatility constraint* $\frac{1}{2}\sigma(\boldsymbol{\theta})^2 \equiv \frac{1}{2}\boldsymbol{\theta}^T\boldsymbol{\Sigma}\boldsymbol{\theta} \le \frac{1}{2}\sigma_0^2$, *i.e.,*

$$\max_{\boldsymbol{\theta}\in\triangle} \boldsymbol{\mu}^T\boldsymbol{\theta}, \quad s.t.\ \frac{1}{2}\boldsymbol{\theta}^T\boldsymbol{\Sigma}\boldsymbol{\theta} \le \frac{1}{2}\sigma_0^2,$$

where the decision set is the simplex $\triangle = \{\boldsymbol{\theta} \in \mathbb{R}^n \mid \boldsymbol{\theta} \ge \mathbf{0}, \mathbf{1}^T\boldsymbol{\theta} = 1\}$. *The Lagrange multiplier for the volatility constraint can be viewed as the parameter* φ *that enters the parametric optimization problem (74).*

Next, let $\partial_x \alpha$ be the total differential of the function $\alpha(x, \tau, \varphi)$, where $\varphi = \varphi(x, \tau)$, i.e.,

$$\partial_x \alpha(x, \tau, \varphi) = \alpha'_x(x, \tau, \varphi) + \alpha'_\varphi(x, \tau, \varphi)\, \partial_x \varphi.$$

Here, α'_x and α'_φ are partial derivatives of α with respect to x- and φ- variables, respectively.

Kilianová and Ševčovič [48, Theorem 4.2] recently established the relationship between the transformed function φ and the value function V. They reported that an increasing value function $V(x, t)$ in the x-variable is a solution to the HJB equation (66) if and only if the transformed function $\varphi(x, \tau) = -\partial_x^2 V(x, t)/\partial_x V(x, t)$, $t = T - \tau$, is a solution to the Cauchy problem for the quasilinear parabolic PDE:

$$\partial_\tau \varphi - \partial_x^2 \alpha(\cdot, \varphi) = -\partial_x \left(\alpha(\cdot, \varphi)\varphi\right), \tag{75}$$

$$\varphi(x, 0) = \varphi_0(x) \equiv -u''(x)/u'(x), \quad (x, \tau) \in \mathbb{R} \times (0, T). \tag{76}$$

Note that the Cauchy problem for the quasilinear parabolic PDE (75) is equivalent to the nonlinear parabolic equation (25) in one-dimensional space. This is obtainable after some shift/perturbation in the main operator of the transformed equation (75). This nonlinear parabolic equation (25) presented in an abstract setting can be viewed as nonlinear PIDE in some sense.

7.2.2. Properties of the Value Function as a Diffusion Function

In this subsection, we present the qualitative properties of the value function and sufficient conditions on the decision set $\triangle$ and functions μ and σ that guarantee higher smoothness of the value function α. Let $C^{k,1}(\mathcal{D})$ be the space consisting of all k-differentiable functions defined on the domain $\mathcal{D} \subset \mathbb{R}^{d+2}$, whose k-th derivative is globally Lipschitz continuous. The next proposition shows (under certain assumptions) that the value function α belongs to $C^{0,1}(\mathcal{D})$, where $\mathcal{D} = \mathbb{R}^d \times (0, T) \times (\varphi_{min}, \infty)$. The following proposition and its proof are contained in our recent paper ([4, Theorem 3]).

Proposition 9. *[4, Proposition 1] Let $\triangle \subset \mathbb{R}^n$ be a given compact decision set. Assume that the functions $\mu(x, t, \boldsymbol{\theta})$ and $\sigma(x, t, \boldsymbol{\theta})^2$ are globally Lipschitz continuous in $x \in \mathbb{R}^d, t \in [0, T]$ and $\boldsymbol{\theta} \in \triangle$ variables and there exist positive constants $\omega, L > 0$ such that $\omega \le \frac{1}{2}\sigma(x, t, \boldsymbol{\theta})^2 \le L$ for any $x \in \mathbb{R}^d, t \in [0, T]$, and $\boldsymbol{\theta} \in \triangle$.*

Then, $\alpha \in C^{0,1}(\mathcal{D})$. Moreover, the function α is strictly increasing, and

$$0 < \omega \le \frac{\alpha(x, \tau, \varphi_2) - \alpha(x, \tau, \varphi_1)}{\varphi_2 - \varphi_1} \le L, \quad \textit{for any } (x, \tau, \varphi_i) \in \mathcal{D}, \tag{77}$$

i.e., $\omega \le \alpha'_\varphi(x, \tau, \varphi) \le L$, and

$$|\nabla_x \alpha(x, \tau, \varphi)| \le p(x, \tau) + L_0 |\varphi|, \tag{78}$$

for a.e., $(x, \tau, \varphi) \in \mathcal{D}$*, where* $p(x,\tau) := \max_{\boldsymbol{\theta}\in\triangle} |\nabla_x \mu(x,t,\boldsymbol{\theta})|$ *and* $L_0 := \max_{\boldsymbol{\theta}\in\triangle, t\in[0,T], x\in\mathbb{R}^d} |\nabla_x \sigma^2(x,t,\theta)|$*, where* $t = T - \tau$*.*

Proof. First, let $\alpha^{\boldsymbol{\theta}}(x,\tau,\varphi) := -\mu(x,t,\boldsymbol{\theta}) + \frac{\varphi}{2}\sigma(x,t,\boldsymbol{\theta})^2$, where $t = T - \tau$. Then,

$$\alpha(x,\tau,\varphi) = \min_{\boldsymbol{\theta}\in\triangle} \alpha^{\boldsymbol{\theta}}(x,\tau,\varphi).$$

For any given $\boldsymbol{\theta} \in \triangle$, the function $\alpha^{\boldsymbol{\theta}}(x,\tau,\varphi)$ is globally Lipschitz continuous in all variables. Therefore, the minimal function α is globally Lipschitz continuous. Moreover, the function $\alpha^{\boldsymbol{\theta}}(x,\tau,\varphi)$ satisfies the inequality (77) for any $\boldsymbol{\theta} \in \triangle$, and so does the minimal function α.

Next, we prove the inequality (78). Let $x_1, x_2 \in \mathbb{R}^d$ such that $x_2 = x_1 + he^i$, where $e^i, i = 1, \cdots, d$, is the standard normal vector, i.e., $e^i = (0,0,...,0,1,0,...,0)^T$. Then, we have

$$\begin{aligned}
&\alpha^{\boldsymbol{\theta}}(x_1,\tau,\varphi) - \alpha^{\boldsymbol{\theta}}(x_2,\tau,\varphi) = -(\mu(x_1,\tau,\boldsymbol{\theta}) - \mu(x_2,\tau,\boldsymbol{\theta})) + \frac{\varphi}{2}(\sigma(x_1,\tau,\boldsymbol{\theta})^2 - \sigma(x_2,\tau,\boldsymbol{\theta})^2) \\
&= \int_0^h (-\partial_{x_i}\mu(x_1 + \xi e^i, \tau, \boldsymbol{\theta}))d\xi + \int_0^h \frac{\varphi}{2}\partial_{x_i}\sigma^2(x_1 + \xi e^i, \tau, \boldsymbol{\theta})d\xi \\
&\le \int_0^h |\partial_{x_i}\mu(x_1 + \xi e^i, \tau, \boldsymbol{\theta}))|d\xi + \int_0^h \frac{|\varphi|}{2}|\partial_{x_i}\sigma^2(x_1 + \xi e^i, \tau, \boldsymbol{\theta})|d\xi \\
&\le \max_{\boldsymbol{\theta}\in\triangle, 0\le\xi\le h} |\partial_{x_i}\mu(x_1 + \xi e^i, \tau, \boldsymbol{\theta})|\, h + \max_{\boldsymbol{\theta}\in\triangle, x\in\mathbb{R}^d} |\partial_{x_i}\sigma^2(x,\tau,\theta)| \frac{|\varphi|}{2} h.
\end{aligned}$$

Hence,

$$\alpha^{\boldsymbol{\theta}}(x_1,\tau,\varphi) \le \alpha^{\boldsymbol{\theta}}(x_2,\tau,\varphi) + \max_{\boldsymbol{\theta}\in\triangle, 0\le\xi\le h} |\partial_{x_i}\mu(x_1 + \xi e^i, \tau, \boldsymbol{\theta})|\, h + \max_{\boldsymbol{\theta}\in\triangle, x\in\mathbb{R}^d} |\partial_{x_i}\sigma^2(x,\tau,\theta)||\varphi| h.$$

We note that $x_2 - x_1 = he^i$ so that $|x_2 - x_1| = h$. Taking minimum over $\boldsymbol{\theta} \in \triangle$, we obtain

$$\alpha(x_1,\tau,\varphi) \le \alpha(x_2,\tau,\varphi) + \max_{\boldsymbol{\theta}\in\triangle, 0\le\xi\le h} |\partial_{x_i}\mu(x_1 + \xi e^i, \tau, \boldsymbol{\theta})|\, h + \max_{\boldsymbol{\theta}\in\triangle, x\in\mathbb{R}^d} |\partial_{x_i}\sigma^2(x,\tau,\theta)||\varphi| h.$$

By exchanging the role of x_1 and x_2 and taking the limit as $x_2 \to x_1$, i.e., $h \to 0$, we obtain inequality (78), as stated.

□

According to Proposition 9, the value function α given in (74) satisfies the assumptions of Theorem 4 provided that the functions

$$p(x,\tau) = \max_{\boldsymbol{\theta}\in\triangle} |\nabla_x \mu(x,\tau,\boldsymbol{\theta})|, \quad \text{and } h(x,\tau) = \alpha(x,\tau,0) = -\max_{\boldsymbol{\theta}\in\triangle} \mu(x,\tau,\boldsymbol{\theta})$$

belong to the Banach space $L^\infty((0,T);H)$.

The next result was proved in [50]. It provides sufficient conditions imposed on the decision set $\triangle$ and functions μ and σ that guarantee higher smoothness of the value function α. Its proof is based on the classical envelope theorem due to Milgrom and Segal [56] and the result on the Lipschitz continuity of the minimizer $\hat{\boldsymbol{\theta}} = \hat{\boldsymbol{\theta}}(x, \tau, \varphi)$ belonging to a convex compact set $\triangle$ according to Klatte [57].

Theorem 8. *[50, Theorem 1] Suppose that $\triangle \subset \mathbb{R}^n$ is a compact convex set and the functions $\mu(x, t, \boldsymbol{\theta})$ and $\sigma(x, t, \boldsymbol{\theta})^2$ are $C^{1,1}$ smooth such that the objective function $f(x, t, \varphi, \boldsymbol{\theta}) := -\mu(x, t, \boldsymbol{\theta}) + \frac{\varphi}{2}\sigma(x, t, \boldsymbol{\theta})^2$ is strictly convex in the variable $\boldsymbol{\theta} \in \triangle$ for any $\varphi \in (\varphi_{min}, \infty)$. Then, the function α belongs to the space $C^{1,1}(\mathcal{D})$.*

7.2.3. Point-Wise a-Priori Estimates of Solution with Their Existence and Uniqueness

In this section, we present a-priori estimates on a solution φ to the Cauchy problem (75). We will assume that the function μ is independent of time $t \in [0, T]$, and σ is independent of $x \in \mathbb{R}$ and $t \in [0, T]$, i.e.,

$$\mu = \mu(x, \boldsymbol{\theta}), \qquad \sigma = \sigma(\boldsymbol{\theta}).$$

Then, the value function $\alpha = \alpha(x, \varphi)$ is also independent of the $\tau = T - t$ variable. Next, we prove a-priori estimates for the transformed function $\psi = \psi(x, \tau)$ defined as $\psi(x, \tau) = \alpha(x, \varphi(x, \tau))$. Since the function α is strictly increasing in the φ variable, there exists an inverse function $\beta(x, \psi)$ such that $\alpha(x, \beta(x, \psi)) = \psi$. Therefore, the function $\varphi(x, \tau)$ is a solution to (75) if and only if the function $\psi(x, \tau)$ is a solution to the following linear parabolic PDE:

$$-\partial_\tau \psi + a(x, \tau)\partial_x^2 \psi + b(x, \tau)\partial_x \psi + c(x, \tau)\psi = 0,$$

where

$$a(x, \tau) = \alpha'_\varphi(x, \varphi(x, \tau)), b(x, \tau) = -\alpha'_\varphi(x, \varphi(x, \tau))\varphi(x, \tau) - \alpha(x, \varphi(x, \tau)),$$
$$c(x, \tau) = \alpha'_x(x, \varphi(x, \tau)).$$

Note that $0 < \omega \leq a(x, \tau) \leq L$. Next, suppose that the function $c(x, \tau)$ is bounded from above by a constant $\lambda \geq 0$. Then, the function $\psi_\lambda(x, \tau) = \psi(x, \tau)e^{-\lambda\tau}$ is a solution to the linear PDE:

$$\mathcal{L}[\psi_\lambda] = 0, \qquad \text{where } \mathcal{L}[\psi_\lambda] \equiv \partial_\tau \psi_\lambda - a(x, \tau)\partial_x^2 \psi_\lambda - b(x, \tau)\partial_x \psi_\lambda - c_\lambda(x, \tau)\psi_\lambda,$$

where $c_\lambda(x, \tau) = c(x, \tau) - \lambda$ is nonpositive, i.e., $c_\lambda(x, \tau) \leq 0$ for all x, τ. Let $\underline{\psi} \leq 0$ be a constant. Then, $\mathcal{L}[\psi_\lambda - \underline{\psi}] = c_\lambda(x, \tau)\underline{\psi} \geq 0$. We employ the maximum principle for parabolic equations on unbounded domains according to Meyer and Needham [58, Theorem 3.4] to obtain $\psi_\lambda(x,\tau)-\underline{\psi}\geq 0$ for all x,τ, provided that $\psi_\lambda(x,0)-\underline{\psi}=\psi(x,0)-\underline{\psi}\geq 0$ for all x. In other words, $\underline{\psi}$ is a subsolution. Similarly, if $\overline{\psi}\geq 0$ is a given constant, then $\mathcal{L}[\psi_\lambda-\overline{\psi}]=c_\lambda(x,\tau)\overline{\psi}\leq 0$ and $\psi_\lambda(x,\tau)-\overline{\psi}\leq 0$ for all x,τ, provided that $\psi(x,0)-\overline{\psi}\leq 0$ for all x, i.e., $\overline{\psi}$ is a supersolution. Consequently, we have the following implication:

$$\underline{\psi}\leq\psi(x,0)\leq\overline{\psi}\quad\Longrightarrow\quad\underline{\psi}e^{\lambda\tau}\leq\psi(x,\tau)\leq\overline{\psi}e^{\lambda\tau}\quad\text{for all } x\in\mathbb{R},\tau\in[0,T].$$

In terms of the solution φ to the Cauchy problem (75)–(76), we obtain the following a-priori estimate:

$$\underline{\psi}e^{\lambda\tau}\leq\alpha(x,\varphi(x,\tau))\leq\overline{\psi}e^{\lambda\tau}\quad\text{for all } x\in\mathbb{R},\tau\in[0,T],\tag{79}$$

where

$$\underline{\psi}=\min\{0,\inf_{x\in\mathbb{R}}\alpha(x,\varphi(x,0))\},\qquad\overline{\psi}=\max\{0,\sup_{x\in\mathbb{R}}\alpha(x,\varphi(x,0))\}.\tag{80}$$

Now, we can apply the general Theorem 4 on existence and uniqueness of a solution.

Theorem 9. *[4, Theorem 5] Let the decision set $\triangle\subset\mathbb{R}^n$ be compact and $u:\mathbb{R}\to\mathbb{R}$ be an increasing utility function such that $\varphi_0(x)=-u''(x)/u'(x)$ belongs to the space $L^2(\mathbb{R})\cap L^\infty(\mathbb{R})$. Suppose that the drift $\mu(x,\boldsymbol{\theta})$ and volatility function $\sigma^2(\boldsymbol{\theta})>0$ are C^1 continuous in the x and $\boldsymbol{\theta}$ variables and the value function $\alpha(x,\varphi)$ given in (74) satisfies $p\in L^2(\mathbb{R})\cap L^\infty(\mathbb{R})$, $h\in L^\infty(\mathbb{R})$, and $\partial_x^2 h\in L^2(\mathbb{R})$, where*

$$p(x)=\max_{\boldsymbol{\theta}\in\triangle}|\partial_x\mu(x,\boldsymbol{\theta})|,\quad h(x)=-\max_{\boldsymbol{\theta}\in\triangle}\mu(x,\boldsymbol{\theta}).$$

Then, for any $T>0$, there exists a unique solution φ to the Cauchy problem

$$\partial_\tau\varphi-\partial_x^2\alpha(\cdot,\varphi)=-\partial_x\left(\alpha(\cdot,\varphi)\varphi\right),\quad\varphi(x,0)=\varphi_0(x),\quad(x,\tau)\in\mathbb{R}\times(0,T),\tag{81}$$

satisfying $\varphi\in C([0,T];H)\cap L^2((0,T);V)\cap L^\infty((0,T)\times\mathbb{R})$.

Proof. Since $\sigma^2(\boldsymbol{\theta})>0$ and $\triangle$ is a compact set, there exist constants $0<\omega\leq L$ such that $0<\omega\leq\sigma^2(\boldsymbol{\theta})\leq L$ for all $\boldsymbol{\theta}\in\triangle$. It follows from Proposition 9 that

$$\omega|\varphi|\leq|\alpha(x,\varphi)-\alpha(x,0)|\leq L|\varphi|.\tag{82}$$

Since $\varphi_0,h\in L^\infty(\mathbb{R})$ and $h(x)=\alpha(x,0)$, we obtain $M:=\sup_{x\in\mathbb{R}}|\alpha(x,\varphi_0(x))|<\infty$.

Now, let us define the shift diffusion function by $\tilde{\alpha}(x,\varphi) = \alpha(x,\varphi) - \alpha(x,0)$. Note that $\alpha(x,0) = \min_{\boldsymbol{\theta}\in\triangle} -\mu(x,\boldsymbol{\theta}) = h(x)$. Then, equation (81) is equivalent to

$$\partial_\tau \varphi + A\tilde{\alpha}(\cdot,\varphi) = \tilde{\alpha}(\cdot,\varphi) + \partial_x^2 h - \partial_x\left(\alpha(\cdot,\varphi)\varphi\right),$$

where $A = I - \partial_x^2$.

Next, let $g_0(\varphi) = \tilde{\alpha}(\cdot,\varphi) + \partial_x^2 h$ and $g_1(\varphi) = -w(\alpha(\cdot,\varphi))\varphi$. Here, $w : \mathbb{R} \to \mathbb{R}$ is a suitable cutoff function

$$w(\alpha) = \begin{cases} \underline{\psi} e^{\lambda T}, & \text{if } \alpha \le \underline{\psi} e^{\lambda T}, \\ \alpha, & \text{if } \underline{\psi} e^{\lambda T} < \alpha < \overline{\psi} e^{\lambda T}, \\ \overline{\psi} e^{\lambda T}, & \text{if } \alpha \ge \overline{\psi} e^{\lambda T}, \end{cases}$$

where $\overline{\psi} = M, \underline{\psi} = -M$. Then, the functions $g_0, g_1 : H \to H$ are globally Lipschitz continuous.

Note that the diffusion function $\tilde{\alpha}$ satisfies the assumptions of Theorem 4 with $\tilde{h}(x) = \tilde{\alpha}(x,0) \equiv 0$. Now, applying Theorems 4 and 5, we obtain the existence and uniqueness of a solution $\varphi \in C([0,T];H) \cap L^2((0,T);V)$ to the Cauchy problem (28). The solution φ satisfies the point-wise estimate (79). Thus, $w(\alpha(x,\varphi(x,\tau))) = \alpha(x,\varphi(x,\tau))$, and φ is a solution to the Cauchy problem (81).

Finally, from (82), we deduce the $L^\infty((0,T)\times\mathbb{R})$ estimate for the solution φ since $\sup_{x\in\mathbb{R}} |\alpha(x,\varphi(x,\tau))| \le Me^{\lambda\tau}$, where $\lambda = \sup_{x\in\mathbb{R}} p(x)$. Moreover, $\varphi \in L^\infty((0,T)\times\mathbb{R})$, and

$$\sup_{x\in\mathbb{R},\tau\in[0,T]} |\varphi(x,\tau)| \le \omega^{-1}(Me^{\lambda T} + \max_{x\in\mathbb{R}} |h(x)|).$$

□

7.2.4. Application to Stochastic Dynamic Optimal Portfolio Selection Problem

This subsection presents examples of the stochastic process (69). We also present examples of the corresponding utility function. For the stochastic process (69), we can consider a portfolio optimization problem with regular cash inflow ($\varepsilon > 0$)/outflow ($\varepsilon < 0$) to a portfolio representing pension fund savings. Here, we consider Slovak pension fund savings according to Kilianová and Ševčovič [49]). It is well-known that in a stylized financial market, the stochastic process $\{y_t\}_{t\ge0}$ driven by the stochastic differential equation

$$dy_t = (\varepsilon(y_t) + \boldsymbol{\mu}^T\boldsymbol{\theta} y_t)dt + \sigma(\boldsymbol{\theta}) y_t dW_t, \tag{83}$$

represents a stochastic evolution of the value of a synthesized portfolio y_t consisting of n-assets with weights $\boldsymbol{\theta} = (\theta_1, \cdots, \theta_n)^T$, mean returns $\boldsymbol{\mu} =$

$(\mu_1, \cdots, \mu_n)^T$, and an $n \times n$ positive definite covariance matrix $\mathbf{\Sigma}$, i.e., $\sigma(\boldsymbol{\theta})^2 = \boldsymbol{\theta}^T \mathbf{\Sigma} \boldsymbol{\theta}$.

In this study, we assume that the value of the inflow/outflow rate $\varepsilon = \varepsilon(y)$ also depends on the y such that it is characterized by the following

$$\varepsilon(y) = \begin{cases} 0, & \text{if } 0 < y < y_-, \\ C, & \text{if } y_- < y, \end{cases}$$

where C is a constant. It represents a realistic pension saving model in which there is no inflow/outflow provided that the value y of the portfolio is very small. Based on the logarithmic transformation $x = \ln y$ and Itô's lemma, the stochastic process $\{x_t\}$ satisfies (69), where $\mu(x, \boldsymbol{\theta}) = \boldsymbol{\mu}^T \boldsymbol{\theta} - \frac{1}{2}\sigma(\boldsymbol{\theta})^2 + \varepsilon(e^x)e^{-x}$.

Recently, Kilianová and Trnovská [59] established a further generalization of the drift and volatility functions arising from the so-called worst-case portfolio optimization problem. In such a case, the volatility function is given by

$$\sigma(\boldsymbol{\theta})^2 = \max_{\mathbf{\Sigma} \in \mathcal{K}} \boldsymbol{\theta}^T \mathbf{\Sigma} \boldsymbol{\theta},$$

where $\mathcal{K}$ is a bounded uncertainty convex set of positive definite covariance matrices. In general, only a part of the covariance matrix can be calculated precisely, whereas entries are not precisely determined. For example, if only the diagonal d is known, we have $\mathcal{K} = \{\mathbf{\Sigma} \succ 0,\ diag(\mathbf{\Sigma}) = d\}$. Furthermore, the drift function is given by

$$\mu(x, \boldsymbol{\theta}) = \min_{\boldsymbol{\mu} \in \mathcal{E}} \boldsymbol{\mu}^T \boldsymbol{\theta} - \frac{1}{2}\sigma(\boldsymbol{\theta})^2 + \varepsilon(e^x)e^{-x},$$

where $\mathcal{E}$ is a given bounded uncertainty convex set of mean returns.

Remark 8. *Consider a class of utility functions characterized by a pair of exponential functions:*

$$u(x) = \begin{cases} -e^{-a_0 x} - c^*, & x \le x^*, \\ -(a_0/a_1)e^{-a_1 x + (a_1 - a_0)x^*}, & x > x^*, \end{cases} \tag{84}$$

where $c^* = e^{-a_0 x^*}(a_0 - a_1)/a_1$ *and* $a_0, a_1 \in \mathbb{R} > 0$ *are given constants. Here,* $x^* \in \mathbb{R}$ *is a point at which risk aversion changes. We observe that* u *is an increasing* C^1 *function with a jump in the second derivative at the point* x^*.

If $a_0 > a_1 > 0$, *then the utility function* u *is called the decreasing absolute risk aversion (DARA) function, representing an investor with a non-constant decreasing risk aversion. It means that the higher the wealth of investors, the lower their risk aversion, and hence the higher exposition of the portfolio to more risky assets. According to Post, Fang and Kopa [60], the piece-wise exponential DARA utility function plays an essential role in the analysis of decreasing absolute risk aversion stochastic dominance introduced by Vickson [61] (see also [48]). It is worth noting that the coefficients of absolute risk aversion of the above utility functions* $-u''(x)/u'(x)$ *is equal to* a_0 *if* $x \le x^*$ *or to* a_1 *if* $x > x^*$.

The piece-wise constant function $\varphi_0(x) = -u''(x)/u'(x)$ *should be truncated outside of some interval* $(-\gamma, \gamma)$*, where* γ *is large enough. Then,* $\varphi_0 \in L^2(\mathbb{R}) \cap L^\infty(\mathbb{R})$*. Therefore, the underlying utility function is modified by linear functions for* $x < -\gamma$ *and* $x > \gamma$.

Another simple example of a convex-concave utility function is the function $u(x) = \arctan(x)$*. Then,* $\varphi_0(x) = -u''(x)/u'(x) = 2x/(1+x^2)$*. Clearly,* $\varphi_0 \in H = L^2(\mathbb{R})$*. It is worth noting that the individual's reduction in marginal utility arising from a loss is absolutely greater than the marginal utility from a financial gain. In the domain of gain (*$x > x^*$*), the utility function is concave, indicating that investors show risk aversion in this domain. On the contrary, investors become risk seekers when dealing with losses, i.e., the utility function is convex for* $x \le x^*$.

7.2.5. Numerical Examples

This subsection presents examples of the value function $\alpha = \alpha(\varphi)$ to the parametric optimization problem with different decision sets. First, we consider a simple decision set $\triangle = \{\boldsymbol{\theta} \in \mathbb{R}^2, \boldsymbol{\theta} \ge 0, \mathbf{1}^T\boldsymbol{\theta} = 1\}, n = 2, \mu(\boldsymbol{\theta}) = \boldsymbol{\mu}^T\boldsymbol{\theta}, \sigma^2(\boldsymbol{\theta}) = \boldsymbol{\theta}^T\boldsymbol{\Sigma}\boldsymbol{\theta}$, where $\boldsymbol{\Sigma}$ is a positive definite covariance matrix, and $\boldsymbol{\mu}$ is a positive vector of mean return. The value function $\alpha = \alpha(\varphi)$ can be explicitly expressed as follows:

$$\alpha(\varphi) = \begin{cases} E^-\varphi + D^-, & \text{if } 0 < \varphi \le \varphi_*^-, \\ A - \frac{B}{\varphi} + C\varphi, & \text{if } \varphi_*^- < \varphi < \varphi_*^+, \\ E^+\varphi + D^+, & \text{if } \varphi_*^+ \le \varphi. \end{cases}$$

Here, $(\varphi_*^-, \varphi_*^+)$ is the maximal interval where the optimal value $\hat{\boldsymbol{\theta}}(\varphi) \in \triangle$ of the function $\boldsymbol{\theta} \mapsto -\boldsymbol{\mu}^T\boldsymbol{\theta} + \frac{\varphi}{2}\boldsymbol{\theta}^T\boldsymbol{\Sigma}\boldsymbol{\theta}$ is strictly positive ($\hat{\boldsymbol{\theta}}(\varphi) > 0$) for $\varphi \in (\varphi_*^-, \varphi_*^+)$, and $C, E^\pm > 0$, $B \ge 0$, $A, D^\pm$ are constants explicitly depending on the covariance matrix $\boldsymbol{\Sigma}$ and the vector of mean return $\boldsymbol{\mu}$ such that the function α is C^1 continuous at $\varphi_*^\pm$, i.e., $E^\pm = B/(\varphi_*^\pm)^2 + C$ and $D^\pm = A - B/\varphi_*^\pm + C\varphi^\pm - E^\pm\varphi_*^\pm$. It is clear that α is only $C^{1,1}$ continuous function with two points $\varphi_*^\pm$ of discontinuity of the second derivative α''.

Furthermore, consider a decision set consisting of finite number of points. Then, the value function $\alpha(\varphi)$ corresponding to such a decision set is only piece-wise linear. In other words, if $\hat{\triangle} = \{\boldsymbol{\theta}^1, \cdots, \boldsymbol{\theta}^k\} \subset \{\boldsymbol{\theta} \in \mathbb{R}^2, \boldsymbol{\theta} \ge 0, \mathbf{1}^T\boldsymbol{\theta} = 1\}$, then $\alpha(\varphi) = \min_{i=1,\cdots,k} \alpha^i(\varphi)$, where $\alpha^i(\varphi) = E^i\varphi + D^i$ is a linear function with slope $E^i = (1/2)(\boldsymbol{\theta}^i)^T\boldsymbol{\Sigma}\boldsymbol{\theta}^i > 0$ and intercept $D^i = -\boldsymbol{\mu}^T\boldsymbol{\theta}^i$.

Figure 4 a) shows the graph of the value function α corresponding to the Slovak pension fund system with the two types of decision sets. According to

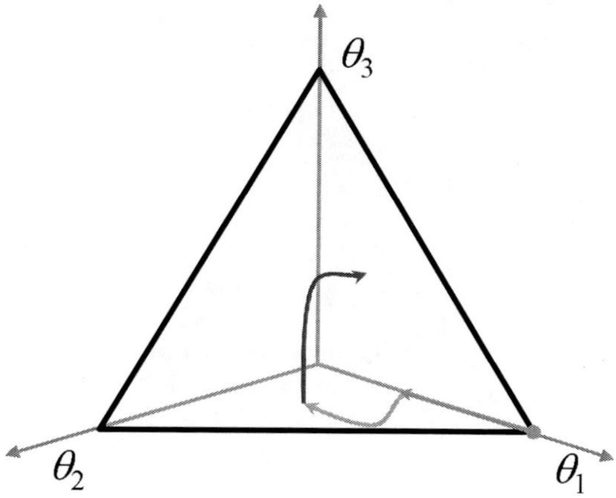

Figure 3. The path $\hat{\boldsymbol{\theta}}(\varphi)$ as a function of φ.

the data set obtained from [62], the portfolio consists of the stock index (with a high mean return $\mu_s = 0.10$ and high volatility $\sigma_s = 0.3$) and bonds (with mean return $\mu_b = 0.03$ and very low volatility $\sigma_s = 0.01$). The returns on stocks index and bonds are negatively correlated $\varrho = -0.15$. $\boldsymbol{\mu} = (\mu_s, \mu_b)^T = (0.1, 0.05)^T$. Then $\boldsymbol{\Sigma}_{11} = \sigma_2^2, \boldsymbol{\Sigma}_{22} = \sigma_b^2, \boldsymbol{\Sigma}_{12} = \boldsymbol{\Sigma}_{21} = \varrho\sigma_s\sigma_b$. As shown in Figure 4 a), the solid blue line corresponds to the convex compact decision set $\triangle = \{\boldsymbol{\theta} \in \mathbb{R}^2, \boldsymbol{\theta} \geq 0, \mathbf{1}^T\boldsymbol{\theta} = 1\}$. The piece-wise linear value function α (red dotted line) corresponds to the discrete decision set $\hat{\triangle} = \{\boldsymbol{\theta}^1, \boldsymbol{\theta}^2, \boldsymbol{\theta}^3\} \subset \triangle$. It represents the Slovak pension fund system consisting of three funds: growth funds with $\boldsymbol{\theta}^1 = (0.8, 0.2)^T$ (80% of stocks and 20% of bonds), balanced funds with $\boldsymbol{\theta}^2 = (0.5, 0.5)^T$ (equal proportion of stocks and bonds), and conservative funds with $\boldsymbol{\theta}^3 = (0, 1)^T$ (only bonds) (cf. [62]). Figure 4 b) shows the graph of the second derivative $\alpha''_\varphi(\varphi)$ of the value function $\alpha(\varphi)$ corresponding to the convex compact decision set $\triangle$. It has the first point of discontinuity φ_*^- close to the value 2. For $n > 2$, the number of discontinuities of α''_φ increases (cf. [49]). Figure 5 shows another example of the value function and its second derivative for the portfolio consisting of five stocks (BASF, Bayer, Degussa-Huls, FMC, and Schering) entering the German DAX30 stock index. The covariance matrix $\boldsymbol{\Sigma}$ is taken from [63]. We set the vector of yields $\boldsymbol{\mu} = (0.03, 0.02, 0.04, 0.01, 0.01)^T$.

The minimizer $\hat{\boldsymbol{\theta}}(\varphi)$ of the convex optimization problem

$$\alpha(\varphi) = \min_{\boldsymbol{\theta}\in\triangle}\{-\boldsymbol{\mu}^T\boldsymbol{\theta} + \frac{\varphi}{2}\boldsymbol{\theta}^T\Sigma\boldsymbol{\theta}\}$$

when considered as a function of the risk aversion parameter φ is only Lipschitz

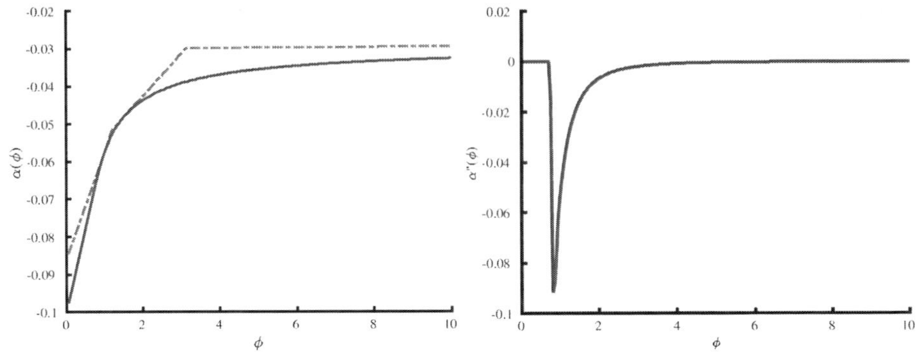

a) b)

Figure 4. a) A graph of the value function α, b) its second derivative $\alpha''(\varphi)$ for the portfolio consisting of the stocks index and bonds (cf. [62]) for the convex compact decision set $\triangle$. The dotted line in a) corresponds to the discrete decision set $\hat{\triangle} = \{\boldsymbol{\theta}^1, \boldsymbol{\theta}^2, \boldsymbol{\theta}^3\} \subset \triangle$. Source: our computation is based on the method from [4].

continuous in φ. According to Millgrom–Segal envelope theorem, the derivative $\alpha'(\varphi)$ is given by $\alpha'(\varphi) = \frac{1}{2}\hat{\boldsymbol{\theta}}(\varphi)^T \Sigma \hat{\boldsymbol{\theta}}(\varphi)$. Figure 3 shows the path $\hat{\boldsymbol{\theta}}(\varphi)$ as a function of φ when increasing φ from $\varphi = 0$ to $\varphi \to \infty$. For small values of φ, only one asset with maximal mean return is active, i.e., $\theta_1 > 0, \theta_2 = \theta_3 = 0$. For intermediate values of φ, two assets are active $\theta_1 > 0, \theta_2 > 0, \theta_3 = 0$. Furthermore, for larger values of φ, all three assets are active, i.e., $\theta_1 > 0, \theta_2 > 0, \theta_3 > 0$. The path $\varphi \mapsto \hat{\boldsymbol{\theta}}(\varphi)$ has a discontinuity in the first derivative when it leaves a lower-dimensional object (vertex, edge) and enters a higher-dimensional object volume.

The advantage of the Riccati transformation of the original HJB is twofold. First, the diffusion function α can be calculated in advance as a result of quadratic optimization problem when the vector $\boldsymbol{\mu}$ and the covariance matrix $\boldsymbol{\Sigma}$ are given or semidefinite programming problem when they belong to a uncertainity set of returns and covariance matrices (cf. [59]). Figure 6 shows the vector of optimal weights $\boldsymbol{\theta}$, as a function of the parameter φ, obtained as the optimal solution to the quadratic optimization problem with the covariance matrix from [63] corresponding to the five assets (BASF, Bayer, Degussa–Huls, FMC, Scheringfrom) entering the DAX30 index from 2008. There are more nontrivial weights θ_i when the parameter φ increases.

In contrast to the fully nonlinear character of the original HJB equation (66), the transformed equation (75) represents a quasilinear parabolic equa-

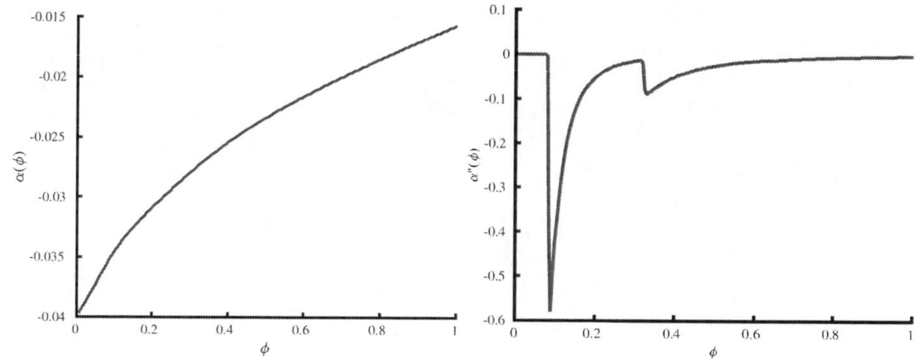

a) b)

Figure 5. a) A graph of the value function $\alpha(\varphi)$, and b) the second derivative $\alpha''_{\varphi}(\varphi)$ corresponding to five stocks (BASF, Bayer, Degussa–Huls, FMC Scheringfrom) entering DAX30 index. Source: our computation is based on the method from [4].

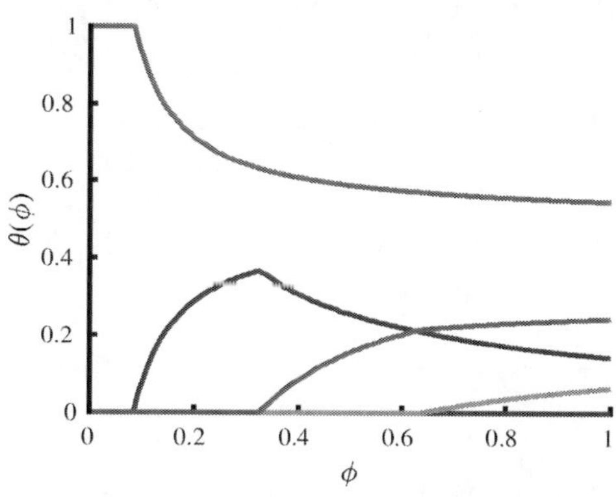

Figure 6. The optimal vector $\boldsymbol{\theta} = (\theta_1, \cdots, \theta_n)^T$ as a function of φ for the German DAX30 index. Source: our computation is based on the method from [4, 49].

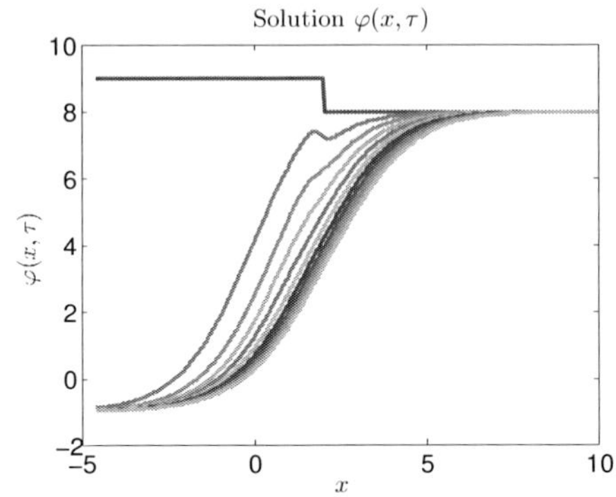

Figure 7. A solution $\varphi(x,\tau)$ for the DARA utility function with $a_0 = 9$, $a_1 = 8$, $x^* = 2$. Source: our computations based on the numerical method from [4].

tion in divergence form. Thus, efficient numerical schemes can be constructed for this class of equations. In our computational experiments, we employ the finite volume discretization scheme proposed and investigated by Kilianová and Ševčovič [48, 49, 50]). Figure 7 shows the results of a time-dependent sequence of profiles $\varphi(x,\tau)$ for a constant initial condition $\varphi_0 \equiv 9$. This figure also shows the solution profiles for the initial condition φ_0 attaining four decreasing values $\{9, 8, 7, 6\}$. It represents the DARA utility function. The function $\varphi(x,\tau)$ is increasing in the x variable and decreasing in the $\tau = T - t$ variable. Therefore, the optimal vector $\boldsymbol{\theta}(x,\tau)$ contains a more diversified portfolio of assets when x increases and the time $t \to T$ (see Figure 6). Furthermore, it is reasonable to invest in an asset with the highest expected return when the account value x is low, whereas an investor has to diversify the portfolio when x is large and time t is approaching the terminal maturity T.

Conclusion

This review chapter presents the analysis of solutions to nonlinear and nonlocal PDEs arising from the financial market. Specifically, we studied and analyzed linear and nonlinear PIDEs arising from option pricing and portfolio selection problem. For option pricing, we investigated the systematic relationships between the corresponding PIDEs and Black–Scholes models that study market illiquidity when the underlying asset price follows a Lévy stochastic process with jumps. We employ the theory of abstract semilinear parabolic equations in Bessel potential spaces to establish the qualitative properties of solutions to nonlocal linear and nonlinear PIDEs for a general class of the so-called admissible Lévy measures satisfying suitable growth conditions at infinity and origin in the multidimensional space. We considered a general shift function arising from nonlinear option pricing models, which takes into account a large trader stock-trading strategy with the underlying asset price following the Lévy process. Then, for the portfolio allocation problem, we presented the qualitative properties results to the fully nonlinear HJB equation arising from the stochastic dynamic optimization problem in Sobolev spaces using the theory of monotone operator technique. Such an HJB equation, presented in abstract setting, arises from portfolio management problem, where the goal of an investor is to maximize the conditional expected terminal utility of a portfolio. We also presented a stable, convergent, and consistent numerical scheme that approximates the solution of such a PIDE. Several numerical simulations were performed to demonstrate the influence of a large trader and the intensity of jumps on the option price.

Acknowledgments

Support of the Slovak Research and Development Agency under the project APVV-20-0311 (C.U.) is kindly acknowledged. The research was also supported by the VEGA 1/0611/21 grant (D.Š.).

References

[1] Cruz, J., Ševčovič, D. 2018. Option Pricing in Illiquid Markets with Jumps. *Applied Mathematical Finance*, (25), 389–409.

[2] Cruz, J., and Ševčovič, D. 2020. On solutions of a partial integro-differential equation in Bessel potential spaces with applications in option pricing models. *Jpn. J. Ind. Appl. Math.*, (37), 697–721.

[3] Ševčovič D, and Udeani C. I. 2021. Multidimensional Linear and Nonlinear Partial Integro-Differential Equation in Bessel Potential Spaces with Applications in Option Pricing. *Mathematics*, 9(13) 1463.

[4] Udeani, C., Ševčovič, D. 2021. Application of maximal monotone operator method for solving Hamilton–Jacobi–Bellman equation arising from optimal portfolio selection problem. *Japan Journal of Industrial and Applied Mathematics, Springer*, (5), 1–21.

[5] Cruz, J. 2021. "*Option Pricing in Iliquid markets with jumps.*" PhD. diss., ISEG, Lisbon School of Economics and Management, University of Lisbon.

[6] Kwok, Y. K. 2008. *Mathematical models of financial derivatives*, second ed.; Springer Finance, Springer, Berlin, xvi+530.

[7] Avellaneda, M., Levy, A., and Paras, A. 1995. Pricing and hedging derivative securities in markets with uncertain volatilities. *Applied Mathematical Finance*, (2), 73–88.

[8] Schönbucher, P.J., and Wilmott, P. 2000. The feedback effect of hedging in illiquid markets. *SIAM J. Appl. Math.*, (61), 232–272.

[9] Frey, R., and Stremme, A. 1997. Market volatility and feedback effects from dynamic hedging. *Math. Finance*, 7, 351–374.

[10] Frey, R., and Patie, P. 2002. Risk management for derivatives in illiquid markets: a simulation study. In *Advances in finance and stochastics*; Springer, Berlin, 137–159.

[11] Jandačka, M., and Ševčovič, D. 2005. On the risk-adjusted pricing-methodology-based valuation of vanilla options and explanation of the volatility smile. *J. Appl. Math.*, 235–258.

[12] Frey, Rüdiger. 1998. Perfect option hedging for a large trader *Finance Stochastic*, 1(2), 305–325.

[13] Frey, Rüdiger. 1998. Market Illiquidity as a Source of Model Risk in Dynamic Hedging *Finance Stochastic*, 1(2), 305–325.

[14] Jarrow, Robert A. 1994. Derivative Securities Markets, Market Manipulation and Option Pricing Theory. *Journal of Finance and Quantitative Analysis*, (29), 241–261.

[15] Platen, E., and Schweizer, M. 1998. On Feedback Effects from Hedging Strategies. *Mathematical Finance*, (8), 67–84.

[16] Schönbucher, Philipp J., and Wilmott, P. 2000. The feedback effect of hedging in illiquid markets *ASIAM J. Appl. Math.*, 1(61), 232–272.

[17] Sircar, K. Ronnie, and Papanicolaou, George. 1998. General Black-Scholes models accounting for increased market volatility from hedging strategies. *Applied Mathematical Finance*. 1(5), 45–82.

[18] Barndorff–Nielsen, O. E., and Levendorskiĭ, S. Z. 2001. Feller processes of normal inverse Gaussian type. *Quant. Finance*, 318–331.

[19] Carr, P., Madan, and Dilip, B. 1999. Option valuation using the fast Fourier transform. *Journal of Computational Finance*, 2(4), 61–73.

[20] Cont, R., and Voltchkova, E. 2005. Integro-differential equations for option prices in exponential Lévy models. *Finance and Stochastics*, 3(9), 299–325.

[21] Eberlein, E., Keller, U., and Prause, K. 1998. New Insights into Smile, Mispricing, and Value at Risk: The hyperbolic Model. *Journal of Business*, 71, 371–405.

[22] Merton, R. C. 1976. Option pricing when underlying stock returns are discontinuous. *J. Financial Economics*, (3), 125–144.

[23] Madan, D. B., Carr, P., and Chang, E. C. 1998. The Variance Gamma Process and Option Pricing. *European Finance Review*, (2), 79–105.

[24] Nualart, D., and Schoutens, W. 2001. Backward stochastic differential equations and Feynman-Kac formula for Lévy processes, with applications in finance. *Bernoulli. Official Journal of the Bernoulli Society for Mathematical Statistics and Probability*, 7(5), 761–776.

[25] Abels, H., and Kassmann, M. 2009. The Cauchy problem and the martingale problem for integro-differential operators with non-smooth kernels. *Osaka J. Math.*, (46), 661–683.

[26] Florescu, I., and Mariani, M. C. 2010. Solutions to integro-differential parabolic problems arising in the pricing of financial options in a Lévy market. *Electronic Journal of Differential Equations (EJDE)[electronic only]*, 1–10.

[27] Aboodh, K., Farah, R., Almardy, I., and Almostafa, F. 2016. Solution of partial integro-differential equations by using Aboodh and double Aboodh transforms methods. *Global Journal of Pure and Applied Mathematics*, (13), 4347–4360.

[28] Arregui, I., Salvador, B., Ševčovič, D., and Vázquéz, C. 2019. Mathematical analysis of a nonlinear PDE model for European options with counterparty risk. *Comptes Rendus Mathematique*, (357), 252–257.

[29] Yüzbaşı, Şuayip. 2016. Improved Bessel collocation method for linear Volterra integro-differential equations with piecewise intervals and application of a Volterra population model. *Applied Mathematical Modelling*, (40), 5349–5363.

[30] Mikulevičius, R., and Pragarauskas, H. 1992. On the Cauchy problem for certain integro-differential operators in Sobolev and Hölder spaces. *Lithuanian Mathematical Journal*, (32), 238–264.

[31] Ishii, H. 1996. Viscosity solutions of nonlinear partial differential equations. *Sugaku Expositions*, 9, 135–152.

[32] Burzoni, M., Ignazio, V., Reppen, A. M., and Soner, H. M. 2020. Viscosity solutions for controlled McKean–Vlasov jump-diffusions. *SIAM Journal on Control and Optimization*, (58), 1676–1699.

[33] Barles, G., Buckdahn, R., and Pardoux, E. 1997. Backward stochastic differential equations and integral-partial differential equations. *Stochastics: An International Journal of Probability and Stochastic Processes*, (60), 57–83.

[34] SenGupta, I., Mariani, M. C., and Amster, P. 2012. Solutions to integro-differential problems arising on pricing options in a Lévy market. *Acta Appl. Math.*, (118), 237–249.

[35] Cont, R., and Voltchkova, E. 2005. Integro-differential equations for option prices in exponential Lévy models. *Finance and Stochastics*, (9), 299–325.

[36] Palatucci, G., Kuusi, T. 2017. *Recent developments in nonlocal theory*, De Gruyter Open.

[37] Cont, R., and Tankov, P. 2004. Financial Modeling with Jump Processes. *Journal of Business*.

[38] Arregui, I., Salvador, B., Ševčovič, D., and Vázquéz, C. 2018. Total value adjustment for European options with two stochastic factors. Mathematical model, analysis and numerical simulation. *Computers and Mathematics with Applications*, (76), 725–740.

[39] Henry, Daniel. 1981. *Geometric theory of semilinear parabolic equations*. Vol. 840, *Lecture Notes in Mathematics*, Springer-Verlag, Berlin-New York, iv+348.

[40] Stein, Elias. M. 1970. Singular integrals and differentiability properties of functions. *Princeton Mathematical Series*, No. 30, Princeton University Press, Princeton, N.J. xiv+290.

[41] Brezis, Häim. 2010. *Functional analysis, Sobolev spaces and partial differential equations*. Springer Science & Business Media.

[42] Barbu, Viorel. 2010. *Nonlinear differential equations of monotone types in Banach spaces*. Springer Science & Business Media.

[43] Showalter, Ralph Edwin. 2013. Monotone operators in Banach space and nonlinear partial differential equations. *American Mathematical Soc.*, (49).

[44] Ševčovič, D., Stehlíkov, B., and Mikula, K. 2011. *Analytical and numerical methods for pricing financial derivatives*. Nova Science Publishers, Inc., Hauppauge, (1)–309.

[45] Leland, H. E. 1985. Option pricing and replication with transactions costs. *The Journal of Finance*, (40), 1283–1301.

[46] Ševčovič, D., and Žitňanská, M. 2016. Analysis of the nonlinear option pricing model under variable transaction costs. *Asia-Pac. Financ. Mark.*, (23), 153–174.

[47] Barles, G., and Soner, H. M. 1998. Option pricing with transaction costs and a nonlinear Black–Scholes equation. *Finance and Stochastics*.

[48] Kilianová, S., and Ševčovič, D. 2018. Expected Utility Maximization and Conditional Value-at-Risk Deviation-based Sharpe Ratio in Dynamic Stochastic Portfolio Optimization. *Kybernetika*, 54(6), 1167-1183.

[49] Kilianová, S., and Ševčovič, D. 2013. A Transformation Method for Solving the Hamilton-Jacobi-Bellman Equation for a Constrained Dynamic Stochastic Optimal Allocation Problem. *ANZIAM Journal*, 55, 14–38.

[50] Kilianová, S., and Ševčovič, D. 2019. Dynamic intertemporal utility optimization by means of Riccati transformation of Hamilton-Jacobi Bellman equation. *Japan Journal of Industrial and Applied Mathematics*, (36(2), 497–517.

[51] Bertsekas, Dimitri P. 1976. *Dynamic programming and stochastic control.* (Academic Press, New York.

[52] Macová, Z, and Ševčovič, D. 2010. Weakly nonlinear analysis of the Hamilton-Jacobi-Bellman equation arising from pension savings management. *Int. J. Numer. Anal. Model.*, 7(4), 619–638.

[53] Federico, S., Gassiat, and P., Gozzi, F. 2015. Utility maximization with current utility on the wealth:regularity of solutions to the HJB equation. *Finance Stoch.*, 19, 415–448.

[54] Ishimura, N., and Ševčovič, D. 2013. On traveling wave solutions to a Hamilton-Jacobi-Bellman equation with inequality constraints. *Japan J. Ind. Appl. Math.*, 30(1), 51–67.

[55] Abe, R., and Ishimura, R. 2008. Existence of solutions for the nonlinear partial differential equation arising in the optimal investment problem. *Proc. Japan Acad., Ser. A*, 84(1), 11–14.

[56] Milgrom, P., and Segal, I. 2002. Envelope theorems for arbitrary choice sets. *Econometrica*, 70(2), 583–601.

[57] Klatte, D. 1985. On the Lipschitz behavior of optimal solutions in parametric problems of quadratic optimization and linear complementarity. *Optimization: A Journal of Mathematical Programming and Operations Research*, 16(6), 819–831.

[58] Meyer, J. C., Needham, D. J.. Extended weak maximum principles for parabolic partial differential inequalities on unbounded domains. *Proc. R. Soc. Lond. Ser. A Math. Phys. Eng. Sci.*, 470(2014), 20140079.

[59] Kilianová, S., and Trnovská, M. 2016. Robust Portfolio Optimization via solution to the Hamilton-Jacobi-Bellman Equation. *Int. Journal of Computer Mathematics*, (93), 725–734.

[60] Post, T., Fang, Y., and Kopa, M. 2015. Linear Tests for DARA Stochastic Dominance. *Management Science*, (61), 1615–1629.

[61] Vickson Raymond G. 1975. Stochastic Dominance for Decreasing Absolute Risk Aversion. *Journal of Financial and Quantitative Analysis*, (10), 799–811.

[62] Kilianová, S., Melicherčík, I., and Ševčovič, D. 2006. Dynamic Accumulation Model for the Second Pillar of the Slovak Pension System, Finance a uver - *Czech Journal of Economics and Finance*, 56(11-12), 506–521.

[63] Deelstra, G., Diallo, I., and Vanmaele, M. 2008. Bounds for Asian basket options. *Journal of Computational and Applied Mathematics*, 218, 215–228.

Index